W9-CSB-020

REVIEWS OF INFRARED AND MILLIMETER WAVES

VOLUME 2

Optically Pumped Far-Infrared Lasers

REVIEWS OF INFRARED AND MILLIMETER WAVES

VOLUME 2

Optically Pumped Far-Infrared Lasers

EDITED BY KENNETH J. BUTTON
National Magnet Laboratory
Massachusetts Institute of Technology
Cambridge, Massachusetts

M. INGUSCIO AND F. STRUMIA
Institute of Physics
Pisa, Italy

PLENUM PRESS • NEW YORK AND LONDON

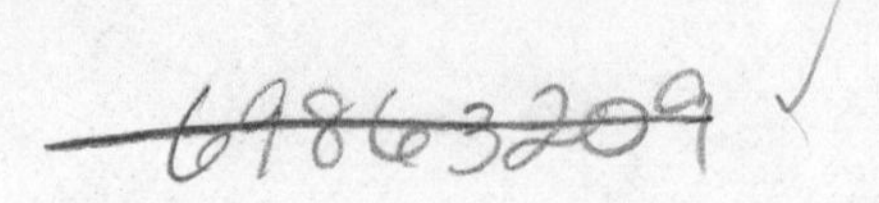

Library of Congress Cataloging in Publication Data

Main entry under title:

Optically pumped far-infrared lasers.

(Reviews of infrared and millimeter waves; v. 2)
Includes bibliographical references and index.
1. Far infra-red lasers—Addresses, essays, lectures. I. Button, Kenneth J. II. Inguscio, M. III. Strumia, F. IV. Series.
TK7876.R48 1983 vol. 2 [TA1696] 621.3813s 83-19268
ISBN 0-306-41487 [621.36′2]

A Division of Plenum Publishing Corporation
233 Spring Street, New York, N.Y. 10013

Printed in the United States of America

PREFACE

This book represents a compendium of the twenty most useful far-infrared (or submillimeter) lasers. In the case of each laser described here, we have been fortunate to have the author who is the pioneer and acknowledged authority describe the principles of operation and to prepare the list of emission lines.

Until these lasers were developed during the past decade, the submillimeter range of the spectrum has been almost barren due to lack of sources of radiation. The lasers described here remain the only practical, powerful source of radiation between the wavelengths of one millimeter and ten micrometers. Many hundreds of emission lines have been listed here, some providing hundreds of kilowatts of peak power in pulsed operation, others providing many tens of milliwatts of continuous power. The hundreds of wavelengths of the emission lines are so closely spaced in the wavelength range between one millimeter and one-tenth millimeter that this source of radiation can be considered to be step-tunable. Of course, laboratory scientists still depend upon the black body source (mercury vapor lamp) and the Fourier transform spectrometer to provide continuous spectra, but for this we must deal with true energy starvation at the sub-microwatt level.

This critical review can be expected to serve as a handbook for decades in the future because it contains descriptions of fundamental principles and listings of fundamental physical data. The manuscripts have been prepared only after the research and development of optically pumped lasers has matured to the point where the data can be considered to be sufficiently complete to be relied upon for some years to come.

Indeed, the compendium of many more optically pumped molecular gas lasers remains to be compiled. We hope that this information can be provided in an additional book in this series eventually.

We are grateful to the authors for their hard work, tangible assistance and patience. Only they know how difficult it was to bring this treatise to fruition.

K. J. Button
M. Inguscio
F. Strumia

CONTENTS

FAR-INFRARED LASER LINES IN $^{12}CH_3F$ AND $^{13}CH_3F$

T. Y. Chang

Bell Telephone Laboratories
Holmdel, New Jersey 07733
USA

INTRODUCTION

CH_3F (methyl flouride) is a relatively simple symmetric top molecule with a strong and well understood infrared absorption band in the 9 to 10-μm region. Since the discovery of close coincidences between two of its absorption lines and two CO_2 laser lines (Chang et al., 1969), it has played more key roles in the development of optically pumped far infrared (FIR) lasers than any other molecule. With this molecule, many new ideas including the very first optically pumped Fir laser (Chang and Bridges, 1970a, b), the generation of 1-MW FIR pulses (Evans et al., 1975), the beneficial effects of some complex buffer-gas molecules (Chang and Lin, 1976), the generation of synchronously mode-locked subnanosecond FIR pulses (Lee et al., 1979), the generation of continuously tunable FIR Raman-laser output (Mathieu and Izatt, 1981), and the usefulness of several novel FIR laser structures have been demonstrated. The extremely wide ranges of FIR output power from milliwatts (cw) to hundreds of kilowatts (pulsed) that are available for the 496-μm line of $^{12}CH_3F$ and the 1,222-μm line of $^{13}CH_3F$ are still unique among the more than a thousand optically pumped FIR laser lines known today. The CH_3F laser is also the most extensively studied and clearly understood optically pumped FIR laser both theoretically and experimentally. We summarize in the following the important experimental facts and parameters and spectroscopic information for all the CH_3F laser frequencies.

FIR Laser Lines

Important data on optically pumped FIR laser lines in $^{12}CH_3F$ and $^{13}CH_3F$ are tabulated in Tables Ia and Ib respectively. The notations used and their meanings are explained in footnotes a through g. Many of the $^{12}CH_3F$ laser lines are pumped off resonance by a pulsed high-power CO_2 laser. For these laser lines, the dominant FIR emission frequency is expected to be Raman shifted from the related rotational transition by approximately the same amount as the pump frequency offset. The observed FIR Raman shift is given in column five along with the calculated pump frequency offset (in parentheses). Similar Raman shifted FIR lines have been obtained in $^{13}CH_3F$, but no details have been given (Biron et al., 1980). No Raman shift is expected for a cascade line. In column one, the upward and downward arrows indicate the observed continuous tuning ranges of the FIR Raman emission which is discussed in some detail in the following section. It is apparent from Tables Ia and Ib that the entire submillimeter-wave range is well covered by the two isotopic species of CH_3F with their pulsed laser lines.

Spectroscopic Data and Laser Line Assignments

The spectra of CH_3F have been studied in great detail by conventional infrared spectroscopy (Smith and Mills, 1963), tunable diode-laser spectroscopy (Sattler and Simonis, 1977), microwave spectroscopy (Winton and Gordy, 1970), laser Stark spectroscopy (Freund et al., 1974), FIR laser frequency measurements (Chang and Bridges, 1970 a,b; Kramer and Weiss, 1976, Bava et al., 1977), and by correlation among FIR laser lines and pump frequencies (Chang and McGee, 1971). The relevant spectroscopic constants for $^{12}CH_3F$

Table Ia $^{12}CH_3F$ Laser Lines

$^{12}CH_3F$ FIR EMISSION					IR LINE	CO_2 PUMP LASER				FIR	$^{12}CH_3F$	
$\lambda(\mu m)^a$	ν(GHz)	R.P.[b]	ν_3:RO(J)[c]	$\delta\nu$ (GHz)[d]	V_3'':PQR(J,K)[e]	LINE[f]	cm^{-1}	MODE[g]	W/J	W/J	torr	REF
157.4	1905	(||)	1~RO(37)	(+2)	0~P(39,K)	10R6	966.25	PLS	~E7W	~E3W	150	h
171.4	1749	(||)	1~RO(34)	(-10)	0~P(36.K)	10R16	973.29	PLS	~3E7W	~E3W	~100	h
176.0	1703	||	1~RO(33)	-1.5(-4)	0~P(35.K)	10R20	975.93	PLS	~E6W	5E3W	40-50	i
180.7	1659	||	1~RO(32)	+4.5(-0.6)	0:P(34,K)	10R24	978.47	PLS	~E6W	6E3W	40-50	i
187.2	1601	||	1~RO(31)	-4.2(0.5)	0:P(33,K)	10R28	980.91	PLS	~E6W	5E3W	40-50	i
192.1	1561	(⊥)	1~RO(30)	(-3)	0~Q(31,K)	9P30	1037.43	PLS	~E7W	~E3W	~100	h,j
192.78	1555.1	||	1:RO(30)	(~0)	0:P(32,3)	10R32	983.25	CHP	55W	5E-3W	0.3	k
192.9	1554	||	1:RO(30)	-0.9(-0.9)	0:P(32,K)	10R32	983.25	PLS	~E6W	6E3W	20	i
195			1~RO(30)	(-2)	0~R(30,K)	9R40	1090.03	PLS	~E6W	~E1W		j
196		||				88I9R10	1091.02	CW	E1W		>0.055	ℓ
198.5	1510	(||)	1:RO(29)	(-5)	0~P(31.K)	10R36	985.49	PLS	~E7W	~E3W	~100	h
199.14	1505.4		1:RO(29)	CASCADE	0:P(32,3)	10R32	983.25	CHP	55W			k
200		(||)	1~RO(29)	(-3)	0:R(29,K)	9R38	1089.00	PLS	~E6W	~E1W		j

(continued)

Table Ia $^{12}CH_3F$ Laser Lines (Continued)

206.7	1450	||	1~RO(28)	-5(-4)	0~R(28.K)	9R36	1087.95	PLS	~6E5W	5E2W	15-18	i
214.8	1396	||	1~RO(27)	-10(-5)	0~R(27,K)	9R34	1086.87	PLS	~7E5W	2E3W	15-18	i,j
221.1	1356	(||)	1:RO(26)	(0)	0:R(26,K)	~9R32	1085.94	PLS	~E-1J	~E-4J	25	m
222.0	1351	||	1~RO(26)	-5(-6)	0~R(26,K)	9R32	1085.77	PLS	8E5W	7E3W	15-18	i
224		(||)	1~RO(26)	~(-16)	0~R(26,K)	~9R32	1085.3	PLS	~E-1J	~E-4J	25	m
224		(||)	0~RO(25)	~(+16)	0~R(25,K)	~9R32	1085.3	PLS	~E-1J	~E-4J	25	m
226.3	1324	(||)	0~RO(25)	(0)	0:R(25,K)	~9R30	1084.85	PLS	~E-1J	~E-4J	25	m
229.5	1306	(||)	1~RO(25)	(0)	0:R(25,K)	~9R30	1084.85	PLS	~E-1J	~E-4J	25	m
231.0	1298	||	1~RO(25)	-8(-8)	0~R(25,K)	9R30	1084.64	PLS	~9E5W	6E3W	15-18	i
232		(||)	1~RO(25)	~(-16)	0~R(25,K)	~9R30	1084.2	PLS	~E-1J	~E-4J	25	m
232		(||)	0~RO(24)	~(+16)	0~R(24,K)	~9R30	1084.2	PLS	~E-1J	~E-4J	25	m
235.4	1274	(||)	0~RO(24)	(0)	0:R(24,K)	~9R28	1083.74	PLS	~E-1J	~E-4J	25	m
238.7	1256	(||)	1:RO(24)	(0)	0:R(24,K)	~9R28	1083.74	PLS	~E-1J	~E-4J	25	m
239.0	1254	(⊥)	1~RO(24)	(-3)	0~Q(25,K)	9P26	1041.28	PLS	~E7W	~E3W	~100	h
241.0	1244	||	1~RO(24)	-12(-9)	0~R(24.K)	9R28	1083.48	PLS	~E6W	2E3W	15-18	i
242		(||)	1~RO(24)	~(-17)	0~R(24,K)	~9R28	1083.2	PLS	~E-1J	~E-4J	25	m

242		(‖)	0~RO(23)	~(+17)	0~R(23,K)	~9R28	1083.2	PLS	~E-1J	~E-4J	25	m
245.2	1223	(‖)	0:R0(23)	(0)	0:R(23,K)	~9R26	1082.61	PLS	~E-1J	~E-4J	25	m
248.6	1206	(‖)	1:RO(23)	(0)	0:R(23,K)	~9R26	1082.61	PLS	~E-1J	~E-4J	25	m
250.6	1196	‖	1~RO(23)	-10(-10)	0~R(23,K)	9R26	1082.30	PLS	~1.1E6W	4E3W	15-18	i
251.91	1190.1	‖	2:RO(23)	(~0)	1:P(25,11?)	10R34	984.38	CHP	83W	1.7E-4W	125	k
252		(‖)	1~RO(23)	~(-17)	0~R(23,K)	~9R26	1082.0	PLS	~E-1J	~E-4J	25	m
252		(‖)	0~RO(22)	~(+17)	0~R(22.K)	~9R26	1082.0	PLS	~E-1J	~E-4J	25	m
255.8	1172	(‖)	0:RO(22)	(0)	0:R(22,K)	~9R24	1081.47	PLS	~E-1J	~E-4J	25	m
259.3	1156	(‖)	1:RO(22)	(0)	0:R(22,K)	~9R24	1081.47	PLS	~E-1J	~E-4J	25	m
261.8	1145	‖	1~RO(22)	-11(-12)	0~R(22,K)	9R24	1081.09	PLS	~1.2E6W	6E3W	15-18	i
263		(‖)	1~RO(22)	~(-18)	0~R(22,K)	~9R24	1081.9	PLS	~E-1J	~E-4J	25	m
263		(‖)	0~RO(21)	~(+18)	0~R(21,K)	~9R24	1081.9	PLS	~E-1J	~E-4J	25	m
267.4	1121	(‖)	0:RO(21)	(0)	0:R(21,K)	~9R22	1080.29	PLS	~E-1J	~E-4J	25	m
271.0	1106	(‖)	1:RO(21)	(0)	0:R(21,K)	~9R22	1080.29	PLS	~E-1J	~E-4J	25	m
275.1	1090	‖	1~RO(21)	-16(-14)	0~R(21,K)	9R22	1079.85	PLS	~1.3E6W	3E3W	15-18	i
276		(‖)	1~RO(21)	~(-18)	0~R(21,K)	~9R22	1079.7	PLS	~E-1J	~E-4J	25	m

(continued)

Table Ia $^{12}CH_3F$ Laser Lines (Continued)

276		(||)	0~RO(20)	~(+18)	0~R(20,K)	~9R22	1079.7	PLS	~E-1J	~E-4J	25	m
280.1	1070	(||)	0:RO(20)	(0)	0:R(20,K)	~9R20	1079.09	PLS	~E-1J	~E-4J	25	m
283.9	1056	(||)	1:RO(20)	(0)	0:R(20,K)	~9R20	1079.09	PLS	~E-1J	~E-4J	25	m
288.2	1040	||	1~RO(20)	-16(-16)	0~R(20,K)	9R20	1078.59	PLS	~1.4E6W	1E3W	15-18	i
289		(||)	1~RO(20)	~(-18)	0~R(20,K)	~9R20	1078.5	PLS	~E-1J	~E-4J	25	m
289		(||)	0~RO(19)	~(+18)	0~R(19,K)	~9R20	1078.5	PLS	~E-1J	~E-4J	25	m
293.9	1020	(||)	0:RO(19)	(0)	0:R(19,K)	~9R18	1077.87	PLS	~E-1J	~E-4J	25	m
298.1	1006	(||)	1:RO(19)	(0)	0:R(19,K)	~9R18	1077.87	PLS	~E-1J	~E-4J	25	m
298.6	1004	||	1:RO(20)	CASCADE	0~R(20,K)	9R20	1078.59	PLS	~1.4E6W		15-18	i
303.4	988	||	1~RO(19)	-18(-18)	0~R(19,K)	9R18	1077.30	PLS	~1.4E6W	5E2W	15-18	i
304		(||)	1~RO(19)	~(-19)	0~R(19,K)	~9R18	1077.2	PLS	~E-1J	~E-4J	25	m
304		(||)	0~RO(18)	~(+19)	0~R(18,K)	~9R18	1077.2	PLS	~E-1J	~E-4J	25	m
309.4	969	(||)	0:RO(18)	(0)	0:R(18,K)	~9R16	1076.63	PLS	~E-1J	~E-4J	25	m
313.7	956	(||)	1:RO(18)	(0)	0:R(18,K)	~9R16	1076.63	PLS	~E-1J	~E-4J	25	m
314.1	955	||	1:RO(19)	CASCADE	0~R(19,K)	9R18	1077.30	PLS	~1.4E6W		15-18	i
320		(||)	1~RO(18)	~(-19)	0~R(18,K)	~9R16	1076.0	PLS	~E-1J	~E-4J	25	m

320		(||)	0~RO(17)	~(+19)	0~R(17,K)	~9R16	1076.0	PLS	~E-1J	~E-4J	25	m
320.1	937	||	0~RO(17)	+18(+18)	0~R(17,K)	9R16	1075.99	PLS	~1.3E6W	3E3W	15-18	i
326.6	918	(||)	0:RO(17)	(0)	0:R(17,K)	~9R16	1075.37	PLS	~E-1J	~E-4J	25	m
331.0	906	(||)	1:RO(17)	(0)	0:R(17,K)	~9R16	1075.37	PLS	~E-1J	~E-4J	25	m
338		(||)	1~RO(17)	~(-20)	0~R(17,K)	~9R14	1074.7	PLS	~E-1J	~E-4J	25	m
338		(||)	0~RO(16)	~(+20)	0~R(16,K)	~9R14	1074.7	PLS	~E-1J	~E-4J	25	m
339.5	883	||	0~RO(16)	+16(+17)	0~R(16,K)	9R14	1074.65	PLS	~1.2E6W	2E3W	15-18	i
345.8	867	(||)	0:RO(16)	(0)	0:R(16,K)	~9R14	1074.07	PLS	~E-1J	~E-4J	25	m
360.1	833	||	0~RO(15)	+16(+15)	0~R(15,K)	9R12	1073.28	PLS	~1.1E6W	1E3W	15-18	i
372.68	804.4	||	1:RO(15)	(~0)	0:P(17,8)	9P50	1016.72	CHP	14W	5.8E-3W	0.095	k
384.7	779	(||)	0~RO(14)	+13(+13)	0~R(15,K)	9R10	1071.88	PLS	~E7W	~E3W	~15	h
397.51	754.2	(||)	1:RO(14)	CASCADE	0:P(17,8)	9P50	1016.72	CHP	14W			k
411.4	729	(||)	0~RO(13)	+14(+12)	0~R(13,K)	9R8	1070.46	PLS	~E7W	~E3W	~15	h
419	715		0:RO(13)	CASCADE	0:Q(12,2)	9P20	1046.85	PLS	0.65J	~E-6J	0.25	n,o
451.901	663.403	⊥	0:RO(12)	(~0)	0:Q(12,1)	9P20	1046.85	QSW	1E3W	~E-2W	~0.08	p
451.923	663.371	⊥	0:RO(12)	(~0)	0:Q(12,2)	9P20	1046.85	QSW	1.5E3W	~E-2W	~0.08	p

(continued)

Table Ia $^{12}CH_3F$ Laser Lines (Continued)

452	663	⊥	0:RO(12)	(~0)	0:Q(12,K)	9P20	1046.85	PLS	65J	1E-2J	2	q
494	607	||	1:RO(11)	(~0?)	0:R(11,?)	88I9P22	1067.36	CW	10W			l
496.072	604.333	⊥	1:RO(11)	(~0)	0:Q(12,1)	9P20	1046.85	QSW	1E3W	~0.1W	0.1	p
496.1	604	⊥	1:RO(11)	(~0)	0:Q(12,K)	9P20	1046.85	PLS	17J	6E5W	2-3	r
								PLS	65J	1.3E-2J	4	q
496.1008	604.2975	⊥	1:RO(11)	(~0)	0:Q(12,2)	9P20	1046.85	CW	30W	2.2E-2W	<0.1	k,s,t
								QSW	1.5E3W	~0.1W	0.1	p
541	554	⊥	1:RO(10)	CASCADE	0:Q(12,K)	9P20	1046.85	PLS	0.8J	4E-4J	1	u
541.113	554.029	⊥	1:RO(10)	CASCADE	0:Q(12,1)	9P20	1046.85	QSW	1E3W	~E-2W	~0.08	p
541.147	553.994	⊥	1:RO(10)	CASCADE	0:Q(12,2)	9P20	1046.85	QSW	1.5E3W	~E-2W	~0.08	p
595	504		1:RO(9)	CASCADE	0:Q(12,K)	9P20	1046.85	PLS	8J	~10W	<2	n
992	302	⊥	1:RO(5)	(~0)	0:Q(6,2)	S9P15	1048.14	CW	2.5W	~E-3W	0.08	v

Table Ib $^{13}CH_3F$ Laser Lines

$^{13}CH_3F$ FIR EMISSION					IR LINE	CO_2 PUMP LASER				FIR	$^{13}CH_3F$	
$\lambda(\mu m)$[a]	ν(GHz)	R.P.[b]	ν_3:RO(J)[c]	$\delta\nu$ (GHz)[d]	V_3'':PQR(J,K)[e]	LINE[f]	cm^{-1}	MODE[g]	W/J	W/J	torr	REF
388	773		2~RO(15)		0:R(4)+1~R(15)	9P32	1035.47	PLS	8E7W	1E4W	4	w
412	728		2:RO(14)	CASCADE	0:R(4)+1~R(15)	9P32	1035.47	PLS	8E7W	1E3W	4	w
862	348		0:RO(6)	CASCADE	0:R(4,K)	9P32	1035.47	PLS	1J	1.5E-4W	1.4	x
1006	298		0:RO(5)	CASCADE	0:R(4,K)	9P32	1035.47	PLS	8E7W	3E3W	2	w,x
1207	248	(||)	0:RO(4)	(~0)	0:R(4,K)	9P32	1035.47	PLS	8E7W	5E3W	2	w,x
1221.9	245.3	||	1:RO(4)	(~0)	0:R(4,K)	9P32	1035.47	PLS	8E7W	3E3W	2	w,x
				(~0)	0:R(4,K)	9P32	1035.47	CW	30	1E-2W	<0.1	k,s

a. Arrows indicate observed continuous tuning ranges when a high-pressure, continuously tunable CO_2 laser is used as the pump source (Mathieu and Izatt, 1981).

b. R.P. = relative polarization of the FIR output w.r.t. the pump polarization. Parentheses imply expected but not experimentally verified.

c. An entry such as 1:RO(30) means it is a J = 31 → 30 rotational transition in the v_3 = 1 vibrational state. The symbol ~ indicates that the FIR emission is Raman in nature and that the rotational transition given is the nearest resonance line. The K value of each transition would be the same as that for the IR pump line given in column six.

d. The observed frequency offset of the FIR output from the nearest resonance line. The numbers in parentheses are the calculated frequency offset of the pump signal from the nearest resonance line.

e. An entry such as 0:R(29,K) means that several K components of the R(29) transition of the v_3 = 0→1 band of CH_3F are simultaneously excited by the pump radiation. The symbol ~ indicates off-resonant pumping (c.f. footnote C).

f. An entry such as ~9R32 means that the CO_2 laser is tuned more than ~2GHz away from the center of the R(32) line of the 9.4-μm band. The notation 88I indicates isotope $C^{18}O_2$. A letter S is used to denote the first sequence band.

g. PLS = pulsed; CHP = chopped; QSW = Q-switched.

h. Biron et al., 1979b.

i. Biron et al., 1979a.

j. Fetterman et al., 1972. Some of the CO_2 laser lines appear to be erroneously identified in this reference. Data presented in the present table have been corrected after comparing with the data of Biron et al.(1979a,b), Smith and Mills (1963), and Freund et al (1974).

k. Chang and McGee, 1971.

l. Davies and Jones, 1980. The original interpretation of the 196-μm line in this reference has not been included in the present table due to its inconsistency with the known spectroscopic data for CH_3F.

m. Mathieu and Izatt, 1981.

n. Brown et al., 1974.

o. Temkin et al., 1975.

p. Chang and Bridges, 1970 a,b.

q. Evans et al., 1975.

r. Brown et al., 1977.

s. Hodges et al., 1976.

t. Bava et al., 1977; Kramer and Weiss, 1976.

u. Lipton and Nicholson, 1978.

v. Danielewicz and Weiss, 1978.

w. Peebles et al., 1980.

x. Hacker et al., 1976

and $^{13}CH_3F$ are given in Tables IIa and IIb respectively.

Shown in Figure 1a is a simplified energy-level diagram for the 496-μm line and its related cascade lines. The details of the K manifolds are illustrated in Figure 1b. The Q(12,2) line of CH_3F is shown to be located approximately 43 MHz above the center of the 9P(20) line of CO_2 laser. The Q(12,1) line, on the other hand, lies below the CO_2 laser line center at about the same distance. These two CH_3F lines are within the tuning range of most CO_2 lasers. For $K \geq 3$, the 9P(20) line of the CO_2 laser is significantly off resonant and optical pumping by this CO_2 laser line gives rise to two distinct FIR spectral subcomponents: one characteristic to the molecule (denoted by 3') and the other Raman in nature (denoted by 3"). The overall FIR gain spectrum of CH_3F when excited by a CO_2 laser tuned to the center of its 9P(20) line is shown in Figure 2 for three different pump intensities (Chang, 1977b; Drozdowicz, 1979a,b). It can be seen that the dynamic Stark effect causes K = 1 and K = 2 components to split and saturate as the pump intensity is increased. The dynamic Stark shift of Raman subcomponents (e.g. 3" and 4") is also evident. The characteristic (K') and Raman (K") subcomponents exhibit the same strength when the population in the ν_3 vibrational state is neglibibly small. However, as soon as the population builds up in the ν_3 state, self-absorption will cause both the gain and the FIR saturation intensity for the characteristic subcomponent to drop significantly. The Raman subcomponent is affected significantly by self-absorption only when the pump offset is comparable or smaller than the linewidth or the dynamic Stark splitting (Chang, 1977; Osche, 1978).

For the first refilling transition in the ground vibrational state (e.g. 452-μm line in Figure 1a), self-absorption is particularly severe. Within the self-absorption region (covering all K components and allowing for dynamic Stark splitting), the FIR laser relies on actual population inversion to operate and the saturated FIR output intensity remains relatively low. As the pump offset

Table IIa. Spectroscopic Constants of $^{12}CH_3F$ [a]

State	0	ν_3	$2\nu_3$
E_v	0	1,048.61077(±6x10^{-5})cm^{-1}	2,081.3803(±4x10^{-4})cm^{-1}
B	25,536.1466(±0.0015) MHz	25,197.57(±0.03) MHz	24,870.6(±0.2) MHz
$A-A_o$	0	-294.1(±0.6) MHz	-583 (±7) MHz
D_J	59.9 (±0.2) kHz	55.5(±1.2) kHz	
D_{JK}	440.3 (±0.4) kHz	575 (±63) kHz	
$D_K - K_{Ko}$	0	-145 (±74) kHz	
μ	1,8585 (±0.0005) D	1,9054(±0.0006) D	1,9519 (±0.0020) D

Table IIb. Spectroscopic Constants of $^{13}CH_3F$[a]

E_v	0	1,02749319($7x10^{-5}$) cm^{-1}
B	24,862.36(0.01) MHz	24,542.1(0.4) MHz
$A-A_o$	0	-288.8 (0.4) MHz
D_J	56.6 (8.5) kHz	56 (12) kHz
D_{JK}	407. (35) kHz	464 (24) kHz
$D_K - D_{KO}$	0	-55 (30) kHz
	1.8579(0.0006)D	1.9039(0.0006) D

a. Freund et al., 1974, and References cited therein.

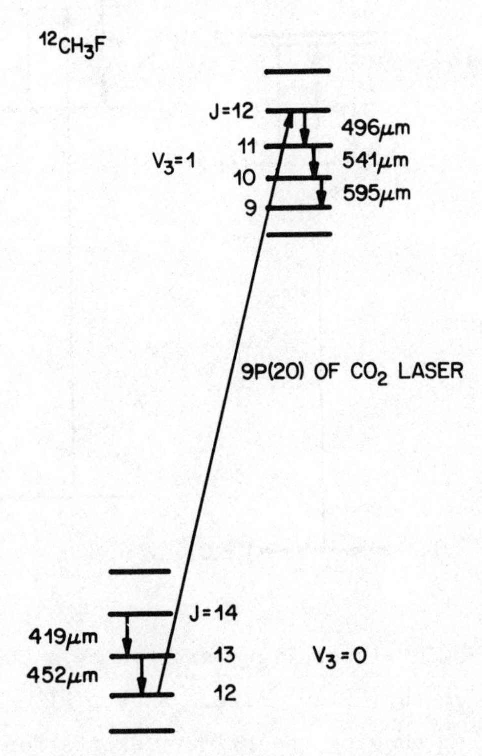

Figure 1 (a). The $^{12}CH_3F$ laser lines excited by a CO_2 laser operating on the 9P(20) line. A simplified energy-level diagram showing the 496-μm laser transition and related cascade transitions.

becomes large enough for the Raman subcomponent to move out of the self-absorption region, the Raman subcomponent becomes strong and dominant, provided that the pump intensity is high enough to generate sufficient Raman gain.

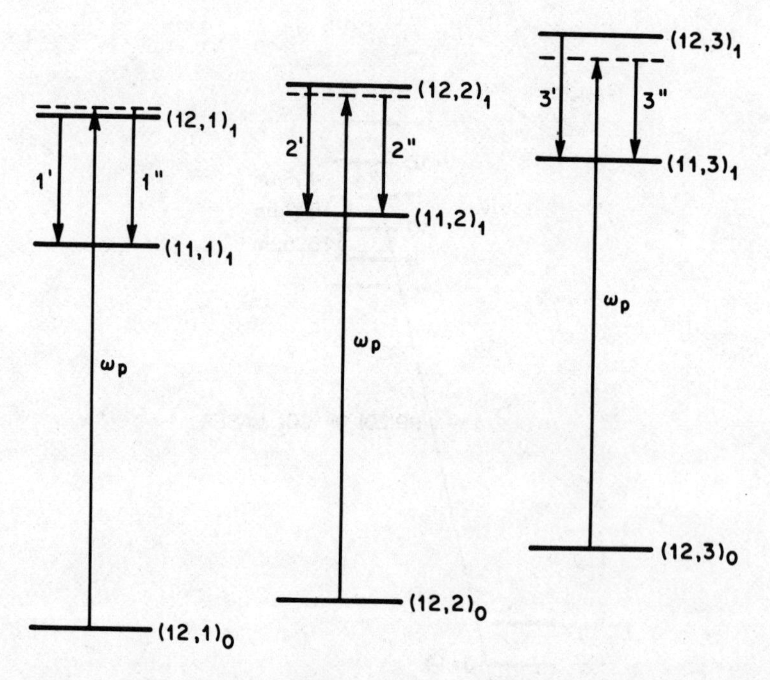

Figure 1 (b). The $^{12}CH_3F$ laser lines excited by a CO_2 laser operating on the 9P(20) line. A partial energy-level diagram showing the primary excitation processes in the K = 1, 2 and 3 manifolds that give rise to the characteristic (1', 2', and 3') and Raman like (1", 2", and 3") subcomponents.

At the same time, the increasing difficulty of achieving actual population inversion as the pump offset is increased causes the characteristic FIR subcomponent to be suppressed eventually. This phenomenon is useful for the generation of a continuously tunable FIR output.

Continuous tuning of the FIR Raman signal between the

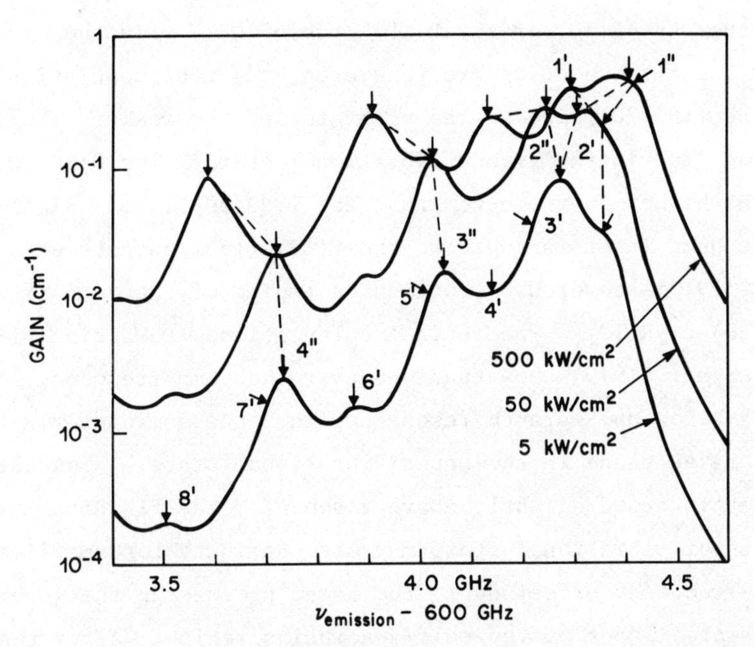

Figure 2. Calculated gain spectra near 496 μm for pump intensities of 5, 50, and 500 kW/cm^2. Short arrows indicate the positions of various subcomponents and the dashed lines indicate the shift of some of these components with the pump intensity. Positions of some subcomponents are not indicated for higher curves to avoid cluttering of the figure (from T. Y. Chang, 1977b).

characteristic rotational transition frequencies has been demonstrated by using a pump intensity of $8MW/cm^2$ obtained from a continuously tunable high-pressure CO_2 TE laser (Mathieu and Izatt, 1981). The excitation processes as the pump frequency is tuned from the R(19) to the R(20) line of CH_3F are illustrated by means of energy-level diagrams in Figure 3. Illustrated in Figure 3a are the situations when the pump frequency is tuned to resonance with either the R(20) or the R(19) line of CH_3F. In the first case, two FIR laser frequencies are generated by actual population inversion: the 1:RO(20) line in the ν_3 state and the weaker 0:RO(20) refilling line in the ground vibrational state. The 0:RO(20) line is 14-GHz higher in frequency than the 1:RO(20) line. Similarly, when the pump laser is tuned to the R(19) line, outputs at 0:RO(20) and 0:RO(19) is covered by continuous tuning of the pump frequency from R(20) to R(19). Theoretical calculations (DeMartino, 1978; Temkin et al., 1979) show that whenever the pump frequency is tuned slightly below an R-branch resonance line, the dominant FIR Raman process takes place in the upper vibrational state. When the pump frequency is tuned slightly above resonance, the FIR Raman process in the upper vibrational state may also dominate for small values of pump-frequency offset until the Raman process in the lower vibrational state is out of the self-absorption region. After that, the Raman process in the lower vibrational state becomes dominant. As a rule of thumb, when the pump frequency is far off resonance, the virtual level of the dominant Raman process lies below the nearest real level. These situations are illustrated in Figure 3b. The upper half of the 36-GHz range between 1:RO(20) and 0:RO(19) can therefore be identified as 1~RO(20), while most of the lower half of the range can be identified as 0~RO(19). In Table I, these continuous tuning ranges are indicated by downward and upward arrows. Experimentally, the FIR output is found to be depressed near the resonance frequencies for a range of approximately 7-GHz due to self-absorption.

Narrow dips in the FIR output have also been observed experimentally midway between pump resonances. No explanation has yet been given on this phenomenon. We conjecture that this narrow dip is caused by two-photon self-absorption which occurs exactly midway between pump resonances as is illustrated in the right-hand part of Figure 3c. The narrowness of the dip is probably related to the

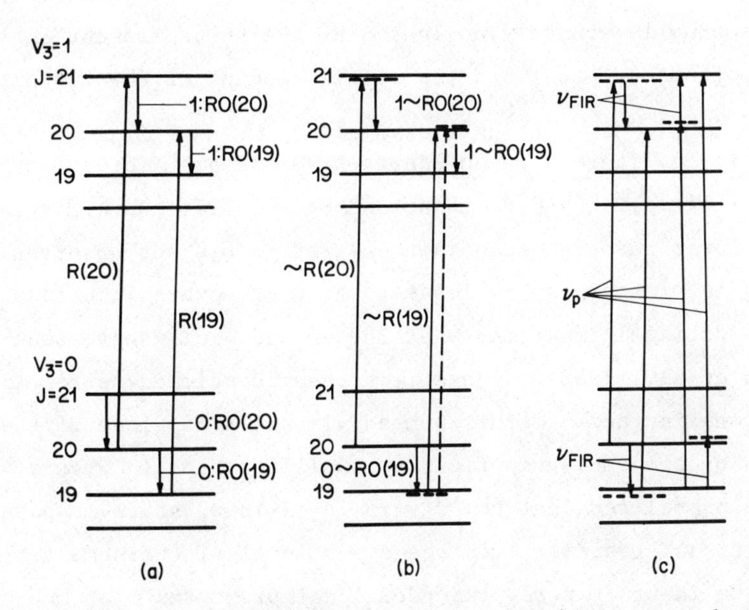

Figure 3. Energy-level diagrams (not to scale) showing the excitation processes as the pump frequency is tuned between the R(19) and R(20) lines of CH_3F. (a) Resonant pumping with the pump frequency tuned to either R(20) or R(19) absorption line. 0:RO(20) is higher than 1:RO(20) by ~14 GHz and the latter higher than 0:RO(19) by ~36 GHz. (b) The dominant processes when the pump is tuned either slightly below R(20) or slightly above R(19). When the pump frequency is only very slightly (<4 GHz) above R(19) 0~RO(19) is still within the self-absorption region of 0:RO(19), the secondary process indicated by dashed arrows may compete effectively. (c) The situation when the pump frequency ν_p is tuned exactly halfway between R(19) and R(20). Processes in upper and lower vibrational states contribute about equally to the generation of the same FIR Raman signal. The sum of ν_p and ν_{FIR} also coincides with J=19 to J=21 two photon transition as is illustrated on the right.

fact that for far-off resonance pumping in the R-branch, the Raman gain spectrum is strongly dominated by the K=0 component due both to the Boltzmann factor and the matrix element of the transitions involved.

Shown in Figure 4 is an energy-level diagram for the principal FIR laser lines in $^{13}CH_3F$ (Chang and McGee, 1971; Hacker et al., 1976). Lower cascade lines in the ν_3 state are not observed due probably to greater cavity losses. At pump powers in excess of 30 MW, additional laser lines at 388 μm and 412 μm have been observed (Peebles et al., 1980). These have been identified as rotational transitions in the $2\nu_3$ state pumped via the R(15) line of the $2\nu_3$-ν_3 hot band by the same pump field as is illustrated in Figure 4. It has been postulated that the J = 15 level in ν_3 state is postulated by collisional transfer from the J = 5 level of the same vibrational state. We identify here, two additional processes that may contribute significantly to the overall pumping process. These are indicated by dashed arrows in Figure 4. One of them is two-photon pumping of the Q(13) transition involving the CO_2 pump and the 1006-μm FIR laser signal. The other one is a three-photon pumping process involving two CO_2 pump photons and a 1006-μm photon. In both cases, the 1006-μm signal is parametrically amplified.

Laser Designs and Performance Studies

In the low power regine (<1W), the CH_3F laser has been operated using either a Fabry-Perot cavity (Chang and Bridges, 1970; Chang, 1974) or waveguide cavity (Hodges et al., 1976b and 1977). While the FIR signal is most commonly coupled out through a hole, the use of metal-mesh-dielectric output mirror (Danielewicz et al., 1975; Weitz et al., 1978) and Michelson output coupler (Duxbury and Herman, 1978) for better beam quality has also been investigated. The performance of the CH_3F laser and its optimization have been studied both experimentally and theoretically (Henningsen and Jensen, 1975; DeTemple and Danielewicz, 1976; Tucker, 1976; Hodges et al., 1976a;

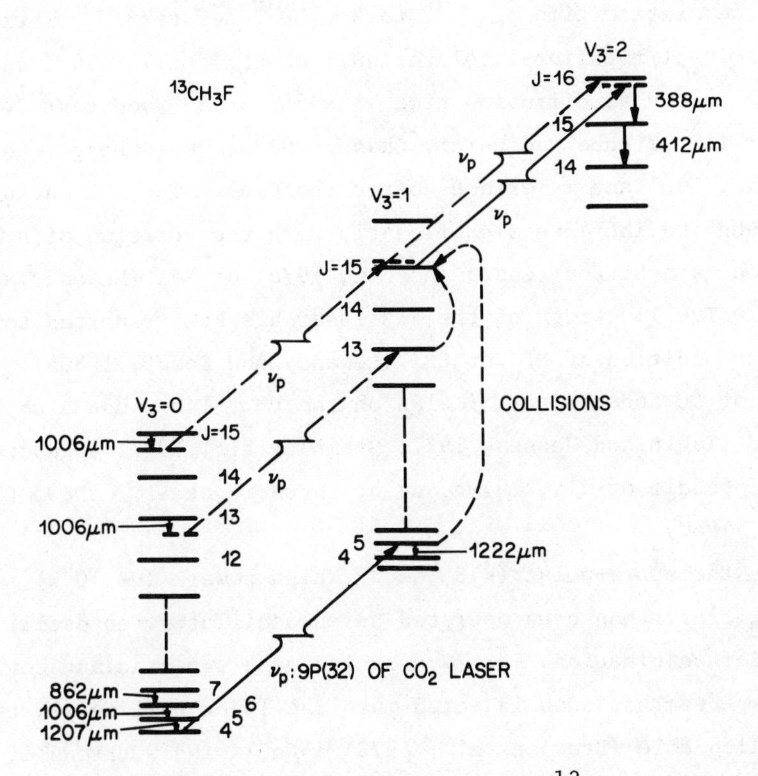

Figure 4. A partial energy-level diagram for $^{13}CH_3F$ showing optical pumping by a CO_2 laser tuned to the 9P(32) line. The main process leads to laser outputs at 1222 μm and its associated refill-cascade wavelengths. Various pumping pathways that many contribute to the generation of 388-μm and 412-μm signals in the $2\nu_3$ vibrational state are also illustrated.

Weiss, 1976; Chang, 1977; Koepf and Smith, 1978; Inguscio et al., 1979c; Walzer and Tacke, 1980). Some of the relaxation times relevant to the performance of the CH_3F laser are as follows:

rotational relaxation time $\tau_R \sim 10^{-8}$ sec.torr/p, vibrational-vibratioanl relaxation time $\tau_{VV} \sim 2$ to 9×10^{-6} sec.torr/p, vibrational-translational/rotational relaxation time $\tau_{VT} \sim 1.7 \times 10^{-3}$ sec.torr/p, and the molecular diffusion time $\tau_D \sim 5.6 \times 10^{-3}$ sec.torr^{-1}.cm^{-1}.pr, where r is the tube radius(cf. Chang, 1977a, Rosenmberger and DeTemple, 1981 and references cited therein). The FIR output has been found to increase significantly with the addition of a buffer gas such as n-Hexane (Chang and Lin, 1976) or SF_6 (Lawandy and Koepf, 1980b). The linewidth of the FIR laser has been reported to decrease with the addition of SF_6 or CS_2 (Lawandy and Koepf, 1980a). The effect of DC and RF Stark fields on the CH_3F laser has also been studied (Tobin and Jensen, 1977; Stein et al., 1977; Inguscio et al., 1977; Inguscio et al., 1979a and b; Inguscio et al., 1980; Rak and Dyubko, 1980).

In the short-pulse ($<10^{-6}$sec.), high-power ($10- 10^6$W) regime, the CH_3F laser has been operated as an oscillator, an oscillator-amplifier combination, or as a nonresonant superradiant device. The pump beam has been injected into the laser collinearly through a coupling hole (Brown et al., 1972) a NaCl window (Brown et al., 1973), a KCl window (Evans et al., 1975) a Ge Brewster window and a Si output etalon (Semet and Luhmann, 1976), a Ge input etalon (Brown et al., 1977), a NaCl Brewster window (Plant et al., 1974), or by reflection off a specially coated crystal-quartz plate (Drozdowicz et al., 1977). The pump beam has also been injected off axis at a very small angle (Sharp et al., 1975), transversely with cylindrical focusing (Brown et al., 1974), or at an intermediate angle into a parallel-plate light guide (Cohn et al., 1975; Allen et al., 1979). The FIR output coupler can be a hole in a mirror, a single mesh, a mesh Fabry-Perot, or a Mylar beam splitter. It can also be a Fox-Smith mode selector to obtain a narrow oscillation width (Evans et al., 1977; Drozdowicz et al., 1977). Although the conventional Fabry-Perot cavity is most commonly used in CH_3F oscillators, the advantanges of unstable resonators (Weber et al.,

1978; Ewanizky et al., 1979) and a distributed feedback waveguide (Affolter and Kneubuhl, 1981) for transverse and longitudinal mode purity have also been demonstrated. Self pulsing of a thightly coupled system of IR-FIR lasers due to passive Q-switching (Koepf, 1977; Dyrna and Tacke, 1979), and the gain and stability of a regenerative FIR amplifier (Galantowicz, 1977) have also been studied.

The understanding of pulsed CH_3F lasers has been gained through rate-equation analysis (Temkin and Cohn, 1976; Pichamuthu and Sinha, 1979), theoretical studies of the dynamic Stark effect (Panock and Temkin, 1977; Chang, 1977b; Drozdowicz et al., 1979a), and experimental verification of the gain spectra (Drozdowicz et al., 1979b). A detailed study of superradiance in CH_3F has also been reported (Rosenberger and DeTemple, 1981). Finally, the generation of subnanosecond FIR pulsed by synchronous, mode-locked pumping of a CH_3F laser has been demonstrated (Lee et al., 1979; Lemley and Nurmikko, 1979). A general review of pulsed optically pumped FIR lasers has been given by DeTemple (1979).

Suggestions for Further Investigation

The entire FIR region can be conveniently covered by quasi-continuously tunable (R-branch pumping) or step tunable (Q-Branch pumping) FIR lasers as discussed above. The spectral coverage can be made faily complete by using several different molecules. To make such a laser a practical laboratory tool, further development work on the high-pressure CO_2 laser is necessary. Further studies to control the spectral purity of the FIR output are also necessary.

References

Affolter, E. and Kneubuhl, F. K., 1981: IEEE J. Quantum Electron., QE-17, 1115-1122.

Allen, L., Dodel, G., and Magyar, G., 1979: Opt. Commun., 28 , 383-388.

Bava, E., DeMarchi, A., Godone, A., Benedetti, R., Inguscio, M.,

Minguzzi, P., Strumia, F., and Tonelli, M., 1977: Opt. Commun., 21 46 - 48.

Biron, D. G., Temkin, R. J., Lax, B., and Danly, B. G., 1979a: Opt. Lett., 4, 381-383.

Biron, D. G., Temkin, R. J., Lax, B., and Danly, B. G., 1979b: In "Digest of the Fourth International Conference on Infrared and Millimeter Waves and Their Applications", (S. Perkowitz, editor), pp. 213-214, IEEE Cat. No. 79 CH 1384-7 MTT.

Biron, D. G., Danly, B. G., Lax, B., and Temkin, R. J., 1980: J. Opt. Soc. Am., 70, 674-675.

Brown, F., Silver, E., Chase, C. E., Button, K. J., and Lax, B., 1972: IEEE J. Quantum Electron., QE-8, 499-500.

Brown, F., Horman, S. R., Palevsky, A., and Button, K. J., 1973: Opt. Commun., 9, 28-30.

Brown, F., Kronheim, S., and Silver, E., 1974: Appl. Phys. Lett., 25, 394-396.

Brown, F., Hislop, P. D., and Tarpinian, J. O., 1977: IEEE J.Quantum Electron., QE13, 445-446.

Chang, T. Y., Wang, C. H., and Cheo, P. K., 1969: Appl. Phys. Lett., 15, 157-159.

Chang, T. Y., and Bridges, T. J., 1970a: Opt. Commun., 1, 423-426.

Chang, T. Y., and Bridges, T. J., 1970b: In "Proc. Symp. Submillimeter Waves", (J. Fox, editor), pp. 93-98, Polytechnic Press, New York.

Chang, T. Y., and McGee, J. D., 1971: Appl. Phys. Lett., 19, 103-105.

Chang, T. Y., 1974: IEEE Trans. Microwave Theory Tech., MTT-22, 983-988.

Chang, T. Y. and Lin, C., 1976: J. Opt. Soc. Am., 66, 362-369.

Chang, T. Y., 1977a: In "Nonlinear Infrared Generation", (Y. R. Shen, editor), pp.215-272, Springer-Verlag, Berlin.

Chang, T. Y., 1977b: IEEE J. Quantum Electron., QE-13, 937-942.

Cohn, D. R., Fuse, T., Button, K. J., Lax, B., and Drozdowicz, Z., 1975: Appl. Phys. Lett., 27, 280-282.

Danielewicz, E. J., Plant, T. K., and DeTemple, T. A., 1975: Opt. Commun., 13, 366-369.

Danielewicz, E. J., and Weiss, C. O., 1978: Opt. Commun., 27, 98-100.

Davies, P. B., and Jones, H., 1980: Appl. Phys., 22, 53-55.

DeMartino, A., Frey, R., and Pradere, F., 1978: Opt. Commun., 27, 262-266.

DeTemple, T. A., and Danielewicz, E. J., 1976: IEEE J. Quantum Electron., QE-12, 40-47.

DeTemple, T. A., 1979: In "Infrared and Millimeter Waves", Vol. 1, (K. J. Button, editor), pp. 129-184, Academic Press, New York.

Drozdowicz, Z., Woskoboinikow, P., Isobe, K., Cohn, D. R., Temkin, R.J., Button, K. J., and Waldman, J., 1977: IEEE J. Quantum Electron, QE-13, 413-417.

Drozdowicz, Z., Temkin, R. J., and Lax, B., 1979a: IEEE J. Quantum Electron., QE-15, 170-178.

Drozdowicz, Z., Temkin, R. J., and Lax, B., 1979b: IEEE J. Quantum Electron., QE-15, 865-869.

Duxbury, G., and Herman, H., 1978: J. Phys. E. Sci. Instrum., 11, 1-2.

Dyrna, P., and Tacke, M., 1979: Appl. Phys. Lett., 35, 908-909.

Evans, D. E., Sharp, L.E., James, J. W., and Peebles, W. A., 1975: Appl. Phys. Lett., 26, 630-632.

Evans, D. E., Sharp, L.E., Peebles, W. A., and Taylor, G., 1977: IEEE J. Quantum Electron., QE-13, 54-58.

Ewanizky, T. F., Bayha, W. T., and Rohde, R. S., 1979: IEEE J. Quantum Electron., QE-15, 538-540.

Fetterman, H.R., Schlossberg, H.R., and Waldman, J., 1972: Opt. Commun., 6, 156-159.

Freund, S. M., Duxbury, G., Romheld, M., Tiedje, J. T., and Oka, T., 1974: J. Mol. Spectrosc., 52, 38-57.

Galantowicz, T. A., 1977: IEEE J. Quantum Electron, QE-13, 459-461.

Hacker, M. P., Drozdowicz, Z., Cohn, D. R., Isobe, K., and Temkin, R. J., 1976: Phys. Lett., 57A, 328-330.

Henningsen, J. O., and Jensen, H. G., 1975: IEEE J. Quantum Electron, QE-11, 248-252.

Hodges, D. T., Tucker, J. R., and Hartwick, T. S., 1976a: Infrared Phys., 16, 175-182.

Hodges, D. T., Foote, F. B., and Reel, R. D., 1976b: Appl. Phys. Lett., 29, 662-664.

Hodges, D. T., Foote, F. B., and Reel, R. D., 1977: IEEE J. Quantum Electron., QE-13, 491-494.

Inguscio, M., Minguzzi, P., and Tonelli, M., 1977: Opt. Commun., 21 208-210.

Inguscio, M., Strumia, F., Evenson, K. M., Jennings, D. A., Scalabrin, A., and Stein, S. R., 1979a: Opt. Lett., 4, 9-11.

Inguscio, M., Minguzzi, P., Moretti, A., Strumia, F., and Tonelli, M., 1979b: Appl. Phys., 18, 261-270.

Inguscio, M., Moretti, A., and Strumia, F., 1979c: Opt. Commun., 30, 355-360.

Inguscio, M., Moretti, A., Strumia, F., 1980: IEEE J. Quantum Electron., QE-16, 955-964.

Koepf, G. A., 1977: Appl. Phys. Lett., 31, 272-273.

Koepf, G. A., and Smith, K., 1978: IEEE J. Quantum Electron., QE-14, 333-338.

Kramer, G., and Weiss, C. O., 1976: Appl. Phys., 10, 187-188.

Lawandy, N. M., and Koepf, G. A., 1980a: J. Chem. Phys., 73, 1162-1164.

Lawandy, N. M., and Koepf, G. A., 1980b: Opt. Lett., 5, 336-338.

Lee, S. H., Petuchowski, S. J., Rosenberger, A.T., and DeTemple, T. A., 1979: Opt. Lett., 4, 6-8.

Lemley, W., and Nurmikko, A. V., 1979: Appl. Phys. Lett., 35, 33-35.

Lipton, K. S., and Nicholson, J. P., 1978: Opt. Commun., 24, 321-326.

Mathieu, P., and Izatt, J. R., 1981: Opt. Lett., 6, 369-371.

Osche, G. R., 1978: J. Opt. Soc. Am., 68, 1293-1298.

Panock, R. L., and Temkin, R. J., 1977: IEEE J. Quantum Electron., QE-13, 425-434.

Peebles, W. A., Brower, D.L., Luhmann, N. C., Jr., and Danielewicz, E. J., 1980: IEEE J. Quantum Electron., QE-16, 505-508.

Pichamuthu, J. P., and Sinha, U. N., 1979: IEEE J. Quantum Electron., QE-15, 501-505.

Plant, T. K., Newman, L. A., Danielewicz, E. J., DeTemple, T. A., and Coleman, P. D., 1974: IEEE Trans. Microwave Theory Tech. MTT-22, 988-990.

Rak, V. G., and Dyubko, S. F., 1980: Sov. J. Quantum Electron., 10, 703-708.

Rosenberger, A. T., and DeTemple, T. A., 1981: Phys. Rev. A, 24, 868-882.

Sattler, J. P., and Simonis, G. J., 1977: IEEE J. Quantum Electron., QE-13, 461-465.

Semet, A., and Luhmann, N. C., Jr., 1976: Appl. Phys. Lett., 28, 659-661.

Sharp, L. E., Peebles, W. A., James, B. W., and Evans, D. E., 1975: Opt. Commun., 14, 215-218.

Smith, W. L., and Mills, I. M., 1963: J. Mol. Spectrosco., 11, 11-38.

Stein, S. R., Risley, A. S., Van de Stadt, H., and Strumia, F., 1977: Appl. Opt., 16, 1893-1896.

Temkin, R. J., Cohn, D. R., Drozdowicz, Z., and Brown, F., 1975: Oot. Commun., 14, 314-317.

Temkin, R. J., and Cohn, D. R., 1976: Opt. Commun., 16, 213-217.

Temkin, R. J., Biron, D. G., Danly, B. G., and Lax, B., 1979: In "Digest of Fourth International Conference on Infrared and Millimeter Waves and Their Applications", (S. Perkowitz, editor), pp. 232-233, IEEE Cat. No. 79 CH 1384-7 MTT.

Tobin, M. S., and Jensen, R.E., 1977: IEEE J. Quantum Electron., QE-13, 481, 484.

Tucker, J. R., 1976: Opt. Commun., 16, 209-212.

Walzer, K., and Tacke, M., 1980: IEEE J. Quantum Electron., QE-16, 255,258.

Weber, B. A., Simonis, G. J., and Kulpa, S. M., 1978: Opt. Lett., 3, 229, 231.

Weiss, C. O., 1976: IEEE J. Quantum Electron., QE-12, 580-584.

Weitz, D. A., Skocpol, W. J., and Tinkham, M., 1978: Opt. Lett., 3, 13-15

Winton, R. S., and Gordy, W., 1970: Phys. Lett., 32A, 219-220.

SUBMILLIMETER LASER LINES IN DEUTERATED METHYL FLUORIDE, CD_3F

G. Duxbury

Department of Natural Philosophy
University of Strathclyde
107 Rottenrow, Glasgow G4 ONG, Scotland

INTRODUCTION

Although methyl fluoride CH_3F, was the first molecule to be used in a submillimeter optically pumped laser (1), it was not until comparatively recently that Tobin, Sattler and Wood (2,3) reported laser action in deuterated methyl fluoride, CD_3F. Among the lines they discovered two are amongst the strongest so far observed in lasers of this type.

The bands of CD_3F which lie in the 10μm region have been studied using laser Stark and interferometric spectroscopy by Duxbury et al. (4,5,6). Some near coincidences have also been studied using optical-optical double resonance by Duxbury and Kato (7). It is therefore possible to use the spectroscopic information in order to assign some of the observed emission lines.

Submillimetre laser lines

The laser lines were obtained using a grating tuned laser producing ca. 30 watts cw on the strongest 10μm P branch lines. For most of the reported lines the laser was operated in an electrically chopped mode, with a pulse length of greater than one millisecond, although the strongest lines operated continuously. The observed lines (2,3) are listed in Table 1. The wavelengths were measured using a metallic mesh Febry-Perot interferometer, and are estimated to be accurate to ± 0.5μm, by calibration using CH_3OH lines of known frequency. Some of the best coincidences, for example that with the 10 R[48] lines, occur near the extremes of the CO_2 laser tuning range where the available pump power is relatively small. One unusual feature of this laser is that the optimum operating pressures are much higher than those of CH_3F, and in fact are well above the pressure at which laser action ceases in $^{12}CH_3F$ and in $^{13}CH_3F$. The 206μm line still operates at pressures up to 750 mtorr, whereas the study of the pressure dependence of the power output from the 496μm line of CH_3F by de Temple and Danielwicz (8) showed that oscillation ceased at pressures of 75 mtorr, with the optimum operating pressure being ca. 35 mtorr. This has been attributed to the rather different energy level patterns in CH_3F and CD_3F, and is discussed more fully in the next section.

Spectroscopic measurements and assignments

The ν_3, ν_5 and ν_6 bands have been studied in detail by Duxbury et al (4,5,6), and the vibrational origins and rotational constants are given in Table 2. The analysis of these bands is complicated by the Coriolis interaction between the ν_3 and ν_6 states (5), between the ν_2 and ν_5 states (6), and possibly between the ν_3 and the ν_5 states. The lines in the ν_3 band appear to be unperturbed when studied by the laser Stark method (4), which favours the observation of transitions with low values of J and K. However, recent diode laser spectra obtained by Sattler (9) have shown that lines of this band with $J' > 13$ are obviously perturbed at $K = 7$ and 8, even though the band appears to be a simple parallel band at medium resolution. Oscilloscope traces of these spectra are shown in Figure 1.

TABLE 1

CD_3F laser emission lines in order of decreasing wavelength

wavelength $\lambda/\mu m$		CO_2 pumpline	relative polarisation	CO_2 mode [a]	rel.FIR signal [b]	pressure/ m torr
1485	(10)	10P46		C	W	120
1450	(20)	9R36		C	VW	110
384.7		10P28		C	W	150
368.4		10R48	⊥	C	W	310
349		9P34		C	W	200
336.6		19P50	⊥	C	W	130
323.3		10P8		C	W	120
265	(5)	9P52	∥	C	W	200
247.5		9R10	∥	CW	S	190 [c]
247.3		10P46	∥	C	W	120
206		9P16	∥	CW	VS	270 [c]
201.5		10P12	∥	CW	M	310
200		9P28	30° from ∥	C	W	310
172.8		10R6	30° from ∥	C	W	310
155.6		10P10	∥	C	W	310

(a) CW-true, C chopped mode of CO_2 laser with pulse > 1 ms

(b) VS > 1 mW S ≃ 1 mW M ~ 100 μW W ~ 1 to 50 μW WW < 1 μW

(c) corresponds to optimum pressure

The medium resolution study of the ν_6 band showed intensity perturbations due to the ν_3:ν_6 interaction (5), and a recent study of the ν_2 band has shown strong ν_2:ν_5 interactions (6). This latter study has been extended by Caldow and Halonen (10), who have shown that the only lines of the ν_5 band which are power saturated sufficiently to give Lamb dip Stark spectra are those in which there is a very large mixing of the levels of the ν_2 and the ν_5 states. Further light has been shed on the interactions

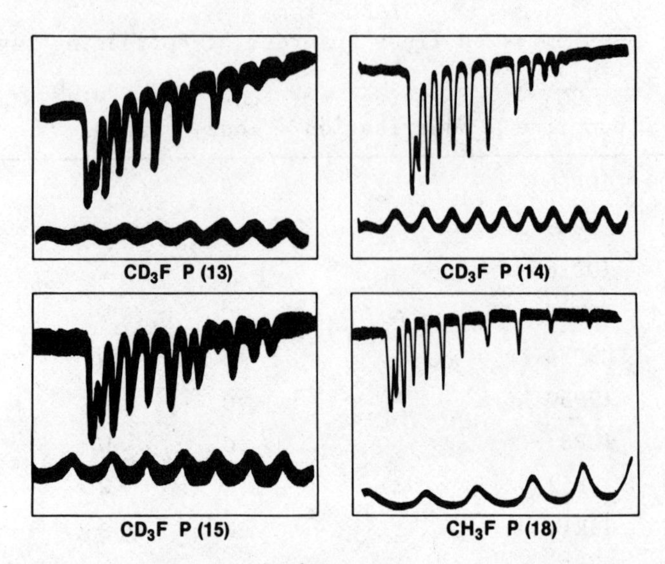

Figure 1
A comparison of diode laser spectra of perturbed P-branch transitions of CD_3F, and an analogous unperturbed transition in CH_3F. The lower calibration channel spectra were derived from Ge etalons of length 7.62 cm (CD_3F) and 2.54 cm (CH_3F).

by the recent work of Duxbury and Kato (7) which has enabled perturbed levels to be characterised by the possession of abnormal dipole moments. It therefore appears that Sattler's speculation about the high pressure operation of the CD_3F laser is plausible, namely that the complicated Coriolis perturbations which occur in this system of closely spaced vibration-rotation bands results in very efficient vibrational relaxation, and helps to remove the vibrational bottleneck effect.

Duxbury and Kato (7) also studied the near coincidences of the $^{13}C^{16}O_2$, $^{12}C^{18}O_2$ and $^{13}C^{18}O_2$ lasers with CD_3F. The resulting OODR spectra have been used to predict the wavelengths of lines pumpable with these lasers, which are summarised in Table 3.

Similar arguments to those given above apply to the discussion of the $^{13}CD_3F$ laser, which appears elsewhere in this volume. It is also probable that the use of waveguide lasers, as

TABLE 2

Molecular Constants of the ν_3, ν_5 and ν_6 bands of CD_3F in cm^{-1}

	Ground State	ν_3	ν_5	ν_6
ν_0	-	992.29882(19)	1072.35093(11)	911.49606(8)
A	2.561	2.559808(20)	2.54663(3)	2.56722
B	0.682134	0.6803298(26)	0.681137(4)	0.6762117(13)
$D_J \times 10^{-6}$	1.29 (2)	1.13(2)	*	0.96(2)
$D_{JK} \times 10^{-6}$	7.395 (4)	6.80(23)	*	10.80(69)
$DK-D_{K_o} \times 10^{-6}$	0	1.52(14)	*	-3.95(88)
$[A\xi_z]$			-0.31805(3)	0.61950(8)
$n_J \times 10^{-6}$			-4.130(7)	-7.1
$q \times 10^{-3}$			8.5804	4.7306
$\|r\| \times 10^{-3}$			1.96	

TABLE 3

Observed and calculated submillimetre laser emission lines in CD_3F, using OODR data (7)

Laser	Laser Line	Band	Emission Line $J'K''$	$J'K''$	λ observed (μm)	λ calculated (μm)
$^{12}C^{16}O_2$	10R48	ν_3	20,6	19,6	368.4	367
$^{12}C^{16}O_2$	9P46		22,5	21,5		334
$^{12}C^{16}O_2$	9P28		37,11	36.11	200	199
$^{13}C^{16}O_2$	9R24		33,9	32,9		223
$^{13}C^{16}O_2$	9R26		34,9	33,9		216
$^{13}C^{18}O_2$	9P18		16,12	15,12		459
$^{13}C^{18}O_2$	9P18		16,8	15,8		459
$^{13}C^{18}O_2$	9P18		16,1	15,1		459
$^{13}C^{18}O_2$	10P8	ν_6	19,6	18,6		389

described recently by Tobin (11), and of frequency shifted lasers, using efficient acousto-optic modulators, will considerably increase the number of lines of CD_3F available.

REFERENCES

1. T.Y. Chang and T.J. Bridges, "Laser action at 452, 496 and 541μm in optically pumped CH_3F", Opt. Commun. Vol.1, pp.423-426 (1970).

2. M.S. Tobin, J.P. Sattler and G.L. Wood, "Optically pumped CD_3F submillimetre-wave laser", Opt. Lett. 4, pp.384-386 (1979).

3. M.S. Tobin, J.P. Sattler and G.L. Wood, "CD_3F optically pumped near millimetre laser", in Conf. Dig. of 4th Int. Conf. Infrared and Millimetre Waves, IEEE Cat. 79C4 1384-7 MTT, 1979, pp.209-210.

4. G. Duxbury, S.M. Freund and J.W.C. Johns, "Stark spectroscopy with the CO_2 laser; the ν_3 band of CD_3F", J. Mol. Spectrosc. 62, 99-108 (1976).

5. G. Duxbury and S.M. Freund, "Stark spectroscopy, with the CO_2 laser the ν_6 band of CD_3F", J. Mol. Spectrosc. 67, 219-243 (1977).

6. G. Duxbury and G.L. Caldow, "Stark spectroscopy with the CO_2 laser the ν_5 band of CD_3F", J. Mol. Spectrosc. 89, 93-106 (1981).

7. G. Duxbury and H. Kato, "Optical optical double resonance spectra of CH_3F and CD_3F using isotopic CO_2 lasers", Chemical Phys. 66, 161-167 (1982).

8. T.A. De Temple and E.J. Danielewicz, "Continuous-wave CH_3F waveguide laser at 496μm: theory and experiment", IEEE J. of Quantum Electron. QE12, 40-47 (1976).

9. J.P. Sattler, (private communication).

10. G.L. Caldow and L. Halonen, "The spectra and analysis of the ν_2 and ν_5 infrared bands of $^{12}CD_3F$", J. Mol. Phys. 46, 223-237 (1982).

11. M.S. Tobin, "cw Submillimeter wave laser pumped by an rf-excited CO_2 waveguide laser", in Proc. Soc. Photo-Opt. Inst. Eng. 259, 13-17 (1980).

FIR LASER LINES OPTICALLY PUMPED IN METHYL CHLORIDE, $CH_3{}^{35}Cl$ AND $CH_3{}^{37}Cl$

J - C. Deroche and G. Graner

Laboratoire d'Infrarouge - Laboratoire Associé au CNRS -
Université de Paris-Sud, Bâtiment 350
91405 Orsay Cédex, France

Experimental Conditions

Optical pumping of CH_3Cl has been reported under several operating conditions. Fetterman et al.(1972) obtained four FIR laser transitions by pumping methyl chloride with the TEA CO_2 laser. Under these conditions, transitions can be pumped even if the coincidence is not perfect (this is called "off resonant optical pumping", or in short, OROP). Although this provides a large number of pumping effects, OROP experiments are difficult to use for the analysis of CH_3Cl itself since the separation between CO_2 and CH_3Cl transitions may be as high as 1 GHz.

Landman et al.(1969) and Meyer et al.(1973) have analyzed the coincidence CO_2, P(26)-$CH_3{}^{35}Cl$, ${}^RQ_3(6)$ and observed the hyperfine quadrupole structure of the ${}^RQ_3(6)$ transition. Such a study is only possible when one gets rid of the Doppler effect. The study of this coincidence is greatly facilitated by the small frequency separation between the transitions : here, only 17 MHz between CO_2 and the farthermost hyperfine component of ${}^RQ_3(6)$.

Wagner and Zelano (1973) used a pulsed CO_2 laser giving 150 watts in 180 μsec pulses. The CO_2 laser was focused on the coupling hole of a 2-m FIR cavity. A scanning Fabry-Perot was used to measure FIR wavelengths. Four FIR emissions were published but with a wavelength uncertainty of several μm. Plant et al. (1974) observed also similar coincidences with a CW CO_2 laser.

Jennings et al. (1975) reported CW FIR emissions, their FIR laser was of the waveguide design and the published wavelengths may be in error by 10 μm for some lines.

Chang and Mc. Gee (1976) using a CO_2 laser delivering 150 μs pulses of up to 200 W peak power at 120 Hz, measured 20 $CH_3C\ell$ FIR emissions with an accuracy of 0.1 μm. Relative polarization, offset (CO_2 line/absorbing transition), threshold were also determined.

Manita (1979) used a CO_2 laser for which the maximum power per pulse was 10 kW and the pulse duration 0.5-3 μsec. The output power of the observed FIR emissions was measured and found to be in a range 350-1800 watt for $CH_3C\ell$.

Spectroscopy

Methyl-chloride is a symmetric-top molecule with 6 fundamental vibrational modes :

- three A_1 modes, giving parallel fundamentals ν_1, ν_2 and ν_3 at 2968, 1355 and 732 cm^{-1} respectively,

- three E modes, giving perpendicular fundamental ν_4, ν_5 and ν_6 at 3039, 1455 and 1018 cm^{-1} respectively.

The ν_6 perpendicular band is the only fundamental in the CO_2/N_2O laser region.

Using only Chang's measurements (Chang, 1976), Deroche determined the spectroscopic parameters of ν_6 band for both isotopic species $^{37}C\ell$ and $^{35}C\ell$ (Deroche, 1978). But since that time we obtained in Laboratoire d'Infrarouge an absorption IR spectrum of $CH_3C\ell$ in the ν_6 region. It was then possible to improve the accuracy of molecular parameters, because transitions were observed for a wider range of quantum numbers values. We got informations from $^RR_6$ to $^PP_{11}$ branches in $CH_3{}^{35}C\ell$ with J up to 42, and from $^RR_6$ to $^PP_3$ with J up to 51 in $CH_3{}^{37}C\ell$. A simultaneous fit was performed on FIR transitions, IR absorption lines and pumped frequencies determined by Chang and Mc. Gee (1976) taking into account the offset with CO_2 laser lines. The fit used the following energy formula :

$$E(J,K,\ell) = \nu + B\,J(J+1) + (A-B)K^2 - 2A\zeta K\ell + \eta_J\,J(J+1)K\ell + \eta_K\,K^3\ell - D_J\,J^2(J+1)^2 - D_{JK}J(J+1)K^2 - D_K K^4$$

and the ℓ-doubling coupling term was set to :

$$\langle J,\ K=-1,\ \ell=-1 | H_2 | J, K=+1, \ell=1 \rangle = \frac{q}{2}\,J(J+1)$$

For ground state parameters we used A_o values given by Di Lauro (1976) B_o, D_J^o, D_{JK}^o from Sullivan et Frenkel (1971) for $^{35}C\ell$ and from Bensari et al.(1977) for $^{37}C\ell$ respectively. D_K^o was taken from a force field calculation by Duncan (1976). Some ν_6 excited state parameters were also fixed according to Shimizu's values (Shimuzu, 1975) obtained from laser Stark spectroscopy. Several of these spectroscopic constants have been determined by MW spectroscopy, or FIR Fourier Transform spectroscopy (Kraitchman, 1954, Orville-Thomas, 1954, Imachi, 1976, Pesenti, 1974, Sakai, 1978), by Raman diffusion (Jensen, 1981), IR Fourier Transform Spectroscopy (di Lauro, 1980) or spectroscopy in a molecular beam (Dubrulle, 1977) in the ground state. But none of them were fully satisfactory for our data both on $^{35}C\ell$ and $^{37}C\ell$.

All derived parameters are gathered in Table I. With these constants all coincidences are reproduced within 15 MHz for $^{35}C\ell$ as well as for $^{37}C\ell$.

Results

FIR emissions with their assignments are given in Table II. A great number of these emissions have been identified, but some of them remain unexplained. The 281.67 μm line pumped by 9R(14) is a cascade from the 275 μm emission. Some pumped transitions, $^{R}P_{10}(41)$

TABLE I

Molecular Parameters of ν_6 Band of Methyl Chloride

	$CH_3{}^{35}C\ell$ cm^{-1}		$CH_3{}^{37}C\ell$ cm^{-1}	
ν_6	1018.0692(3)		1017.6861(3)	
$A_o - A_6$	− 0.025375(5)		−0.025508(30)	
$B_o - B_6$	0.0016332(3)		0.0016117(2)	
$A\zeta_6$	1.30996(5)		1.30742(7)	
$D_J^o - D_J^6$	$-6.01\ 10^{-10}$	e	$-6.01\ 10^{-10}$	f
$D_{JK}^o - D_{JK}^6$	$-1.57(5)\ 10^{-7}$		$-1.60(9)\ 10^{-7}$	
$D_K^o - D_K^6$	$-2.65\ 10^{-6}$	e	$-2.65\ 10^{-6}$	f
η_J	$1.646(4)\ 10^{-5}$		$1.580(7)\ 10^{-5}$	
η_K	$1.087(12)\ 10^{-4}$		$1.065(16)\ 10^{-4}$	
$q_6/2$	$2.4075(230)\ 10^{-4}$		$2.4075\ 10^{-4}$	f
A_o	5.20536	a	5.20536	a
B_o	0.44340238	b	0.4365730	d
D_J^o	$6.004\ 10^{-7}$	b	$5.860\ 10^{-7}$	d
D_{JK}^o	$6.6212\ 10^{-6}$	b	$6.410\ 10^{-6}$	d
D_K^o	$8.342\ 10^{-5}$	b	$8.353\ 10^{-5}$	c

a - fixed to di Lauro (1976)

b - fixed to Sullivan (1971)

c - fixed to Duncan (1976)

d - fixed to Bensari-Zizi (1977)

e - fixed to Shimizu (1975)

f - fixed to $CH_3{}^{35}C\ell$ value.

TABLE II

						PROPOSED ASSIGNMENTS				IR TRANSITIONS		FIR TRANSITIONS			
CO_2 LASER LINE	FIR WAVELENGTH (μm)	OFFSET (MHz)	POLARIZATION	THRESHOLD PUMP POWER (W)	MAX OUT/IN PRESSURE (mW/W ; mTorr)	ISOTOPE	GROUND LEVEL (J,K)	UPPER LASING LEVEL (J,K)ν	LOWER LASING LEVEL (J,K)ν	OBSERVED (cm^{-1})	CALCULATED (cm^{-1})	OBSERVED (cm^{-1})	CALCULATED (cm^{-1})	REFERENCES	COMMENTS
9R(42)	461.20	-25	⊥	2	0.027/11;140	37	(25,10)	(25,11)ν_6	(24,11)ν_6	1091.0294	1091.0294	21.6825	21.6804	(a), (b)	
9R(38)	275.09	10	‖	6	0.05 /60;310	37	(41,5)	(42,6) ν_6	(41,6)ν_6	1089.0014	1089.0019	36.3517	36.3512	(a)	
9R(16)	378.57	50	⊥	15	0.18 /29;225	35	(30,8)	(30,9) ν_6	(29,9) ν_6	1075.9894	1075.9893	26.4150	26.4173	(a), (b), (c)	
9R(14)	275.00	45	‖	4	1.7 /44;310	37	(41,3)	(42,4) ν_6	(41,4) ν_6	1074.6480	1074.6475	36.3636	36.3598	(a), (b)	
	281.67	45	‖			37		(41,4) ν_6	(40,4) ν_6	1074.6480	1074.6475	35.5025	35.5021	(a)	Cascade from 275μ line
9R(12)	943.97	-30	‖	2	13 /45;100	35	(11,6)	(12,7) ν_6	(11,7) ν_6	1073.2775	1073.2777	10.593	10.5931	(d), (a)	
9R(2)	236.25	-10	⊥	2	0.1 /13;310	37	(49,7)	(49,8) ν_6	(48,8) ν_6	1066.0371	1066.0370	42.3280	42.3216	(a), (b)	
9P(12)	273.7					37	(41,10)	(42,10) GS	(41,10) GS	1053.9235	1053.8667*	36.536		(c)	OROP *
9P(26)	1886.87	20	⊥	14	1.6 /57;100	35	(6,3)	(6,4) ν_6	(5,4) ν_6	1041.2797	1041.2791	5.2998	5.3002	(a)	
9P(30)	250.4					37		(46,) GS	(45,) GS	1037.4341		39.936*		(c)	OROP, GS*
9P(38)	958.25	-50	‖	16	0.52 /40;100	37	(13,3)	(12,4) ν_6	(11,4) ν_6	1029.4405	1029.4405	10.4356	10.434	(a)	
9P(42)	333.96	40	⊥	1.8	44 /47;225	35	(34,1)	(34,2) ν_6	(33,2) ν_6	1025.2992	1025.2991	29.944	29.9469	(a), (b)	
9P(48)	227.15	50	‖	5	0.072/26;140	37	(50,6)	(51,5) ν_6	(50,5) ν_6	1018.9023	1018.9022	44.023	44.0296	(a)	
9P(52)	870.80	30	‖	5	1.3 /16;49	35	(14,1)	(13,2) ν_6	(12,2) ν_6	1014.5189	1014.5195	11.484	11.4809	(a)	
10R(52)	511.90	0	⊥	4	0.065/17;93					993.3764		19.535		(a)	
10R(34)	286.79	-35	⊥	22	0.3 /69;225					984.3820		34.8687		(a)	
10R(26)	568.81	5	‖	70	0.022/97;120					979.7054		17.5805		(a)	
10R(24)	397.6					37	(30,2)	(29,1) ν_6	(28,1) ν_6	978.4723	978.4891*	25.151		(c), (b)	OROP *
10R(18)	349.34	- 5	‖	3	7.4 /89;180	37	(34,2)	(33,1) ν_6	(32,1) ν_6	974.6217	974.6217	28.6254	28.6213	(d), (a), (b)	
10P(4)	230.									957.8005		43.47		(e)	
10P(10)	240.98	-30	‖	12	1.3 /90;260	37	(49,3)	(48,2) ν_6	(47,2) ν_6	952.8798	952.8800	41.4972	41.4907	(d), (a), (b)	
10P(20)	271.29	-35	⊥	11	0.33 /86;460	35	(42,11)	(42,10)ν_6	(41,10)ν_6	944.1928	944.1928	36.861	36.8596	(e), (a)	
10P(34)	261.03	-20	‖	20	0.069/106;310					931.0008		38.3097		(e), (a)	
10P(38,40)	240.									927.0083 / 924.9739		41.66		(e)	uncertainty on CO_2 line
11P(19)	307.65	-50	‖	2	0.57 /23;225	35	(38,11)	(37,10)	(36,10)ν_6	911.3580	911.3580	32.504	32.5068	(a)	(01'1)-(11'0) band of CO_2

a) CHANG AND MC. GEE (1976)
b) MANITA (1979)
c) FETTERMAN ET AL. (1972)
d) JENNINGS ET AL. (1975)
e) WAGNER AND ZELANO (1973)

[9P(12)] and $^{P}P_2(30)$ [10R(24)] are rather far from CO_2 emissions since they are pumped through off resonant optical pumping (OROP). It is to be noted that the 275.09 μm emission is actually pumped by 9R(38) and not 9R(36) as stated by Chang (1976), Gallagher (1977) and Rosenbluh (1976). The remaining unassigned lines take place in excited states and are pumped through hot band transitions but it was not possible to ascertain the states involved, because of the lack of information on excited state as $v_3=v_6=1$ for example. Some lines which have been previously published are not included in Table II as for example those of Jennings (1975) : 968μ[9R(12)], 354μ[10R(18)] and 254μ[10P(10)] which actually correspond to 943μ[9R(12)], 349.34μ[10R(18)] and 240.98μ[10P(10]respectively.

Suggestions for further studies

It is to be noted that no frequency measurements have been done on $CH_3C\ell$ SMM emissions and we have only wavelength measurements. Comparison between observed and calculated FIR transitions shows the necessity to improve FIR data by true frequency measurements for methyl chloride.

This would allow the prediction of more accurate frequencies in the ν_6 band to help finding more possibilities of optical pumping in $CH_3C\ell$ (de Temple,1978).

References

Bensari-Zizi, N., Guelachvili, G., Alamichel, C., 1977 : "Etude de la bande $\nu_2 + \nu_3$ de $CH_3C\ell$ en résonance de Coriolis avec $\nu_3 + \nu_5$," Mol. Phys. 34, 1131.

Chang, T.Y., Mc. Gee, J.D., 1976 : "Millimeter and submillimeter-wave laser action in symmetric top molecules optically pumped via perpendicular absorption bands," IEEE Trans. QE-12, 62.

Deroche, J.C., 1978 : "Assignment of submillimeter laser lines in methyl chloride", J. Mol. Spectrosc. 69, 19.

De Temple, T.A., Lawton, S.A., 1978 : "The identification of candidate transitions for optically pumped far infrared lasers : methyl halides and D_2O," IEEE QE-14, 762.

Di Lauro, C., Guelachvili G., Alamichel, C., 1976 : "Etude des bandes infrarouges en résonance $\nu_2 + \nu_6^{\pm 1}, \nu_5^{\mp 1} + \nu_6^{\mp 1}$ et $\nu_5^{\mp 1} + \nu_6^{\pm 1}$ de $CH_3C\ell$," J. Phys. Paris 37, 355.

Di Lauro, C., Alamichel, C., 1980 : "Rotational analysis of the $\nu_2 + \nu_6^{\pm 1}$, $\nu_5^{\pm 1} + \nu_6^{\pm 1}$, $\nu_5^{\pm 1} + \nu_6^{\mp 1}$ and $2\nu_3 + \nu_6^{\pm 1}$ interacting infrared bands of methyl chloride," J. Mol. Spectrosc. 81, 390-412.

Dubrulle, A., Boucher, D., Burie, J., Demaison, J., 1977 : "The high resolution rotational spectrum of methyl chloride. The spin-rotation and nuclear magnetic shielding tensors," Chem. Phys. Lett. 45, 559.

Duncan, J.L., 1976 :"The centrifugal distortion constant D_K of symmetric top molecules,"J. Mol. Spectrosc. 60, 225.

Fetterman, H.R., Schlossberg H.R., Waldman, J., 1972 :"Submillimeter lasers optically pumped off resonance," Opt. Commun 6, 156.

Gallagher, J.J., Blue, M.D., Bean, B., Perkowitz, S., 1977 :"Tabulation of optically pumped far infrared laser lines and applications to atmospheric transmission," Infrared Physics, 17, 43.

Imachi, M., Tanaka, T., Hirota, E., 1976 : "Microwave spectrum of methyl chloride in the excited vibrational states : Coriolis interaction between the ν_2 and ν_5 states," J. Mol. Spectrosc. 63, 265.

Jennings, D.A., Evenson, K.M., Jimenez, J.J., 1975 : "New CO_2 pumped CW Far-infrared laser lines," IEEE Trans. QE-11, 637.

Jensen, P., Brodersen, S., Guelachvili, G., 1981 : "Determination of A_o for $CH_3{}^{35}C\ell$ and $CH_3{}^{37}C\ell$ from the ν_4 infrared and Raman bands," J. Mol.Spectrosc. 88, 378-393.

Kraitchman, J., Dailey, B.P., 1954 : "Variation in the quadrupole coupling constant with vibrational state in the methyl halides," J. Chem. Phys. 22, 1477.

Landman, A., Marantz, H., Early, V., 1969 : "Light modulation by means of the Stark effect in molecular gases. Application to CO_2 lasers," Appl. Phys. Lett. 15, 357.

Manita, O.F., 1979 : "An optically-pumped CH_3Br and $CH_3C\ell$ submillimeter-wave laser," Izv. Vuz. Radioelektron 22, 83.

Meyer, T.W., Brilando, J.F., Rhodes, C.K., 1973 : "Observation of quadrupole hyperfine structure in the ν_6 ${}^RQ_3(6)$ transition of ${}^{12}CH_3{}^{35}C\ell$," Chem. Phys. Lett. 18, 382.

Orville-Thomas, W.J., Cox, J.T., Gordy, W., 1954 : "Millimeter wave spectra and centrifugal stretching constants of the methyl halides,"J. Chem. Phys. 22 1718.

Pesenti, J., Sergent-Rozey, M., 1974 : "Etude par Transformation de Fourier de CH_3Br et CH_3Cl dans l'infrarouge très lointain," Appl. Opt. 13, 1158.

Plant, T., Newman, L. Danielewicz, E., De Temple, T. Coleman, P., 1974 : "High power optically pumped far infrared lasers," IEEE Trans. Microwave Theory Tech. 22, 988.

Rosenbluh, M., Temkin, R.J., Button K.J., 1976 : "Submillimeter laser wavelength tables," Applied Optics 15, 2635.

Sakai, K., Ichimuka, K., Masumoto, H., Kitagawa, Y., 1978 : "High-resolution spectroscopy of some gaseous molecules with a submillimeter Fourier transform spectrometer," Infrared Physics 18, 577-583.

Shimizu, F., 1975 : "Laser stark spectroscopy of $CH_3{}^{35}Cl$ ν_6 band," J. Phys. Soc.Japan 38, 1106.

Sullivan, T.L., Frenkel, L., 1979 : "Measurement of fourth order distortion constants in symmetric top molecules," J. Mol. Spectrosc. 39, 185.

Wagner, R.J., Zelano, A.J., Ngai, L.H., 1973 : "New submillimeter laser lines in optically pumped gas molecules," Opt. Commun. 8, 46.

FAR-INFRARED LASER LINES OBTAINED BY OPTICAL PUMPING OF THE $CD_3C\ell$ MOLECULE

Georges Graner and Jean-Claude Deroche

Laboratoire d'Infrarouge
Université de Paris-Sud
Bâtiment 350
91405 Orsay Cédex France

A. Optically pumped lines

Only two works are known on optically pumped emissions in $CD_3C\ell$.

In 1975, Dyubko et al. published a list of 16 coincidences with the CO_2 laser, giving rise to 19 FIR laser lines. The more precisely measured of these laser lines are reported also in Dyubko and Fesenko (1978).

In 1978, Duxbury and Herman found a new coincidence with the 9P28 transition of CO_2 producing a line at 791 μm. They also published their assignments for Dyubko's FIR laser lines.

Experimental details

The experimental apparatus used in Dyubko et al. (1975) has been described in Dybko et al., 1972, except for the frequency measurements. The CO_2 laser, operating in the CW regime, is 1.3 m long and yields 4 to 10 Watts. The resonant cavity of the submillimeter laser is 1,2 m long and has a diameter of 56 mm. It can operate both with continuous flow of the gas an in the "sealed-off" regime. The submillimeter radiation is recorded with a point-contact detection of beryllium bronze paired with InSb and with a pyroelectric detector.

The output power was measured with a calorimeter according to Dybko et al. (1972), but Dyubko et al. (1975) only give this power in relative units, from 1 to 250. Duxbury and Herman (1978) tried to improve these figures and state that "the Golay acts as a crude powermeter for the submillimeter output." They determined that 1 unit (Dybko et al., 1975) was equivement to 0.016 mW. We give in Table I their estimation of this power, pointing out 3 laser lines where the proportionality to Dyubko's values is not respected.

The FIR frequencies have been accurately measured to within 2×10^{-6} for 12 laser lines (Dybko et al., 1975, 1978), which means an uncertainty of 1 MHz at 500 GHz. For the remaining 7 laser lines (Dybbo et al., 1975), the wavelength has been obtained from cavity scan to within 2×10^{-3} and it is probably the same for the 791 μm line reported in Duxbury and Herman (1978).

The polarization of the FIR radiation relative to the polarization of the CO_2 laser beam, as measured with a one-dimensional wire grating, is given for all lines measured by Dyubko et al. (1975).

Little is known about the offset between the absorption $CD_3C\ell$ transition and the pumping CO_2 laser. Nevertheless, Duxbury and Herman (1978) mention an offset of approximately -35 MHz for 9P28 and +15 MHz for 9P10. This lack of information is specially unfortunate for the three occurrences (9P6, 9P36, 10R20) when the same CO_2 line gives two submillimeter transitions. It would have been important to know whether they are cascades or accidental coincidences of distinct absorptions.

B. Spectroscopic information

The problem for $CD_3C\ell$ is somewhat similar to the problem for CD_3Br although we have here much less spectroscopic information.

TABLE I - FIR LASER LINES IN $CD_3C\ell$

CO_2 LASER LINE	OFF-SET (MHz) (a)	FIR WAVELENGTH (μm) (b)	FIR FREQUENCY (MHz) (c)	POLAR-IZA-TION (c)	FIR POWER (mW) (d)	PROPOSED ASSIGMENT				COMMENTS
						Species	GROUND LEVEL (J,K)	UPPER LASING LEVEL	LOWER LASING LEVEL	
9R34	-	(383.2846)	782166.7	//	0.96	35	(35,1)	(36,0) ν_5	(35,0) ν_5	
9R28	-	224	-	//	0.016	-	-	-	-	J≈61 or 62
9P6	-	(698.5554)	429160.6	⊥	0.096	37	(20,1)	(20,1) ν_5	(19,0) ν_5	CASCADE
9P6	-	(735.1298)	407808.9	⊥	0.032	37		(19,0) ν_5	(18,0) ν_5	CASCADE
9P10	+ 15	(443.2646)	676328.5	//	2.4-4	35	(30,7)	(31,7) ν_2	(30,7) ν_2	
9P12	-	(1239.480)	241869.6	//	0.24	35	(10,3)	(11,2) ν_5	(10,2) ν_5	
9P14	-	(1990.757)	150592.2	//	0.48	37	(8,1)	(7,0) ν_5	(6,0) ν_5	
9P16	-	288	-	//	0.048	-	-	-	-	J≈47 or 48
9P24	-	(293.6480)	1020924.7	⊥	4-0.48	-	-	-	-	J≈46 or 47
9P28	- 35	791	-		(2.4)	35	(17,3)	(18,3) ν_2	(17,3) ν_2	
9P32	-	245	-	⊥	0.080	-	-	-	-	J≈57
9P34	-	(883.598)	339286.0	//	0.48	-	-	-	-	J≈15 cannot belong to or ν_5 of $CD_3^{35}C\ell$[2] or $CD_3^{37}C\ell$

TABLE I - FIR LASER LINES IN $CD_3C\ell$ (Continued)

CO_2 LASER LINE	OFF-SET (MHz) (a)	FIR WAVELENGTH (µm) (b)	FIR FREQUENCY (MHz) (c)	POLAR-IZA-TION (c)	FIR POWER (mW) (d)	PROPOSED ASSIGMENT				COMMENTS
						Species	GROUND LEVEL (J.K.)	UPPER LASING LEVEL	LOWER LASING LEVEL	
9P36	-	(4803102)	624164.3	⊥	0.24	-	-	-	-	J≃28 } cannot belong to ν_2 or ν_5 of $CD_3{}^{35}C\ell$
9P36	-	(519.3032)	577297.5	//	0.080-0.096	-	-	-	-	J≃26 } cannot belong to ν_2 or ν_5 of $CD_3{}^{35}C\ell$
9P38	-	249	-	⊥	0.016	-	-	-	-	J≃55 or 56
10R28	-	318	-	//	0.016	-	-	-	-	J≃43 or 44
10R20	-	(449.7998)	666502.0	//	1.28	35	(32,10)	$(31,10)\nu_2$	$(30,10)\nu_2$	} CASCADE
10R20	-	(464.7568)	645052.4	//	0.64	35		$(30,10)\nu_2$	$(29,10)\nu_2$	} CASCADE
10R18	-	288	-	//	0.016	-	-	-	-	J≃47 or 48
10R14	-	246	-	//	0.080	-	-	-	-	J≃56

(a) From G. Duxbury and H. Herman (1978)

(b) Values within brackets are computed from the frequencies

(c) From S. F.Dyubko, et al., 1975, S.F.Dyubko and L.D. Fesenko, 1978.

(d) FIR powers have been computed by Duxbury and Herman (1978) by multiplying arbitrary units from S.F. Dyubko, et al.,(1975) by 0.016 mW. When two values are given, the first one is obtained this way, the second one comes from (G. Duxbury and H. Herman, 1978). The value between brackets has been measured in G. Duxbury and H. Herman 1978.

Deuterated methyl-chloride is a symmetric-top with six fundamental modes:

- three A_1 modes, giving parallel fundamentals ν_1, ν_2 and ν_3 at 2160, 1029 and 701 cm^{-1}, respectively,
- three E modes, giving perpendicular fundamentals ν_4, ν_5 and ν_6 at 2283, 1060 and 769 cm^{-1}, respectively.

These values are given for the main isotopic variety $^{12}CD_3{}^{35}C\ell$, which represents 75% in natural abundance. The corresponding values for the $^{12}CD_3{}^{37}C\ell$ variety (24%) are very similar, except for ν_3, located 7 cm^{-1} lower. The ^{13}C varieties which account for 1% are barely known.

For $CD_3C\ell$ as for CD_3,Br, the ν_2 and ν_5 bands are the only ones located in the CO_2/N_2O laser region. The last infrared spectrum recorded in this region dates back to 1966 (Jones et al. 1966).

The resolving power was about 0.5 cm^{-1} so that for ν_2 only P and R "lines" without K structure were analyzed. For ν_5, although P and R lines are clearly visible (see Fig. 2 of Jones et al., 1966), only Q branches were analyzed. It was of course impossible at that time to resolve the isotopic structure.

As in the case of $CD_3C\ell$, the important fact is the Coriolis-type interaction between ν_2 and ν_5 which was pointed out and analyzed by di Lauro and Mills (1966), using the spectra from Jones et al. (1966).

Recently two papers giving accurate data were published by Kyusku University group (Imachi et al, 1976; Yamada and Hirota, 1977). Imachi et al. (1976) give a few microwave transitions in the $v_2 = 1$ and $v_5 = 1$ excited states, both for $CD_3{}^{35}C\ell$ and $CD_3{}^{37}C\ell$ up to $J = 3 \leftarrow 2$ as well as direct ℓ-type doubling transitions (i.e. $k\ell = 1 \leftarrow k\ell = 1$) up to $J = 20$. Yamada and Hirota (1977) deal with the Laser Stark spectrum of $CD_3{}^{35}C\ell$ only from which 75 zero field transition frequencies are deduced. These authors are thus able to derive a set of molecular constants for $CD_3{}^{35}C\ell$ including band centers for ν_2 or ν_5. Nevertheless, the relatively low values used for J' (maximum 10) and the K' (maximum 4) has the consequence that the distortion constants published (Yamada and Hirota, 1977) are difficult to trust, specially for extrapolations.

As for $CD_3{}^{37}C\ell$, no vibration-rotation transition is available so that little can be said about the band centers and not much more about other molecular constants.

Duxbury and Herman (1978) give a set of molecular constants both for $CD_3{}^{35}C\ell$ and for $CD_3{}^{37}C\ell$. But these constants depend strongly on their assignment of the FIR emission lines. We have therefore decided to derive our own constants and our own assignments.

C. The $CD_3{}^{35}C\ell$ molecule

Since no high-resolution spectrum of this molecule was available, we decided first to fit the Laser Stark and microwave data (Imachi et al, 1976; Yamada and Hirota, 1977) alone. The following diagonal energy formula was used (with ζ, η_J and η_K equated to zero in ν_2)

$$E(J,k,\ell) = \nu+BJ(J+1)+(A-B)k\ -2A\zeta k\ell+n_J J(J+1)k\ell+\eta_K k^3\ell-D_J J^2(J+1)^2$$
$$-D_{JK}J(J=1)K^2\ =\ D_K k^4$$

The main coupling was the Coriolis interaction

$$\langle v_2 = 1,\ J,\ k|H_1|\ v_5 = 1,\ \ell_5 = \pm 1,\ J,\ K\pm 1\rangle$$
$$= [\pm W_1+W_2(2k\pm 1) + W_3 J(J+1)\pm W_4(2k\pm 1)^2][J(J+1)\ -\ k(k\pm 1)]^{\frac{1}{2}}$$

and the $\ell(2,2)$ type coupling term was written as

$$<J,k,\ell=-1|H_2|J,k+2,\ell=+1> = \tfrac{1}{2}[q + q'J(J=1) + q''J^2(J+1)^2][J(J+1)-k(k+1)]^{\frac{1}{2}}$$
$$[J(J+1) - (k+1)(k+2)]^{\frac{1}{2}}$$

In the first step, we could fit the available data for $J \leq 10$ with only 8 parameters namely: ν_2, A_2, B_2, ν_5, A_5, B_5, $A\zeta_5$ and the coupling term W_1, the distortion constants being constrained to their ground state values.

In this problem, there is a crossing of the Coriolis interacting levels (v_2 = 1,k) and (v_5 = 1, ℓ_5 = -1, k-1) between k = 5 and 6. Therefore, the next logical step would have been to free the second Coriolis term w_2 as was done for CD_3Br. Unfortunately, the available sample contains no transition whatsoever with k larger than 5. On the contrary, there is wealth of information for low k value and specially for $k\ell$ = 1. With such a biased sample, it is no wonder that we were unable to determine a value for W_2. On the reverse, when fitting all data up to J = 20, we were progressively compelled to introduce q, q' and q" as was done by Yamada and Hirota (1977).

In table II, we give the set of eleven constants which reproduce nicely all 94 data (Imachi et al., 1976; Yamada and Hirota, 1977) as well as three new informations: the coincidence of the 9P28 laser line with $^QR_3(17)$, the coincidence of 9P12 with $^PR_3(10)$ and the corresponding F.I.R. laser line $(\nu_5,11,2)\rightarrow(\nu_5,10,2)$.

The fact that this set of constants was not fully satisfactory was clearly revealed when we tried to fit new data (F.I.R. lines and CO_2 laser coincidences) with higher J and k values such as $^QR_7(30)$ or $^PP_{10}(32)$. With 7 new data which could not be seriously contested, we obtained the second set of constants of Table II, where η_J and η_K are now free. Note that the overall quality of the fit seriously deteriorated when going from J < 20 to J < 36 and that no other constant was found really significant.

Extrapolation to new data with J = 45-48 and higher was attempted and abandoned. We feel extremely dangerous to assign

TABLE II

MOLECULAR CONSTANTS USED FOR $CD_3{}^{35}C\ell$

(all in cm^{-1})

GROUND STATE CONSTANTS [(a)]

A_o = 2.618178

B_o = 0.3616485

D_J = 4.27×10^{-7}

D_{DK} = 3.469×10^{-6}

D_K = 3.186×10^{-5}

UPPER STATE CONTANTS

	Set I	Set II
Maximum J Value	20	36
ν_5	1059.9697	1059.9658
A" - A'	15.401×10^{-3}	15.010×10^{-3}
B" - B'	-6.014×10^{-4}	-6.622×10^{-4}
$A\zeta_5$	-0.83672	-0.83836
η_J	-	1.33×10^{-5}
η_K	-	-2.90×10^{-4}
ν_2	1028.6728	1028.6773
A" - A'	-9.145×10^{3}	-7.517×10^{-3}
B" - B'	1.205×10^{3}	1.158×10^{-3}
W_1^{Cor}	0.29199	0.29213
q	3.18×10^{-4}	2.84×10^{-4}
q'	9.04×10^{-7}	11.44×10^{-7}
q"	-2.66×10^{-10}	-6.88×10^{-10}

(a) These ground state constants are taken from M. Imachi et al., 1976. Better values of $A_o = 2.6136 \pm 0.0005$ and $D_K = (1.99 \pm 0.25) \times 10^{-5}$ are found in S. Brodersen, J. Mol. Spectrosc. 71, 312-320 (1978).

new F.I.R. lines of CO_2 laser coincidences on the basis of risky extrapolations and fit this supposed new information with ad hoc constants. It is likely that several of the assignments given in Dyubko et al. (1978) for these high J values are correct but we cannot endorse them at the present time.

D. The $CD_3{}^{37}Cl$ molecule

The situation is much worse here. We have at our disposal only 20 microwave data (Imachi et al., 1976) with one k = 2 value, fifteen k = 1 values and four k = 0 values. This gives of course no information whatsoever on the ν_2 and ν_5 band centers. By fixing these band centers to reasonable values, and most molecular constants to the $CD_3{}^{35}Cl$ values, one can succeed to fit these microwave data.

This allows us to confirm the assignments (Duxbury and Herman, 1978) concerning the F.I.R. lines at 429, 408 and 151 GHz. Therefore we can safely assume that the 9P6 and 9P14 CO_2 lines coincide with ${}^PQ_1(20)$ and ${}^PP_1(8)$, respectively. This gives us some information about the band centers, especially about ν_5.

Fitting this meager sample is not an easy task. If most constants are constrained to the values of $CD^{35}Cl$, the few which are free take quite different values. This suggests, as was already said, that the parameters obtained for $CD^{35}Cl$ are not the true ones and do not express correctly the Coriolis resonance. We shall not give here any molecular parameters for this species.

Extrapolation to high J is still more risky for $CD_3{}^{37}Cl$ as for the main isotopic species and was not attempted.

E. Conclusions and suggestions

Out of the 20 F.I.R. lines found we can safely assign only 6 to the ${}^{35}Cl$ species and 3 to the ${}^{37}Cl$ species. For one transition at 339 GHz, we are sure that it does not occur in the v_2 = 1 or v_5 = 1 states of these two species and for two transitions at 480 and 519 μm, we are sure that it does <u>not</u> occur to $v_2 = 1$ or $v_5 = 1$ of $CD_3{}^{35}C\ell$. For the other 8 lines, we cannot be affirmative although we are convinced that a good proportion of them belong to these two fundamental states.

This conclusion shows clearly that there is a dramatic need for a high resolution IR spectrum of $CD_3C\ell$ which would certainly yield much better molecular parameters and therefore allow further F.I.R. lines assignments.

References

di Lauro, C. and Mills, I.M., 1966: J. Mol. Spectrosc. 21, 386-413.

Duxbury, G. and Herman, H., 1978: J. Phys. B (G.B.) 11, 935-949.

Dyubko, S.F., Svich, V.A. and Fesenko, L.D., 1972: ZheTF Pis. Red. 16, 592-594, English translation in JETP Letters 16, 418-419.

Dyubko, S.F., Fesenko, L.D., Baskakov, O.I. and Svich, V.A., 1975: Zh. Prikl. Spektr. 23, 317-320.

Dyubko, S.F. and Fesenko, L.D., 1978: 3rd International Conference on Submillimeter Waves and Their Applications, Guilford, Univ. of Surrey, March 1978.

Imachi, M., Tanaka, T. and Hirota, E., 1976: J. Mol. Spectrosc. 63, 265-280.

Jones, E.W., Popplewell, R.J.L. and Thompson, H.W., 1966: Spectrochim. Acta 22, 659-667.

Yamada, C. and Hirota, E., 1977: J. Mol. Spectrosc. 64, 31-46.

FAR-INFRARED LASER LINES OBTAINED BY OPTICAL PUMPING OF CF_3Br

J-M. Lourtioz

Institut d'Electronique Fondamentale
Université Paris XI
Bât. 220, 91405 Orsay, France

C. Meyer

Laboratoire d'Infrarouge
Université Paris XI
Bât. 350, 91405 Orsay, France

Introduction

The CF_3Br molecule is the heaviest molecule which has been optically pumped by a CO_2 laser in CW operation (Knight, 1980). Due to its molecular weight (m = 148 a.m.u for the $CF_3{}^{79}Br$ isotope), it presents the main following features : i) an infrared absorption spectrum of very high density, ii) an infrared Doppler absorption linewidth less than 40 MHz, iii) rotational transitions at long wavelengths ($\lambda \gtrsim 1$ mm).

The first detailed analysis of the CF_3Br infrared spectrum near 1100 cm^{-1} has been achieved by Burczyk et al.(1979). The spectrum was recorded on the SISAM spectrometer of the Laboratoire d'Infrarouge d'Orsay (Meyer, 1979). From this study, many CF_3Br transitions were expected to coincide with the different emission lines of the R branch of the 9.4 μm $^{12}C^{16}O_2$ laser.

FIR laser action of the CF_3Br molecule by optical pumping with a $^{12}C^{16}O_2$ laser was first observed in 1979 (Pontnau, 1979). Eight new

FIR transitions have been observed more recently from this molecule owing to the optimization of the optical pumping set-up (Lourtioz, 1981). The FIR wavelengths are determined with a 10^{-2} - 10^{-3} accuracy. However due to the complex rovibrational structure of the CF_3Br infrared spectrum, we just can propose tentative assignments for the FIR lines observed.

II. Spectroscopic data for the CF_3Br molecule :

II.1. Vibrational constants of the CF_3Br molecule - infrared absorption bands in the region 1000 - 1150 cm^{-1} :

The CF_3Br molecule, commercially labelled as freon 13 B1, presents two main isotopic varieties i.e. $^{12}CF_3{}^{79}Br$ and $^{12}CF_3{}^{81}Br$ in equivalent abundance. This prolate symmetric-top molecule belongs to the C_{3v} group. It exhibits six fundamental vibrational modes. Three of them (ν_1, ν_2, ν_3) have the A_1-type symmetry, the others (ν_4, ν_5, ν_6) having the E-type symmetry.

Table I gives the vibrational constants of $CF_3{}^{79}Br$ and $CF_3{}^{81}Br$ which are deduced from Raman and infrared spectroscopy (Edgell,1955, Burczyk,1981). The energy of a given vibrational level may be calculated from the well-known relation (Townes,1955) :

$$\frac{E_v}{hc} = \sum_i \omega_i \left(v_i + \frac{d_i}{2}\right) + \sum_i \sum_j x_{ij} \left(v_i + \frac{d_i}{2}\right) \left(v_j + \frac{d_j}{2}\right) \qquad (1)$$

Expression (1) allows to know all the vibrational bands (fundamental bands, hot bands, overtone - and combination - bands) which lie in the region between 1100 cm^{-1} and 1000 cm^{-1}.

The central frequencies of the main vibrational bands are given in Table II. Most of these bands have the symmetry A_1. The $2\nu_5$ overtone-band which is composed of two allowed sub-bands of different symmetries (A_1+ E) is the exception. However, as generally observed for prolate symmetric top molecules, the A_1-type subband is the most intense and the contribution of the E-type subband may be neglected. In this condition, the selection rules for

Table I
Available Vibrational Constants of $CF_3{}^{79}Br$ and $CF_3{}^{81}Br$

	$CF_3{}^{79}Br$		$CF_3{}^{81}Br$	
ω_1 (A_1)	1084.763	cm^{-1}	1084.521	cm^{-1}
ω_2 (A_1)	762.140	-	761.960	-
ω_3 (A_1) *	352.40	-	352.40	-
ω_4 (E) *	1208.80	-	1208.80	-
ω_5 (E) *	547.37	-	547.37	-
ω_6 (E)	392.66	-	392.66	-
x_{16}	- 1237	-	-1237	-
x_{26}	- 0.83	-	- 0.81	-
x_{25}	0.73	-	0.69	-
x_{23}	5.50	-	5.57	-

Stars (*) labels the constants that are averaged for the two isotopic species.

the different rovibrational transitions of CF_3Br in the spectral range of interest are given by :

$$\Delta J = 0, \quad \pm 1 \quad , \quad \Delta K = 0 \qquad (2)$$

Then all the absorption bands listed In table II must exhibit a typical (PQR) envelope.

Only three of them may be clearly detected. As an illustration, Fig.1, shows a rovibrational spectrum of natural CF_3Br in the region 1050 cm^{-1} - 1150 cm^{-1} which has been recently recorded with the SISAM spectrometer at the Laboratoire d'Infrarouge d'Orsay. The most intense peaks are respectively related to the Q-branches of the ν_1, $(\nu_1 + \nu_6 - \nu_6)$, and $(\nu_2 + \nu_3)$ rovibrational bands of

Table II

Main Vibrational Bands of $CF_3{}^{79}Br$ and $CF_3{}^{81}Br$ in the 1100-1000 cm^{-1} Spectral range.

Vibrational bands	central frequency (cm^{-1}) $CF_3{}^{79}Br$	$CF_3{}^{81}Br$	Relative intensity
$3\nu_3$	1048.2	unknown	-
$\nu_1 + 2\nu_6 - 2\nu_6$	1082.29	1082.042	0.17
$\nu_1 + \nu_6 - \nu_6$	1083.52	1083.284	0.46
$\nu_1 + \nu_3 - \nu_3$	-	-	0.19
$\nu_1 + \nu_5 - \nu_5$	-	-	0.15
ν_1	1084.76	1084.5207	1
$2\nu_5//$	1095.63	unknown	-
$\nu_2 + \nu_3$	1120.07	unknown	-

Only the bands ν_1, $(\nu_1 + \nu_6 - \nu_6)$, $(\nu_1 + 2\nu_6 - 2\nu_6)$ are presently known with a good precision.

$CF_3{}^{79}Br$ and $CF_3{}^{81}Br$(1). A smaller peak located at $\simeq$ 1095 cm^{-1} is suspected to be associated with a Q-branch of the $2\nu_5$ band which has not been analyzed at the present time. In contrast, the $(\nu_1+2\nu_6-2\nu_6)$ hot band, the $3\nu_3$ overtone band and the $(\nu_1+n\nu_3-n\nu_3)$ hot bands are not observed(2). Besides, some of the absorption

(1) As indicated in Table II, the central frequency of a given band changes only slightly with the isotope. The corresponding Q-branches generally appear as a doublet in the spectrum of fig.1.

(2) A first theoretical estimation indicates that the $3\nu_3$ band is very weak. In the same way, the $(\nu_1+\nu_3-\nu_3)$ hot band is expected to be two times less intense than the $(\nu_1+\nu_6-\nu_6)$ hot band, the ν_6 vibrational mode being doubly generate.

bands related to the $^{13}CF_3{}^{79}Br$ - $^{13}CF_3{}^{81}Br$ isotopes which are in a weaker abundance and which we have previously ignored should be also present in the spectrum of Fig.1[(3)].

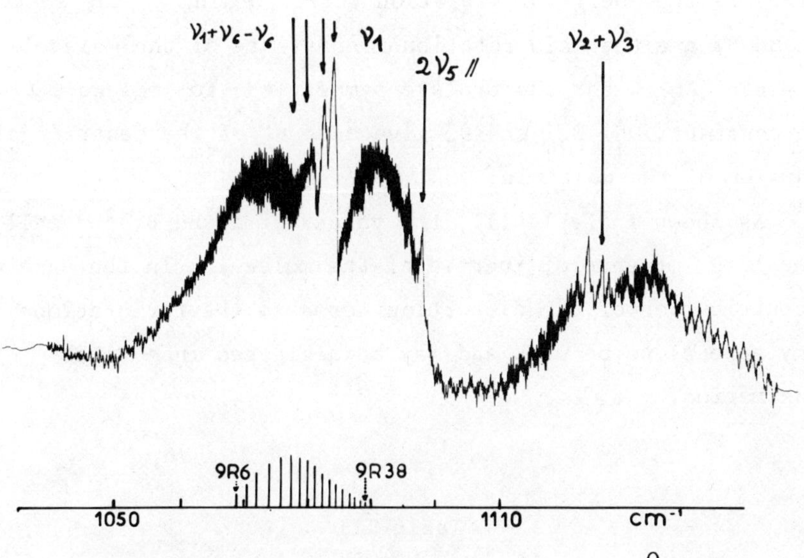

FIGURE 1. The Synthesized Spectrum of $\nu_1 {}^QP_K(89)$

II.2. Rotational constants of the CF_3Br molecule - rotational structure of the IR absorption bands :

Table III gives the rotational constants of $^{12}CF_3{}^{79}Br$ and $^{12}CF_3{}^{81}Br$ respectively for the vibrational ground state and the first excited state ν_1. It must be noted that these constants have been accurately determined from microwave spectroscopy (Sheridan, 1952, Jones, 1976).

The energy of a given rovibrational state (i,K,J) may be written as:

---

(3) According to a recent study of Lombard, the central frequency of the ν_1 fundamental band of $^{13}CF_3{}^{79}Br$ is approximately located at 1058.2 cm^{-1}.

$$\frac{E_{i,K,J}}{hc} = \frac{E_i}{hc} + (A_i - B_i)\,K^2 + B_i\,J(J+1) - D_J^i\,J^2(J+1)^2 - D_{JK}^i\,K^2J^2(J+1)^2 - D_K^i\,K^4 \ldots \quad (3)$$

(E_i/hc) is the energy of vibration corresponding to the state i. (A_i and B_i are the main rotational constants of the molecule in this state (A > B for the prolate symmetric - top molecule CF_3Br). the constants D_J^i, D_{JK}^i and D_K^i give account of the centrifugal distorsion of the molecule.

As shown in Table III, the values of A and B are small due to the large moments of inertia of the molecule. In the same way, the contribution of the distortion terms to the rovibrational energy appears to be weak and may be neglected in a first approximation.

Table III

Available Rotational Constants of CF_3Br

	$CF_3{}^{79}Br$	$CF_3{}^{81}Br$
A_o (cm^{-1})	$190{,}6\ 10^{-3}$	$190{,}6\ 10^{-3}$
B_o (cm^{-1})	$69{,}984\ 10^{-3}$	$69{,}984\ 10^{-3}$
* α_1^A (cm^{-1})	$0{,}44\ 10^{-3}$	$0{,}44\ 10^{-3}$
** α_1^B (cm^{-1})	$0{,}218\ 10^{-3}$	$0{,}217\ 10^{-3}$
α_6^B (cm^{-1})	$0{,}120\ 10^{-3}$	-
D_J (cm^{-1})	0	0
D_{JK} (cm^{-1})	$0{,}42\ 10^{-6}$	-
eqQ (MHz)	619	517

* $\alpha^A = A' - A_o$

** $\alpha^B = B' - B_o$

The wavenumber of a given rovibrational transition may be calculated from the relation (3) :

$$\sigma = \sigma_o + (A' - A'' - B' + B'')K^2 + (B' - B'') J (J+1) + \ldots + \ldots\ldots$$

for Q-branch transitions ($\Delta J = 0$) (4a)

$$\sigma = \sigma_o + (A' - A'' - B' + B'')K^2 + (B' + B'') m + (B' - B'') m^2 + \ldots + \ldots\ldots$$

with $m = J + 1$ for a R-branch transition ($\Delta J = +1$)

$m = -J$ for a P-branch transition ($\Delta J = -1$) (4b)

In relations (4a) and (4b), σ_o is the wavenumber related to the band center, the prime and second index on the constants A and B correspond respectively to the upper and lower states of the transition.

For a detailed analysis of a given branch, it is convenient to gather the transitions ($J'' \rightarrow J'$) related to the same value of J". This set of transitions $\Delta K = 0$ will be labelled as a K-structure (for instance K varies from 0 to J' for a P branch). The remarkably low value of (A-B) leads to very compact K-structures. In contrast, due to the small B-value, the K-structures related to neighbouring J"- values generally overlap.

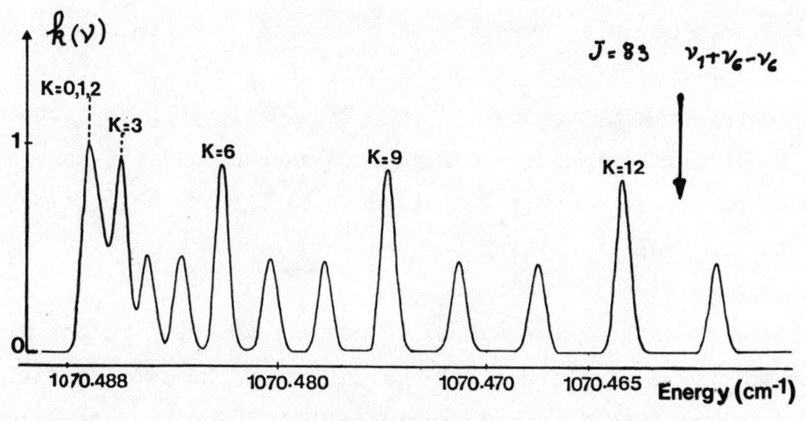

FIGURE 2.

Figure 2 shows the synthetized spectrum corresponding to different $\nu_1\,{}^QP_K(89)$ transitions of $CF_3{}^{81}Br$ with K varying from 0 to 13. The total width of this K-structure (K varying from 0 to 88) is 1.47 cm^{-1}. The transitions with K = 3N, N being an integer, are the most intense. Their intensities I(K,J") vary as following :

$$I(K,J'')\,\alpha\,\frac{(J''+1)^2 - K^2}{J''+1} \times (2J''+1)\exp\left(-\frac{hc}{kT}(B''J''(J''+1)+(A''-B'')K^2)\right) \quad (5)$$

The first term into brackets in expression (5) represents the overall rotational contribution to the oscillator strength of the transition. The second term gives the J",K- dependence of the population of the lower level of the transition. The line intensity slowly decreases with K. For instance, the transitions which correspond respectively to K = 3, K = 48 and K = 87 are in the ratio 1 : 0.25 : 0.001.

If we only take into account the absorption lines with relative intensity larger than 0.25, the so-reduced K-structure ($K = 0 \rightarrow 48$) covers a spectral range of 0.50 cm^{-1}. On the other band, for a given K value the neighbouring J-components of the P-branch are separated by 0.15 cm^{-1}. This emphasizes the overlapping of three consecutive K-structures.

Moreover K-structures which do not belong to the same rovibrational band may also overlap. In the example of Fig. 2, we have indicated the position of the $(\nu_1+\nu_6-\nu_6)\,{}^QP_{K=0}(83)$ transition of $CF_3{}^{81}Br$ within the K-structure $\nu_1\,{}^QP_K(89)$. Unlike the spectrum of fig.2, an integrate spectrum taking into account all the possible K-structures should exhibit a continuous background absorption with a few intense peaks resulting from the exact coincidence of several transitions.

From the discussion above, we may already realize the difficulty to have a complete (J,K)-assignment of the two transitions involved in the IR-FIR pumping cycle of CF_3Br. Besides, an exact assignment of the transitions also needs to take into account the hyperfine structure due to the nuclear spin of bromine. The nuclear spin being I = 3/2, each rotational level splits in four hyperfine levels corresponding respectively to :

$$F = J + 3/2 \quad , \quad F = J + 1/2 \; , \quad F = J - 1/2 \quad , \; F = J - 3/2$$

where F = I + J is the total angular momentum.

The energy of each level becomes :

$$\frac{E}{hc} = \frac{E_{iJK}}{hc} + \Delta W \qquad (6)$$

where ΔW is given by Townes (1955).

The hyperfine structure of the CF_3Br rotational transitions has not been analyzed at the present time and it will be ignored for the IR-FIR assignments. Nevertheless, owing to the high value of the hyperfine constant, i.e. eqQ ≃600 MHz (see table III), it is reasonable to think that the ΔF-splittings are comparable with the IR Doppler linewidth ($\Delta\nu_D \simeq 40$ MHz).

III. Experimental results - Tentative assignments of the IR-FIR transitions :

III.1. Optical pumping set-up - Main features of the CF_3Br laser :

Most of the optical pumping set-up has been described (Lourtioz and Adde, 1979),(Lourtioz, 1980).

- The 1.70 m long CO_2 laser is capable to deliver 20-40 Watts on most of its emission lines. Recently the use of the 171 l/mm grating blazed at 9.3 μm has allowed to increase notably the CO_2 power delivered on the R branch at 9 μm. This high-power level together with a good mode purity and a good frequency stability have been found to be critical for the observation of the weak FIR CF_3Br lines.

- The FIR cavity is a cylindrical metallic waveguide resonator (1.80 m long, 30 mm inner diameter) with external mirrors placed at a few millimeter from the waveguide ends. A single 1 mm - hole in one mirror allows both input-coupling of the pump and output-coupling of the FIR emitted radiation. The FIR beam expands rapidly due to the small size of the coupling hole and is extracted with a 45° off axis mirror having 3 mm diameter hole and 5-10 cm distant from the input mirror (see Fig. 1 of Lourtioz, 1980). This compact high Q resonator is easily operated at long wavelength. For instance, we report in section III.2 the observation of several CF_3Br lines near 2 mm, which are among the longest wavelengths emitted from an FIR optically pumped laser (Knight, 1980). In contrast, the main disadvantage of the low-loss metallic waveguides is that the polarization of the FIR output is not directly related to the pump-beam polarization. This lack of information may be detrimental to the attribution of a given FIR emission line (Chang, 1970).

Figures 3A and 3B emphasize some important features of the CF_3Br laser as compared to other CW optically pumped lasers with lighter molecules (i.e. the 118 μm CH_3OH laser). Figure 3A shows the simultaneous recordings of the 9R8 CO_2 pump power, the CF_3Br absorption and the 824 μm CF_3Br laser power as functions of the CO_2-frequency tuning. Similar results obtained for the 118 μm CH_3OH laser are shown in Fig.3B and are comparable to the results published elsewhere (Inguscio, 1979, Inguscio, 1980).

In both cases, the FIR cavity is tuned at the mode center and the FIR gas absorption is detected with a microphone placed inside the FIR cavity (Busse, 1977). Whereas the CH_3OH absorption curve of Fig. 3B is well related to the Doppler profile of a single absorption line (≃80 MHz Doppler full-width), the CF_3Br absorption curve (Fig.3A) reveals several peaks of different intensities. The CF_3Br laser signal corresponds to the absorption peak of highest intensity. This illustrates in another way the discussion of section II.2 concerning the high density of the rovibrational spectrum

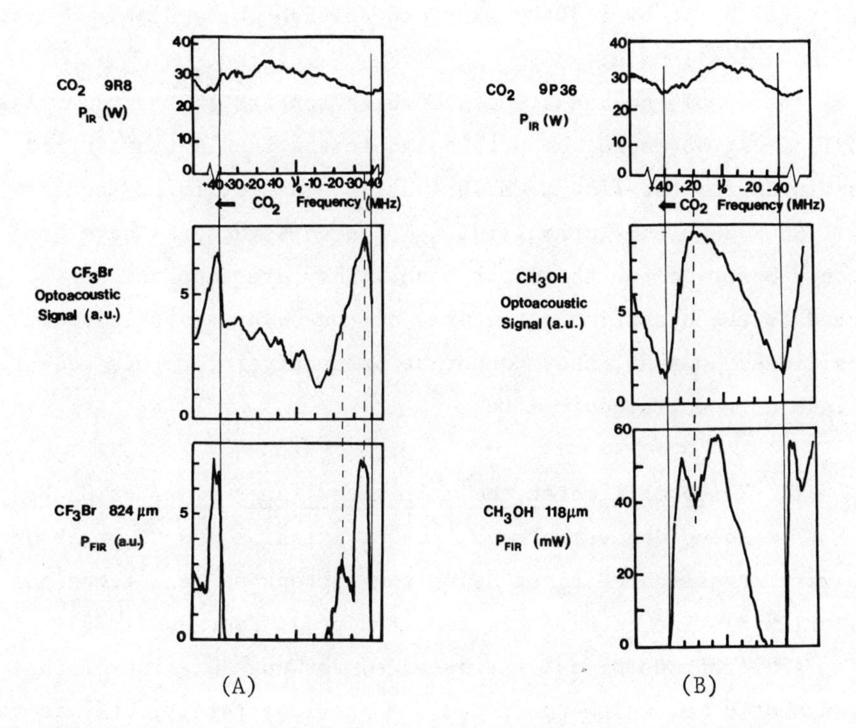

FIGURE 3. Absorbtion features in spectrophone recordings in CF_3Br and CH_3OH

of CF_3Br. Besides due to the small Doppler linewidth of the CF_3Br absorption lines ($\Delta\nu_D \simeq 40$ MHz), the FIR signal is only detected within a narrow CO_2 tuning range (<20 MHz). Thus, the FIR amplitude is very sensitive to small CO_2 frequency detunings and the CF_3Br laser generally exhibits a poor amplitude stability. As another consequence, in the search of CF_3Br lines it has been found convenient to use the following porcedure : the whole CO_2 frequency profile is PZT swept by a 30 Hz sawtooth voltage while slowly scanning the FIR cavity length.

The double-peaked structure which appears in the CF_3Br laser signal of Fig.3B cannot be attributed to the IR-FIR transferred Lamp-dip as for the 118 μm CH_3OH laser (Inguscio, 1979, Inguscio, 1980). Such double - (or multiple -) peaked structures have been commonly observed for the CF_3Br laser. They are suspected to be related to the hyperfine structures of the CF_3Br rovibrational transitions. However, they cannot be presently interpreted due to the lack of spectroscopic data.

III.2. Submillimeter and millimeter laser lines of CF_3Br

The coincidences between $^{12}C^{16}O_2$ emission and CF_3Br absorption were investigated first using the microphone cell technique (Walzer,1978). The results are reported by J. Pontnau (1979).

A microphone signal was detected for any CO_2 line of the R-branch at 9 μm, which confirmed the previous results of infrared spectroscopy (see section II.1). Other coincidences were also found with several CO_2 lines of P-branch at 9 μm.

Up to now, all the FIR emission lines of CF_3Br have been obtained by pumping with the CO_2 lines of the R-branch at 9 μm. No attempt has been done with the P-branch lines at 9 μm since no spectroscopic data is presently available to assign them in this frequency domain.

The ten laser lines observed using natural CF_3Br are summarized in table IV. Most of them are weak lines ($P_{FIR} < 50$ μW).

Table IV CF_3Br Laser Results

CO_2 pump	FIR wavelength measured (μm)	FIR intensity	Optimum Pressure (mTorr)	IR absorption line center (cm^{-1})	J 79/81 calculated		IR absorption transition (K=0) (calculated)	FIR wavelength calculated μm
9R8	823.5 ± 1.5	S	70	1070. 4623	86/87	B	1070.487 ± 0.05	822
9R10	883 ± 2.5	M	50	1071. 8837	80/81	A	1071.841 ± 0.05	884
9R12	1043 ± 12	W	50	1073. 2784	67/68	B	1073.600 ± 0.005	1033
9R16	1526 ± 10	W	50	1075. 9878	46/46	D	1076.147 ± 0.008	1509
9R20	1895 ± 10	W	50	1078. 5906	37/38	B	1078.793 ± 0.005	1903
9R28	1083 ± 4	M	40	1083. 4787	65/66	B	1083.532 ± 0.005	1080
9R34	1151 ± 5	M	50	1086.8697	61/62			
9R34	2140	VW	50	1086.8697	32/33			
9R38	1556 ± 4	M	50	1089. 0011	45/46	C	1089.163 ± 0.005	1560
9R40	1687 ± 5	M	50	1090. 0283	41/42	B	1090.070 ± 0.005	1682

Power scale
- S ≃ 200 μW
- M ≃ 50 μW
- W ≃ 10 μW
- VW ≃ 1 μW

Labels of the bands
- A ν_1 band of $CF_3{}^{79}Br$
- B band of $CF_3{}^{81}Br$
- C $(\nu_1+\nu_6-\nu_6)$ band of $CF_3{}^{79}Br$
- D $(\nu_1+\nu_6-\nu_6)$ band of $CF_3{}^{81}Br$

The line at 823 μm is the exception ($P_{FIR} \simeq 0.1$ mW). The long wavelength of these lines and the simultaneous coincidence of the pump with many absorption transitions are probably the main reasons of the poor optical pumping efficiency. However, it must be noted that no effort has been made to optimize the FIR cavity structure (coupling scheme, waveguide sizes...).

The FIR wavelengths have been measured with an external Fabry-Perot interferometer equipped with two silicon plates. The whole experimental set-up for wavelength measurement has been described by Lourtioz (1980). The displacement of the FP interferometer is calibrated using the CO_2-pump wavelength as a proper etalion. Care is taken to limit the diffraction effects due to the strong divergence of FIR beams. In the experiments, the FIR output beam is nearly collimated using a teflon lens with the focus point close to the virtual FIR source (i.e. the 1 mm coupling hole of the FIR cavity). The accuracy of this type of measurement is typically better than 10^{-3}, which is an order of magnitude better as compared to the conventional FIR wavelength measurements (Lourtioz, 1980). However, in the case of CF_3Br the experimental performances were limited due both to the weakness and the instability of the FIR signals. The FIR wavelengths reported in table IV (column 2) are given with a precision in the range 10^{-2}- 10^{-3}. This precision is still better than that obtained in many FIR wavelength measurements.

Besides, the direct measurement of the FIR frequency using heterodyne methods (Chang, 1970) seems to be only possible for the strong 824 μm line.

III.3. <u>Tentative assignments of the IR-FIR transitions</u> :

As shown in section II.1, only the ν_1, $(\nu_1+ \nu_6- \nu_6)$, $(\nu_1+2\nu_6-2\nu_6)$ rovibrational bands of the two isotopes $CF_3{}^{79}Br$ and $CF_3{}^{81}Br$ are presently known. Thus, the IR-FIR transitions may be only assigned within these bands. The symmetric-top program which has been developed at the Laboratoire d'Infrarouge d'Orsay allows to calculate the transitions with a precision of $\simeq 0.005\ cm^{-1}$ if $J < 80$ and a precision of $0.05\ cm^{-1}$ for higher J values.

Figure 4 recalls the scheme of the pumping cycle together with the labels of the IR and FIR transitions[(4)]. The J value is related to λ_{FIR} through the simple equation :

$$(J + 1) = \frac{1}{\lambda_{FIR}} \times \frac{1}{2B} \qquad (7)$$

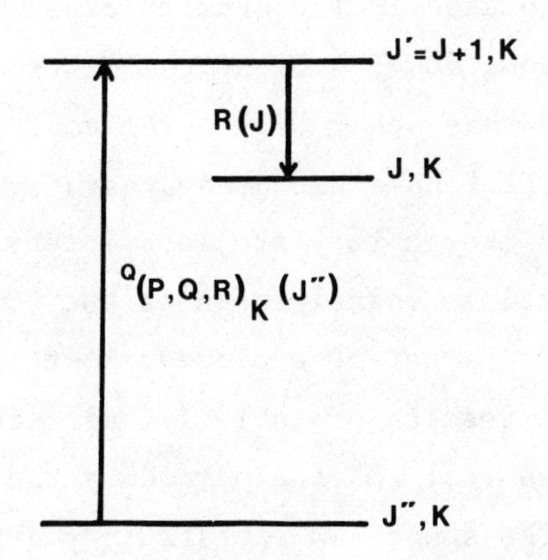

FIGURE 4. : The Scheme of the pumping cycle.

Owing to the precision achieved in the FIR wavelength measurement, the J value for each isotope may be generally determined without any ambiguity. The J values corresponding to the different FIR lines are reported in column 6 of table IV.

The procedure used to assign completely the IR and FIR

---

(4) An error is to be noted (Lourtioz, 1981). The J-labels given respectively in fig. 2 and in table I do not coincide.

transitions is given (Lourtioz, 1981). First, the location of the CO_2 line within the CF_3Br absorption spectrum (see fig.1) allows to know the type of the branch (P,Q,R) which is concerned. Then, the examination of all the K-structures related to the given J"-value allows to determine the rovibrational transitions which coincide with the CO_2-pump. Several coincidences may a priori occur (for instance, one for each isotope). This was not the case for the lines involved.

The only possible assignments of the IR transitions within the ν_1 and $(\nu_1+\nu_6-\nu_6)$ bands of $CF_3{}^{79,81}Br$ are given in column 9 of table IV. The experimental IR frequencies (CO_2 line) are reported in column 5. The calculated IR frequencies given in column 7 correspond to the first transition (K = 0) of the K-structure involved.

We must remind that other $CF_3{}^{79,81}Br$ rovibrational bands mentioned in section II.1 have not been presently investigated for the IR assignments. Although they are less intense than the ν_1 and $(\nu_1+\nu_6-\nu_6)$ bands, possible coincidences of the CO_2-pump with transitions of these bands cannot be excluded. In this condition, we may only consider the results of table IV. as tentative assignments.

The case of the 1151 μm line pumped by the 9R 34 CO_2 line illustrates the remarks above. No satisfactory attribution of the pump transition has been found within the ν_1 and $(\nu_1+\nu_6-\nu_6)$ bands of $CF_3{}^{79,81}Br$. The location of the 9R 34 CO_2 line within the CF_3Br spectrum of fig.1 suggests that we might be concerned with a P-branch transition of one of the two isotopic $2\nu_5$ // bands : as a first estimation[(5)], the CO_2-pump should coincide with a ${}^{2\nu_5}P_K(62)$ transition of $CF_3{}^{79}Br$. It is to be noted that J" = 62 is also the J" value calculated from relation (7).

As another exception the IR transition pumping the 2140 μm line cannot have been assigned. But, the uncertainty of the FIR

(5) The wavenumbers of ${}^{2\nu_5}P_K(J)$ transitions are calculated from relation (4b) by setting A' = A" and B' = B" = B, where B is the rotational constant of the ground state. The band center σ_o is taken from table II..

wavelength measurement for this very weak FIR line (< 1 μW) probably explains this result.

It must also be mentioned that the pump transitions of the 824 μm and 883 μm CF_3Br lines are not completely assigned. The theoretical precision of the symmetric-top model degrades for high J-values (J > 80) and the K-value could not be exactly determined. Nevertheless, the high intensity of the 824 μm transition suggests that it could be pumped by a $K \simeq 0$ transition.

Finally, the absence of FIR transitions for the 9R 22, 9R 24, 9R 26, 9R 30, 9R 32 CO_2-pump can be explained in the following way. Each of these lines coincides with $CF_3{}^{79,81}Br$ absorption transitions with low J-values. Thus, the possible FIR emissions should be at very long wavelength (λ_{FIR}> 2 mm) which probably corresponds to the upper limit of the experimental set-up.

IV. Conclusion and suggestion for further investigations :

Most of the FIR lines of CF_3Br are satisfactorily assigned within the ν_1,(ν_1 +ν_6 - ν_6) bands of the two main isotopes ($^{12}CF_3{}^{79}Br$ and $^{12}CF_3{}^{81}Br$). That corroborates the values of the spectroscopic constant deduced from microwave and infrared spectroscopy.

The final confirmation of these attributions could be obtained in four ways :

1/ The direct frequency-measurement of the strong 824 μm line should help for its complete (J,K) assignment. Moreover, since a high J value is concerned in this case, such a datum should also allow us to refine the symmetric-top model used for CF_3Br.

2/ Experiments with mono isotopic specie (i.e. $CF_3{}^{79}Br$ or $CF_3{}^{81}Br$) should be quite useful. First, the examination of the infrared spectra should provide more information about the weakest absorption bands. Secondly, this should greatly facilitate the IR-FIR attributions.

3/ The cooling of the FIR gas should strongly favor the FIR

emission within the fundamental ν_1 bands. This should also facilitate the IR-FIR attributions.

4/ A better knowledge of the hyperfine structure of the rovibrational bands seems to be necessary. This has been examplified by the results presented in section III.1.

Then FIR emission lines at very long wavelength ($\lambda > 1$ mm) are presently obtained from the CF_3Br molecule. This reveals the molecule as a good FIR candidate as regards to the poor optical pumping efficiency achievable in this wavelength region. New FIR emission lines could be certainly obtained with an improved experimental set-up and by using other pump sources (i.e. isotopic CO_2 sources). Besides, the possible coincidences of some CO_2 emission lines with rovibrational transitions of the $2\nu_5$ overtone bands of $CF_3{}^{79,81}Br$ are to be considered as a favourable situation for the obtention of mid-infrared CF_3Br emissions ($\lambda \simeq 20$ µm) (Tiee,1978).

References

Knight, D.J.E., 1980 : "Ordered list of optical pumped laser lines", NPL Report.

Burczyk, K., Burger, H. and Pinson, P., 1979 : "Rotational analysis of $\nu_1, \nu_1+\nu_6-\nu_6, \nu_1+2\nu_6-2\nu_6$, of $CF_3{}^{79}Br$ and $CF_3{}^{81}Br$ near 1100 cm^{-1}", J. Mol. Spectrosc. 77, 109.

Meyer, C., Pinson, P., 1979 : "Progrés obtenus en spectrométrie infrarouge par l'utilisation du détecteur HgTe-CdTe", Infrared Phys. 16, 335.

Pontnau, J., Lourtioz, J.M. and Meyer, C., 1979 : "Submillimeter laser action of CW optically pumped CF_3Br", I.E.E.E. J. Quantum Electron. QE-15, 1088.

Lourtioz, J.M., Pontnau, J. and Meyer, C., 1981 : "Optically pumped CW CF_3Br FIR laser. New emission lines and tentative assignments", International Journ. of Infrared and Millim. Waves, 2, 525.

Edgell, W.F. and May, C.F., 1955 : "Raman and Infrared spectra of CF_3Br and CF_3I", J. Chem. Phys., 22, 1808.

Burczyk, K., Burger, H., Schultz, P. and Ruoff, A., 1981 : "Vibrational and Rotational Analysis of parallel overtones and Combination bands of CF_3Br", Z. Anorg. Allg. Chem. 474, 74.

Townes, C.H. and Schallow, A.L., 1955 : "Microwave Spectroscopy" Mac Graw Hill Book Company Inc., New-York.

Lombard, J. : "Private Communication", CEN Saclay, FRANCE

Sheridan, J. and Gordy, W., 1952 : "The microwave spectra and molecular structure of trifluoromethyl bromide, iodide and cyanide", J. Chem. Phys., 13, 591.

Jones, H., Kolher, F. and Rudolph,H.D., 1976 : "Precision Infrared spectroscopy in $CF_3$79Br by means of infrared-microwave double resonance", J. Mol. Spectrosc., 63, 205.

Lourtioz, J.M., Adde, R., Bouchon, D. and J. Pontnau, 1979 : "Design and performances of a CW CH_3OH waveguide laser", Revue de Physique Appliquée, 14, 323.

Lourtioz, J.M., Pontnau, J. and Julien,F., 1980 : "Simple and accurate method to measure wavelengths of CW FIR optically pumped lasers", Infrared Phys., 20, 231.

Chang, T.Y., Bridges, T.J., 1970 : "Laser action at 452, 496 and 541 μ in optically pumped CH_3F", Opt-Commun, 1, 423.

Inguscio, M., Moretti, A. and Strumia, F. 1979 : "IR-FIR Transferred Lamb dip spectroscopy in optically pumped molecular lasers", Opt. Commun, 30, 355.

Inguscio, M. Moretti, A. and Strumia, F., 1980 : "New laser lines from optically pumped CH_3OH. Measurements and assignments", Opt. Commun, 32, 87

Busse, G., Beisel, E. and Pfaller, A., 1977 : "Application of the opto-acoustic effect to the operation of optically pumped Far-Infrared gas lasers", Appl. Phys., 12, 387.

Walzer, K., Tacke, M. and Busse, M., 1979 : "Opto-acoustic spectra of some Far-Infrared laser active molecules", Infrared Phys. 19, 175.

Radford, H.E., Peterson, F.R., Jennings, D.A. and Mucha, J.A., 1977 : "Heterodyne measurements of submillimeter laser spectrometer frequencies", I.E.E.E. J. Quantum Electron., 13, 92.

Tiee, J.J., Wittig, C., 1978 : "Optically pumped lasers in the 11-17 μm region", J. Appl. Phys., 49, 61.

FAR-INFRARED LASER LINES OBTAINED BY

OPTICAL PUMPING OF THE CD_3Br MOLECULE

G. Graner, J - C. Deroche, and C. Betrencourt-Stirnemann

Laboratoire d'Infrarouge - Laboratoire Associé au CNRS -
Université de Paris-Sud, Bâtiment 350
91405 Orsay Cédex, France

A) Optically pumped lines

Two independent works are known on optically pumped emissions in CD_3Br.

On the one hand Dyubko (1976) (Dyubko and Fesenko, 1978) obtained 9 coincidences with the CO_2 laser, giving rise to 10 FIR laser lines, which he measured very accurately. On the other hand, Landsberg(1980.a)also published 9 coincidences giving rise to 10 FIR laser lines. Altogether, we had in hand 13 coincidences giving rise to 15 FIR laser lines.

However it can be seen that the FIR line generated by the 9R16 CO_2 line (Dyubko, 1976; Dyubko and Fesenko, 1978) coincides with the FIR line generated by the 9R18 line (Landsberg, 1980.a). Dyubko's experimental value is only a few MHz from the predicted value for this transition. We have therefore assumed a clerical error in Dyubko's papers.

Therefore we have in hand 12 coincidences giving rise to 14 FIR laser lines.

Experimental details

In Dyubko's work (1976) (Dyubko and Fesenko, 1978), the

pumping power was in the range 10-20 W. The output power ranges from 0.15 to 6 mW. The corresponding values are reported in Table II. No information is given about the polarization of the FIR line. From other works of the same group (Dyubko, Fesenko, Baskakov and Svich, 1975), it can be inferred that the FIR frequencies are measured to within 2×10^{-6}, which means an uncertainty of 1 MHz at 500 GHz.

As for Landsberg, he has described his experimental set-up (Landsberg, 1980.a, 1980.b, 1981). For each line, the threshold pumping power is given, ranging from 0.2 to 8 W. The output power has been determined only roughly and is given as Medium, Weak and Very Weak. According to a private communication (Landsberg, 1981), M should correspond to the range 0.01-0.1 mW, W and VW being respectively one and two orders of magnitude smaller. This give output powers much smaller than those of Dyubko. Table I therefore gives both scales.

Note also that Landsberg's laser(1980.a)did not perform well for $\lambda > 600$ µm, which explains why he was not able to find the lines at 692, 851 and 941 µm.

Most of the wavelengths reported (Landsberg, 1980.a) should be accurate to within 0.5 % which should therefore be considered as the standard deviation of the measurements (Landsberg, 1981). Landsberg (1980.a) gives also the relative polarization of the FIR radiation relative to that of the pumping laser,which is very useful information for the assignments.

No value is found in the literature for the offset between the absorbing CD_3Br transition and the pumping CO_2 laser.

B) Spectroscopic Information

Deuterated methyl-bromide is a symmetric-top molecule with 6 fundamental modes :

- three A_1 modes, giving parallel fundamentals ν_1, ν_2 and ν_3 at 2160, 991 and 577 cm^{-1} respectively,

TABLE I - FIR LASER LINES IN CD_3Br

CO_2 LASER LINE	F I R WAVELENGTH (µm) (a)	F I R FREQUENCY (MHz) (b)	POLARIZATION (c)	THRESHOLD PUMP POWER (W) (c)	F I R POWER (d)	PROPOSED ASSIGNMENT: SPECIES	GROUND LEVEL (J,K)	UPPER LASING LEVEL	LOWER LASING LEVEL	CO_2 WAVENUMBER (cm^{-1})	CALCULATED I R TRANSITION (cm^{-1})	CALCULATED F I R TRANSITION (MHz)	CALCULATED F I R TRANSITION (µm)	COMMENTS
9R30	(556.8027)	538417.8	//	1.5	6 M	81	(34,1)	$(35,2)\nu_5$	$(34,2)\nu_5$	1084.6351	.6368	538448	556.77	
9R26	(466.6429)	642445.1			0.2			42 ?	41 ?	1082.2962				
9R18	(441.6738)	678764.4	//	1.5	2 M	79	(43,1)	$(44,0)\nu_5$	$(43,0)\nu_5$	1077.3025	.3028	678728	441.70	(e)
9P14	(366.6252)	817708.3	//	1.0	1 M	79	(52,6)	$(53,5)\nu_5$	$(52,5)\nu_5$	1052.1955	.1959	817662	366.65	
9P18	(431.7360)	694388.4	⊥	2.5	0.6 M	79	(45,2)	$(45,1)\nu_5$	$(44,1)\nu_5$	1048.6608	.6623	694351	431.76	
9P26	(692.0248)	433210.6			0.2	79	(28,3)	$(28,2)\nu_5$	$(27,2)\nu_5$	1041.2791	.2784	433195	692.05	
9P32	297	-	//	2.0	M	81	(67,1)	$(66,2)\nu_5$	$(65,2)\nu_5$	1035.4736	.4731	1011078	296.51	
9P32	(553.8827)	541256.2	⊥	8.0	0.2 W	79	(35,4)	$(35,3)\nu_5$	$(34,3)\nu_5$	1035.4736	.4734	541237	553.90	
9P36	341	-	⊥	4.0	VW	79	(57,5)	$(57,4)\nu_5$	$(56,4)\nu_5$	1031.4774	.4778	878327	341.32	
10R18	(851.3242)	352148.4			0.15			23 ?	22 ?	974.6219				
10R18	(941.4865)	318424.6			0.1			21 ?	20 ?	974.6219				
10R10	(530.1322)	565505.1	//	1.5	5.5 M	79	(38,0)	$(37,0)\nu_2$	$(36,0)\nu_2$	969.1395	.1410	565483	530.15	(f)
10R2	428	-	//	0.2	M	81	(47,0)	$(46,0)\nu_2$	$(45,0)\nu_2$	963.2631	.2634		428.64	(f)
10P18	290	-	//	1.5	M	79	(69,7)	$(68,7)\nu_2$	$(67,7)\nu_2$	945.9802	.9802		290.03	

(a) VALUES WITHOUT BRACKETS COME FROM LANDSBERG (1980.A)THOSE WITHIN BRACKETS ARE COMPUTED FROM THE FREQUENCIES.
(b) FROM DYUBKO (1976) (1978)
(c) FROM LANDSBERG (1980.A)
(d) POWERS FROM DYUBKO (1976) ARE EXPRESSED IN MILLIWATTS. LETTERS M , W AND VW COME FROM LANDSBERG (1980.A)
(e) REPORTED AS 9R16 BY DYUBKO. PROBABLY A CLERICAL ERROR.
(f) THE ASSIGNMENT WITH K = 1 IS ALMOST AS GOOD.

- three E modes, giving perpendicular fundamentals ν_4, ν_5 and ν_6 at 2296, 1055 and 713 cm^{-1} respectively.

The ν_2 and ν_5 bands are the only ones located in the CO_2/N_2O laser region and they give rise to pumped FIR emission lines. The $2\nu_3$ band centered at $\simeq$ 1148 cm^{-1} might have P lines with high J values in coincidence with some CO_2 laser lines but no such coincidence has been found.

The perpendicular band ν_5 of CD_3Br was measured almost simultaneously by Morino and Nakamura (1965) and by Jones et al (1966). In the latter reference ν_2 was also studied. Morino and Hirose (1967) measured several microwave transitions in the excited states v_2=1 and v_5=1 and pointed out the importance of the Coriolis interaction between ν_2 and ν_5. All the information known on CD_3Br up to 1979 was gathered in a recent review by Graner (1981).

Very recently Betrencourt-Stirnemann et al (1981) recorded this spectral region at a resolution of 0.010 cm^{-1} and a wavenumber accuracy of 0.002 cm^{-1}. The reader is referred to Fig.1,2,4,5 of this reference for the general appearance of this band.

These authors assigned about 1600 transitions for each of the isotopic species $CH_3{}^{79}Br$ and $CH_3{}^{81}Br$. They were fitted to a model with 11 free parameters including both the Coriolis interaction and the $\ell(2,2)$ resonance. The standard deviation of the residuals was about 0.004 cm^{-1} for both isotopic species.

The molecular constants (Betrencourt-Stirnemann et al, 1981), are given in columns 2 and 4 of Table II. These constants are quite good for interpolation of frequencies within the range of J and K values used in Betrencourt-Stirnemann (1981), (i.e. $J < 65$ in ν_2 and much smaller values in ν_5) but as usual, they are less reliable for <u>extrapolations</u>. Our experience is that they can be used to interpolate near IR transitions to less than 0.010 cm^{-1} and far IR transitions to less than 20 MHz.

We have attempted to improve these molecular constants by fitting simultaneously the rovibrational data of Betrencourt-Stir-

TABLE II

Molecular Constants of the ν_2 and ν_5 bands of CD_3Br

	$CD_3{}^{79}Br$		$CD_3{}^{81}Br$	
GROUND STATE CONSTANTS				
$A_o (cm^{-1})$	2.6004		2.6004	
$B_o (cm^{-1})$	0.257332		0.256219	
$D''_J (10^{-7} cm^{-1})$	1.965		1.913	
$D''_{JK} (10^{-6} cm^{-1})$	2.12		2.11	
$D''_K (10^{-5} cm^{-1})$	1.72		1.72	
UPPER STATE CONSTANTS	Betrencourt-Stirnemann et al. (1981)	This Work	Betrencourt-Stirnemann et al. (1981)	This Work
$\nu_5 (cm^{-1})$	1055.474	1055.4713	1055.471	1055.4738
$A''-A' (10^{-3} cm^{-1})$	15.26	15.18	15.26	15.15
$B''-B' (10^{-4} cm^{-1})$	- 2.460	- 1.840	- 2.430	- 1.523
$A\zeta_5$, (cm^{-1})	- 0.82021	- 0.82020	- 0.82031	- 0.82008
$\eta_J (10^{-6} cm^{-1})$	- 2.62	- 3.21	- 2.85	- 2.92
$\eta_K (10^{-5} cm^{-1})$	- 6.47	- 6.68	- 6.47	- 6.96
$D''_J-D'_J (10^{-9} cm^{-1})$	0	- 4.0	0	10.0
$D''_{JK}-D'_{JK} (10^{-7} cm^{-1})$	0	- 1.3	0	0
$\nu_2 (cm^{-1})$	991.401	991.4008	991.390	991.3894
$A''-A' (10^{-3} cm^{-1})$	- 8.564	- 8.570	- 8.640	- 8.653
$B''-B' (10^{-4} cm^{-1})$	8.001	6.59	7.952	6.40
$D''_J-D'_J (10^{-9} cm^{-1})$	6.4	5.07	6.4	3.3
$D''_{JK}-D'_{JK} (10^{-7} cm^{-1})$	1.70	1.57	1.32	0.84
$q_5 (10^{-5} cm^{-1})$	13.98	0	14.30	0
$W_1^{Cor} (cm^{-1})$	0.20336	0.21434	0.20243	0.21433
$W_2^{Cor} (10^{-4} cm^{-1})$	0	5.50	0	5.95

nemann (1981) and the information given by the FIR laser lines and by the coincidences with the CO_2 laser. This improvement was possible for the ^{79}Br species and almost impossible for the ^{81}Br, due to the lack of accurate information. The constants obtained are given in columns 3 and 5 of Table II. The quality of these new constants is certainly better for ^{79}Br than for ^{81}Br.

Note that we have found more convenient to constrain to zero the $\ell(2,2)$ constants q_5 and to let free the second order term W_2 in the Coriolis interaction, given by :

$$< v_2=1, J, k \,|H_1'|\, v_5=1, \ell_5=\pm 1, J, k\pm 1>$$
$$= [\pm W_1 + W_2\ (2k\pm 1)]\ [J(J+1)-k(k\pm 1)]^{1/2}$$

This leads to quite different values for (B"-B') for ν_2 and ν_5 in our model as compared to Betrencourt-Stirnemann et al.(1981).

Very little information is known about $^{13}CD_3Br$ (Graner, 1981) and no FIR emission lines has ever been assigned to this compound.

C) Assignment of the FIR lines

Using preliminary spectroscopic information from the authors of Betrencourt-Stirnemann (1981) , Landsberg(1980.a)assigned 7 of the FIR transitions, 4 of them completely and 3 only partially.

We have checked and completed these assignments (Table I). Eleven out of the 14 observed lines are now assigned, although some doubts are still possible for two of them : the two lines at 530 μm and 428 μm that we attributed to K=0 might be also assigned to K=1.

For these eleven lines, we have also given calculated values for the IR and for the FIR transitions. The fact that some calculated FIR frequencies are still 30-40 MHz from the accurate observed values shows that our spectroscopic model is not perfect. Additional resonances such as the $\ell(2,-1)$ interaction should probably be introduced. Nevertheless these constants allow us to reproduce all coincidences with the CO_2 laser within 0.002 cm^{-1}.

The three lines still unassigned do not belong to the $v_2=1$ or the $v_5=1$ states of $CH_3{}^{79}Br$ or $CH_3{}^{81}Br$, unless there is a clerical error. They can belong either to a hot band or to a ^{13}C species. The fact that they are the three weakest lines observed by Dyubko supports this hypothesis.

D) Suggestion for further investigations

Improved values for the FIR frequencies are necessary for at least 4 FIR transitions as well as measurements of offsets (CD_3Br - CO_2 laser) for all coincidences.

Trials with isotopic CO_2 lasers and with N_2O lasers should provide new coincidences.

Spectroscopic work is certainly needed with ^{13}C enriched samples in the (ν_2, ν_5) region. Moreover, spectra of normal CD_3Br in the region of ($\nu_2+\nu_3$,$\nu_5+\nu_3$) and ($\nu_2+\nu_6$, $\nu_5+\nu_6$) would help to understand the hot bands. Both type of studies would help to complete the assignments of FIR lines.

References

Dyubko, S.F., Sept. 1976 : "Optically pumped submillimeter lasers and high resolution spectroscopy", 3rd Symposium on high resolution molecular spectroscopy (Novosibirsk).

Dyubko, S.F. and Fesenko, L.D., March 1978 : "Frequencies of optically pumped submillimeter lasers," 3rd International conference on submillimeter waves and their applications, (Guilford, Univ. of Surrey).

Landsberg, B.M., 1980.a :"New CW optically pumped FIR emissions in HCOOH, D_2CO and CD_3Br," Appl. Phys. 23 , 345-348.

Dyubko, S.F., Fesenko, L.D., Baskakov, O.I. and Svich V.A., 1975, Zh. Prikl. Spektr. 23, 317-320.

Landsberg, B.M.,1980.b : "New CW FIR laser lines from optically pumped ammonia analogues," Appl. Phys. 23, 127-130.

Landsberg, B.M. June 1981, Private communication to the authors.

Morino, Y. and Nakamura K., 1965 : "Vibration-Rotation spectra, the Coriolis coupling constants and the intramolecular force field of symmetric top molecules I. The E-type fundamental bands of methyl and methyl-d_3 halides," Bull. Chem. Soc. Japan 38, 443-458.

Jones, E.W., Popplewell R.J.L. and Thompson H.W., 1966 : "Vibration-rotation bands of methyl bromide-d_3," Spectrochim. Acta 22, 639-646.

Morino, Y. and Hirose, C., 1967 : "Microwave Spectra of Methyl bromide and methyl bromide-d_3. Coriolis resonance between the ν_2 and ν_5 states," J. Mol. Spectrosc. 24, 204-224.

Graner, G., 1981 : "The methyl bromide molecule : A critical consideration of perturbations in spectra," J. Mol. Spectrosc. 90, 394-438.

Betrencourt-Stirnemann C., Paso R., Kauppinen J. and Antilla R., 1981 : "The Coriolis interaction between the fundamentals ν_2 and ν_5 of CD_3Br," J. Mol. Spectrosc. 87, 506-521.

SUBMILLIMETER LASERS IN METHYL IODIDE AND ITS ISOTOPIC SPECIES

E. Arimondo

Istituto di Fisica Sperimentale dell'Università
di Napoli, Italy
Gruppo Nazionale di Struttura della Materia, Pisa, Italy

1. Introduction

FIR emission from CH_3I molecules was observed at first by Dyubko and coworkers[16] as early as 1974, using a medium power cw CO_2 laser. While the number of FIR laser lines emitted by this molecule following cw emission was extended in 1975[17], irradiation by a chopped laser with 200W peak by Chang and McGee[10] increased the number of FIR lines by a factor two. The apparatus of Dyubko et al[17] with cw CO_2 laser irradiation was applied to the deuterated compound CD_3I and the number of FIR lines emitted by this molecule was from the very beginning larger than in the normal compound. No further investigation of FIR emission by this molecule was performed, while it would be expected that more powerful chopped CO_2 lasers would produce an increase in the number of FIR lines in CD_3I. High precision measurement of FIR transitions has allowed resolution of the hyperfine structure due to the interaction of the quadrupole moment of iodine with the gradient of the molecular electric field[19,20,29,38]. The emission from $^{13}CH_3I$ contained in natural abundance was observed in the measurements of Dyubko et al[17] and Chang and McGee[10] as confirmed by the assignment of Arimondo et al[5]. Recently the emission from the $^{13}CD_3I$ molecule was observed[15] and the optical-optical double-resonance was used as an aid to the FIR assignments for this molecule.[30] Isotopic CO_2 lasers have been used for pumping in $^{12}CD_3I$ and $^{13}CD_3I$[31,37]. The optimum parameters for FIR emission from the $^{12}CD_3I$ and $^{12}CD_3I$ molecules was investigated by Dyubko et al[18] and Manila[32]. The earliest spectroscopic analysis of the $^{12}CH_3I$ FIR emission was made by Graner[23] and it is worthwhile pointing out that such an analysis was important in promoting further spectroscopic studies of the FIR-laser molecules.

Methyl iodide, in both its normal and isotopic species, has a low rotational constant, because of the iodide molecular weight ($B \simeq 0.2$ cm^{-1}). That produces characteristic features in their FIR-laser behaviour. Rotational levels with large J numbers are populated at room temperatures, and the FIR lines in the typical region of investigation (50-2000μm) correspond to rotational transitions between high J levels (typically J = 30 to 50). Most of the molecular spectroscopy investigations are restricted to low J number where high-order corrections to the rotational constants and to the intervibration interaction are negligible. Furthermore methyl iodide has vibrational modes of low-frequency thermally populated at room temperature so that hot-band vibrational transitions contribute to the FIR emission. In the end the quadrupole splitting in the vibro-rotational levels of these molecules is large and should be considered in high-resolution analyses. Thus the spectroscopy of the FIR lines in these molecules is far from being complete. The study of the laser-Stark spectroscopy of FIR-lines, widely applied on other molecules, does not provide straightforward information because of the large J number and the quadrupole structure[41]. On the contrary the infrared-radiofrequency intracavity double-resonance, widely applied to methyl-iodide compounds[5,40], has provided spectroscopic assignments of most CH_3I FIR lines[5].

2. Submillimetre Laser lines

The initial observation of laser action in CH_3I was made by Dyubko and coworkers[16,17] using a cw CO_2 laser with pumping power in the 10W range and a mirror cavity for the submillimetre laser. Wavelengths were measured with a $2x10^{-3}$ accuracy and for a few lines heterodyne mixing was applied to obtain the frequencies within $2x10^{-6}$ accuracy, which means an uncertainity of 1 MHz at 500 GHz. In later measurements by the same group[19,20] the heterodyne method has allowed the resolution of the hyperfine structure in the far-infrared emission. In ref.(16) the output power, in mW, of three laser lines was reported, while in ref.(17) the power was measured only in relative units. We have used an averaged conversion factor (45 arbitrary unit = 1 mW) to derive the output power reported in Table I.

The observations on FIR emission CH_3I were extended by Chang and Mc Gee[10] using a chopped CO_2 laser, with 150μs pulses of up to 200W peak power. A FIR mirror cavity was used. The polarization, the threshold power, the pressure of operation, the output power and the wavelengths (with 0.1μm accuracy) were measured and have been reported in Table I. Also the offset frequency of the CO_2 laser for maximum output power has been reported by those authors, however the presence of the hyperfine structure on the infrared absorption line makes this observation hard to interpret.

The strongest lines of FIR emission in CH_3I were measured through heterodyne beating by Kramer and Weiss[29] and Radford et al[38] with 0.3 and 0.5 MHz accuracy respectively, and the measurements reported in the table are the weighted average of the available experimental results.

The observations of FIR emission by the CD_3I molecule were done Dyubko and coworkers[17,20] using the apparatus described above for the CH_3I molecule. Thus the observations reported above are valid also for this molecule, and the same conversion factor has been used to derive the output powers from the relative units reported in ref. (17). (Table 2).

Table I - CH_3I Laser Lines

λ [a]	Pump Line	Pol.	Refs.	Thresh. (b)	FIR(c) Power mW/W	mW	Press. mTorr(d)	Pump	FIR Laser (e)
377.45	9R16	∥	10,16,17	7	5.1/47	0.05	435	${}^{q}R(53,3)2\nu_3 \leftarrow 0$	$(54,3) \rightarrow (53,3); 2\nu_3$
390.53	10P42	∥	17	9	1.8/57	0.03	750	${}^{r}R(53,9)\nu_3+\nu_6 \leftarrow \nu_3$	$(54,10) \rightarrow (53,10); \nu_3+\nu_6$
392.48	9R14	∥	10	24	0.017/40	-	100	?	?
(447.17)(f)	10P18	∥	19,29,38	1.5	23/103	3	225	${}^{r}R(44,5)\nu_6 \leftarrow 0$	$(45,6,93/2) \rightarrow (44,6,91/2); \nu_6$
457.25	" (g)	∥	10	"	-	-	"	"	$(44,6) \rightarrow (43,6)\nu_6$
459.18	10P8	⊥	10	54	0.16/69	-	155	?	?
477.87	9P26	∥	10	58	0.021/82	-	110	?	?
508.37	9P34	∥	10	7	13/66	-	192	${}^{q}P(41,3)2\nu_3 \leftarrow 0$	$(40,3) \rightarrow (39,3); 2\nu_3$
517.33	9P14	∥	10	32	0.38/81	-	435	?	?
525.32	9P4	∥	10	6	0.093/26	-	130	${}^{q}R(38,5)3\nu_3 \leftarrow \nu_3$	$(39,5) \rightarrow (38,5); 3\nu_3$
529.28	10P36	∥	10	20	3.2/60	-	320	$\left[{}^{r}P(39,8)\nu_6 \leftarrow 0\right]$	$(38,9) \rightarrow (37,9); \nu_6$

(continued)

Table I - CH_3I Laser Lines (Continued)

λ[a]	Pump Line	Pol.	Refs.	Thresh. (b)	FIR(c) Power mW/W	FIR(c) Power mW	Press. mTorr	Pump	FIR Laser[e]
542.99	10P26	||	10,17	15	0.002/99	0.02	132	$^{r}R(38,5)\nu_6 \leftarrow 0$[h]	$(39,6) \rightarrow (38,6);\nu_6$
576.17	10P16	||	10,17	16	1.6/58	0.23	320	$^{r}R(34,6)2\nu_6 \leftarrow \nu_6$	$(35,7) \rightarrow (34,7);2\nu_6$
578.90	10R34	||	10	54	0.98/68	-	375	$^{r}R(34,11)\nu_6+\nu_3 \leftarrow \nu_3$	$(35,12) \rightarrow (34,12);\nu_3+\nu_6$
583.87	9P4	||	10	4	0.12/32	-	192	?	?
639.73	9P6	||	10	30	0.17/35	-	79	$^{q}R(31,5)3\nu_3 \leftarrow \nu_3$	$(32,5) \rightarrow (31,5);3\nu_3$
670.99	10P28	||	10	22	0.93/91	-	145	?	?
719.30	10P22	||	10,17	26	0.33/52	-	320	?	?
964.	10P22	||	17	?	?	0.01	-	$\left[^{r}Q(21,8)\nu_3+\nu_6 \leftarrow \nu_3\right]$	$21,9) \rightarrow (20,9);\nu_3+\nu_6$
1063.29	10P38	||	10,17	25	6.5/71	0.34	435	$^{r}R(18,5)\nu_3+\nu_6 \leftarrow \nu_3$	$(19,6) \rightarrow (18,6);\nu_3+\nu_6$
(1253.73)[i]	10P32	||	19,29,38	5	13/98	0.3	170	$^{r}R(15,5)\nu_6 \leftarrow 0$	$(16,6,27/2)\ (15,6,25/2);\nu_6$
(1253.82)[j]			19					"	$(16,6,35/2) \rightarrow (15,6,33/2);\nu_6$

a) These wavelengths in μm are obtained from refs. (10) and (17) except those in brackets computed from the frequencies measurements.

b) Pump power in Watts required to obtain FIR emission in the Chang and Mc Gee experiment of ref. (10).

c) In the first column output power with pulsed CO_2 lasers[10]; second column output power obtained with cw CO_2 lasers, derived from refs. (16,17) as explained in the text.

d) Pressures used in the Chang and Mc Gee experiment[10].

e) The (J,K) numbers of the upper and lower states in the FIR emission are shown, except for the lines measured with heterodyne where (J,K,F) are indicated.

f) Measured frequency ν = 670.4628(3) GHz

g) Collisional line

h) FIR emission of $^{13}CH_3I$ in natural abundance

i) Measured frequency ν = 239.1189 (5) GHz

l) Measured frequency ν = 239.1028 (5) GHz

Observations of FIR laser action in $^{13}CD_3I$ was made by Duxbury and Herman[15] using a cylindrical metallic resonator, where the threshold pump power is greatly reduced. The features of the $^{13}CD_3I$ emission are reported in Table 3. For the 690μm line the output power was measured as function of the $^{13}CD_3I$ pressure for a Q-switched pump laser; a maximum 18 mW output power was obtained at 110 mTorr when a 3W CO_2 laser was Q-switched by a chopper. Table 3 reports also the FIR lines obtained by the same group using isotopic CO_2 lasers (6 to 15 watts power) and measuring the wavelengths with a 5.10^{-3} accuracy.

Dyubko and coworkers[18] have reinvestigated the lasing behaviour of several gases, including CH_3I and CD_3I, as a function of the active-gas pressure. For cw standing--wave pump up to $60W/cm^2$ the gain of the submillimeter laser was measured as function of the pump intensity and the gas pressure. As a function of the active-gas pressure the cw operation of the CH_3I and CD_3I FIR lasers has a maximum in the 40-60 mTorr range, as reported in Table I. The authors observed a gain increase with the pump intensity, limited by saturation effects in the pump absorption transition. For instance for the 447μm line of CH_3I a maximum gain (∼ 2dB/m) was observed at a pressure ∼ 60 mTorr, and saturation effects occur at 30--$40W/cm^2$. Finally the polarization of a few lines of CH_3I and CD_3I was carefully measured.

Manila[32] has made recently a study of the output power in the 447μm and 1253μm lines of CH_3I and the 670μm and 980μm lines of CD_3I, operating a mirror cavity.

Table 2 $^{12}CD_3I$ Laser Lines

$\lambda(\mu m)^a$	$\nu(GHz)^a$	Pump$(cm^{-1})^b$		Ref.Pol.		FIR Power (mW)	Press. (mTorr)	IR line	ν^e_{IR} (cm^{-1})	$\Delta\nu^e_{FIR}$ (MHz)
272.	(1102.)	(12)9P(12)	1053.92	17	$\perp$	0.01		$^rQ(92,0)\nu_5\leftarrow 0$	0.06	2000.
298.0	(1006.)	(18)10R(12)	975.38	31	$\|$	?	82	?	–	–
301.	(996.)	(12)9R(26)	1082.30	17	$\|$	0.11	–	?	–	–
390.	(769.)	(12)9P(26)	1041.28	17	$\|$	$4x10^{-3}$	–	$^pR(63,2)\nu_3+\nu_5\leftarrow\nu_3$	3.11	3847.
(433.10)	692.1955	(12)9P(28)	1039.37	17	$\|$	0.07	–	?	–	–
(444.38)	674.6213	(12)9R(32)	1085.76	17	$\perp$	2.22	–	$^rQ(56,5)\nu_5\leftarrow 0$	0.02	-123.0
448.7	(668.1)	(13)10R(12)	923.11	31	$\|$	?	100	$^qP(57,5)\nu_2\leftarrow 0$	0.13	-1625.
(460.56)	650.9275	(12)9R(12)	1073.28	17	$\perp$	3.33	–	$^rQ(54,3)\nu_5\leftarrow 0$	0.08	-65.6
(487.22)	615.3046	(12)9P(10)	1055.62	17	$\|$	0.27	–	$^pR(50,3)\nu_5\leftarrow 0$	-0.02	+134.4
(490.39)	611.3336	(12)9R(22)	1079.85	17	$\|$	0.89	–	?	–	–
(523.40)	572.7721	(12)10P(38)	927.01	17	$\|$	0.27	–	$^qP(49,13)\nu_2\leftarrow 0$	0.14	+287.7
523.9	(572.2)	(13)10R(18)	927.30	31	$\|$	?	100	$^qP(49,6)\nu_2\leftarrow 0$	0.12	-2380.
540.	(555.)	(12)9R(6)	1069.01	17	$\perp$	0.01	–	?	–	–
555.2	(539.9)	(13)10R(22)	929.99	31	?	?	90	?	–	–
556.2	(539.0)	(13)10R(20)	928.66	31	$\|$	?	100	$^qP(46,2)\nu_2\leftarrow 0$	0.08	-461
(556.87)	538.3473	(12)10P(36)	929.02	17	$\|$	1.11	–	$^qP(46,9)\nu_2\leftarrow 0$	0.11	-650.3

(continued)

Table 2 $^{12}CD_3I$ Laser Lines (Continued)

$\lambda(\mu m)$[a]	ν(GHz)[a]	Pump(cm^{-1})[b]		Ref.	Pol.	FIR Power (mW)	Press. (mTorr)	IR line	$\Delta\nu^{e}_{IR}$ (cm^{-1})	$\Delta\nu^{e}_{FIR}$ (MHz)
557.1	(538.1)	(13)10R(16)	925.92	31	$\parallel$	?	61	?	–	–
567.7[c]	(528.1)	(13)10R(20)	928.66	31	$\parallel$	?	100	$^{q}P(46,2)\nu_2 \leftarrow 0$	0.08	-1118.
(569.47)[c]	526.4344	(12)10P(36)	929.02	17	$\parallel$	0.67	–	$^{q}P(46,9)\nu_2 \leftarrow 0$	0.11	-4.1
(599.55)	500.0292	(12)10R(22)	977.21	17	$\parallel$	3.33	–	$^{q}P(43,4)2\nu_3 \leftarrow 0$	-0.01	-35.1
(614.11)[c]	488.1740	"	"	17	$\parallel$	2.22	–	"	"	-31.3
640.	(468.)	(12)10R(18)	974.62	17	$\perp$	0.02	–	?	–	–
644.	(465.)	(12)10P(16)	947.74	17	$\perp$	0.01	–	$^{q}Q(39,5)\nu_2 \leftarrow 0$	-0.02	-2000.
660.2	(454.1)	(13)10R(26)	932.60	31	$\parallel$	?	100	$^{q}P(39,10)\nu_2 \leftarrow 0$	–	–
(660.58)	453.8306	(12)10P(46)	918.72	17	$\parallel$	0.04	–	?	–	–
(667.23)	449.3075	(12)10P(10)	952.88	17	$\parallel$	0.67	–	$^{p}P(38,10)\nu_3+\nu_5 \leftarrow \nu_3$	2.88	5775.0
(670.09)	447.3886	(12)10R(8)	967.71	17	$\parallel$	0.22	–	?	–	–
(670.11)	447.3751	(12)10R(8)	967.71	17	$\parallel$	0.22	–	?	–	–
(691.12)	433.7782	(12)9R(20)	1078.59	17	$\parallel$	0.07	–	$^{r}R(35,5)\nu_3+\nu_5 \leftarrow \nu_3$	2.27	2041.6
728.1	(411.7)	(18)10R(24)	982.16	31	$\perp$	?	60	$^{p}Q(34,11)\nu_5 \leftarrow 0$	-0.01	400.
(730.32)	410.4927	(12)9R(28)	1083.48	17	$\parallel$	0.22	–	$^{r}P(35,7)\nu_5 \leftarrow 0$	0.01	-24.3
(734.26)	408.2906	(12)9P(22)	1045.02	17	$\parallel$	0.11	–	$^{r}R(33,0)\nu_3+\nu_5 \leftarrow \nu_3$	0.65	+354.6
745.	(402.)	(12)10P(8)	954.54	17	$\parallel$	0.02	–	$^{q}R(32,?)\nu_2+\nu_3 \leftarrow \nu_3$	-1.4	-4000.

$\lambda(\mu m)$ [a]	ν(GHz) [a]	Pump (cm^{-1}) [b]		Ref.	Pol.	FIR Power (mW)	Pressure (m Torr)	IR Line	$\Delta\nu_{IR}$ [e] (cm^{-1})	$\Delta\nu_{FIR}$ [e] (MHz)
(788.48)	380.2149	(12)10P(12)	951.19	17	$\|\|$	0.07	-	$^{q}R(30,?)\nu_2+\nu_3\leftarrow\nu_3$	-1.3	2000.
842.9	(355.7)	(18)10P(34)	939.61	31	?	?	150	-	-	-
895.	(335.)	(12)10P(30)	934.89	17	$\|\|$	0.02	-	$^{q}P(29,?)\nu_2+\nu_6\leftarrow\nu_6$	+1	-700.
(918.61)	326.3544	(12)9R(28)	1083.48	17	$\|\|$	4.44	-	$^{r}R(26,3)\nu_5\leftarrow 0$	+0.01	-10.2
(953.88) [c]	314.2874	"	"	17	$\|\|$	0.67	-	"	"	-9.3
960.2	(312.2)	(18)10P(32)	941.43	31	$\|\|$	?	80	-	-	-
(981.71)	305.3780	(12)10P(22)	942.38	17	$\|\|$	0.04	-	?	-	-
(1005.34)	298.1978	(12)10P(34)	931.00	17	$\|\|$	0.44	80	$^{q}P(26,10)\nu_2+\nu_3\leftarrow\nu_3$	-0.18	431.1
(1099.54)	272.6516	(12)10P(22)	942.38	17	$\|\|$	0.22	-	?	-	-
(1549.50)	193.4763	(12)9R(10)	1071.81	17	$\|\|$	0.22		$^{r}P(17,4)\nu_5\leftarrow 0$	0.02	-9.6

a) Values without brackets from Refs. (17) or (31). Those within are computed from frequencies (or wavelengths)

b) The first number in brackets denotes the CO_2 isotopic species

c) Collisional line.

d) Output power obtained with cw CO_2 laser, ref(17), with same normalization ratio as CH_3I

e) Observed - calculated

TABLE 3 - $^{13}CD_3I$ Laser lines

λ(μm)	ν(GHz)[a]	Pump[b]	Pol.	Ref.	Thresh. Pump (W)	FIR Power (mW/W)	Pressure (mTorr)
554.7	540.5	(13)10R(20)	$\parallel$	(31)	-	-	82
574.6	521.7	(18) 9P(20)	$\perp$	(31)	-	-	78
690.	434.	(12)10P(10)	?	(15)	0.8	4/4	100
745.5	402.1	(13)10R(10)	$\parallel$	(31)	-	-	80
806.	372.	(12)10P(12)	?	(15)	0.9	1/4	60
901.3	332.6	(13)10R(18)	$\parallel$	(31)	-	-	82
929.8	322.4	"	$\parallel$	(31)	-	-	82
1182.2	253.6	(13)10R(22)	$\perp$	(31)	-	-	110

a) Computed from the wavelength measurements

b) The first number in bracket denotes the CO_2 isotopic species

For a 1J pump energy a maximum output energy in the mJ range resulted with a 5 Torr active-gas pressure. The FIR laser pulse duration and the delay time of the output pulse relative to the pump pulse were measured versus the active-gas pressure.

In refs.(7) and (41) the Stark splitting of the 508.37μm line of CH_3I with an applied 1.4 KV/cm electric field was reported; the observed tuning of the M components is in reasonable agreement with an estimate based on a first-order Stark effect corresponding to the dipole moment $\mu = 1.65D$[42].

3. Spectroscopy

3a CH_3I

The absorption in the 10 μm region is produced by the $\nu_6 \leftarrow 0$ band and the $\nu_6+\nu_i \leftarrow \nu_i$ hot bands, the strongest one being $\nu_6+\nu_3 \leftarrow \nu_3$ starting from the ν_3 level at 533 cm^{-1}, with a population one tenth of that in the ground state. The $\nu_6 \leftarrow 0$ band has been analyzed by Matsuura and Overend[33] in conventional high resolution absorption spectroscopy. The CO_2 intracavity infrared-radio--frequency two-photon saturated absorption method has been used by Arimondo and Glorieux[6] to measure with high accuracy a few transitions, while the diode laser method has been applied by Das et al[11] to perform accurate measurements in the 11.8μm region. The ν_6

molecular constants derived in those last investigations differ in the band origin and in the K-dependent A terms. Moreover the molecular constants of Das et al[11] do not reproduce satisfactorily the positions of the CH_3I absorption lines measured in the infrared-radiofrequency experiments with the CO_2 lasers. The diode-laser data[11] and the saturated absorption data[6] have been reanalyzed in a simultaneous fit, introducing in the data of ref. (6) the recent measurements of isotopic CO_2 laser line[22]. For the rotational constants of the ν_6 state a good agreement has been found with the recent microwave results of Dubrulle et al[13] that have made use of the definitions of Amat et al[1] for the constants. The ir data have been reproduced with a high accuracy and the improved molecular constants of the $\nu_6 \leftarrow 0$ band are reported in Table 4. For the band origin and the K-dependent A terms the discrepancy with the constants of Das et al[11] remains large, outside the derived uncertainity, but those constants are fixed by the very precise saturated absorption data. However for what concerns the FIR emission only two CO_2 laser lines coincide with $\nu_6 \leftarrow 0$ absorptions and they were assigned by Graner[23] on the basis of the Matsuura and Overend[33] data. The hyperfine components of the FIR transition on the 10P34 at 1253.8μm were assigned by Dyubko et al[19].

The 9μm absorption is produced by the $2\nu_3 \leftarrow 0$ overtone band, and of course the hot-bands. The $2\nu_3 \leftarrow 0$ absorption was investigated in low-resolution infrared spectroscopy by Jones and Thompson[25].

Table 4 - Molecular constants of $^{12}CH_3I$ ν_6 Band (MHz)

$\nu_0+A_6(1-\zeta_6)^2-B_6+3/4\eta_6^k$	26552193.3(9)
$2\left[A_6(1-\zeta_6)-B_6+\eta_6^k\right]$	231650.3(5)
$(A_6-B_6)-(A_o-B_o)+3/2\eta_6^k$	1069.93(7)
B_6	7477.666(4)
D_6^j	$6.36(1)\times10^{-3}$
D_6^{jk}	$9.91(1)\times10^{-2}$
$D_6^k-\frac{1}{4}\eta_6^k$	1.692(3)
H_6^{jk} (a)	$0.287(101)\times10^{-6}$
H_6^{kj} (a)	$2.78(95)\times10^{-6}$
η_j^6	0.2032(6)
B_o (b)	7501.275 699(21)
D_o^j (b)	$6.3070(2)\times10^{-3}$
D_o^{jk} (b)	$9.876\ 25(33)\times10^{-2}$
$D_o^k-\frac{1}{4}\eta_6^k$	1.592(4)
H_o^j (b)	$5.06(61)\times10^{-9}$
H_o^{jk} (b)	$4.3(19)\times10^{-8}$
H_o^{kj} (b)	$4.506(35)\times10^{-6}$

$$H_6^j = H_o^j$$

a) Fixed to the value of ref.(13)

b) Fixed to the value of ref.(8)

Error limits given in parentheses are standard deviations in the last digit.

Several hot-band absorptions are involved in the FIR emission of this molecule, and their band origins have been reported in Table 5. The $3\nu_3 \leftarrow \nu_3$ position is derived from the double resonance investigation of ref. (5).

The $\nu_6 \leftarrow 0$ band of $^{13}CH_3I$ was investigated under high-resolution by Duncan and Allan[14] (a miscalculation in this paper was corrected in the paper by Arimondo et al[5]). The high resolution investigation of this band through laser-diode absorption has been reported by Das et al[12].

3b CD_3I

The $^{12}CD_3I$ molecule presents strong absorption in the 10μm and 9μm regions because of the $\nu_2 \leftarrow 0$, $2\nu_3 \leftarrow 0$ and $\nu_5 \leftarrow 0$ vibration bands respectively, ν_2 and $2\nu_3$ being parallel bands and ν_5 a perpendicular one. These infrared absorptions were investigated under low resolution by Fenton et al[21], Jones et al[24] and by Morino and Nakamura[36], and under high resolution by Anderson and Overend[2]. The near coincidences of the CO_2 laser lines with the CD_3I absorption lines led Sakai and Katayama[39] to observe level crossing and anticrossing signals in laser-Stark spectroscopy. A laser-Stark and laser-microwave double-resonance study of the ν_2 band was made by Kawaguchi et al[26], and a similar study for the ν_5 band by Kawaguchi and Hirota[27]. Level-crossing and laser-Stark saturation spectroscopy have been applied by Caldow et al[9] to investigate few lines of the ν_2 band. A careful combined analysis of the ν_5, ν_2 and $2\nu_3$ bands including their Coriolis coupling was made by Matsuura and Shimanouchi[35] and a fit of vibro-rotational lines with J as high as 45 was obtained. Accurate spectroscopic constants for these bands are found in their paper. The experimental results for the electric dipole moments of the ground and ν_i state have been discussed in the paper by Caldow et al[9].

The ν_3 and ν_6 vibrational states lying around 500 and 650 cm^{-1} respectively, have a significant population at room temperature and may contribute to the absorption of CO_2 laser lines in the 9 μm and 10 μm regions through hot bands. We derived the band origin of these hot-band absorptions from the experimental results of overtones by Fenton et al[21] and Anderson and Overend[31] (see Table 5). For the ν_3 and ν_6 bands use was made of the recent measurement by a Fourier transform spectrometer by Kauppinen and coworkers[4,28].

The $^{13}CD_3I$ molecule has not been studied in infrared absorption spectroscopy, and the origins of the bands lying in the CO_2 laser region was estimated comparing the isotopic shift of the $^{12}CH_3I$ and $^{13}CH_3I$ compounds (Table 5).

Finally the ν_6 band of $CH_3{}^{129}I$ has been studied through tunable diode laser spectroscopy by Walhen and Tucker[43].

Table 5 - Infrared bands involved in the FIR pump

$^{12}CH_3I$			Ref.
$\nu_6 \leftarrow 0$	885.68 cm^{-1}	-	(6,11)
$\nu_3+\nu_6 \leftarrow \nu_3$	880.55 "	$x_{36} = -5.13\ cm^{-1}$	(25,34)
$2\nu_3 \leftarrow 0$	1059.87 "	$x_{33} = -3.28\ cm^{-1}$	(25)
$3\nu_3 \leftarrow \nu_3$	1046.9 "	-	(5)
$^{13}CH_3I$			Ref.
$\nu_6 \leftarrow 0$	881.92 cm^{-1}	-	(12,14)
$^{12}CD_3I$			Ref.
$\nu_2 \leftarrow 0$	949.36 cm^{-1}	-	(35)
$\nu_5 \leftarrow 0$	1049.31 "	-	(35)
$2\nu_3 \leftarrow 0$	998.99 "	$x_{33} = -1.62$	(35)
$\nu_3+\nu_2 \leftarrow \nu_3$	941.68 "	$x_{23} = -7.7\ cm^{-1}$	(4,28)
$\nu_3+\nu_5 \leftarrow \nu_3$	1025.9 "	$x_{35} = -23.4\ cm^{-1}$	(4,28)
$\nu_6+\nu_2 \leftarrow \nu_6$	946.04 "	$x_{26} = -3.3\ cm^{-1}$	(4,28)
$^{13}CD_3I$			Ref.
$\nu_2 \leftarrow 0$	953.7 cm^{-1}	-	(a)
$\nu_5 \leftarrow 0$	1043.2 cm^{-1}	-	(b)

(a) isotope shift estimated comparing the $^{12}CH_3I$ (25) and $^{13}CH_3I$ (14) bands

(b) isotope shift estimated comparing the $^{12}CH_3I$ (34) and $^{13}CH_3I$ (14) bands

4. FIR assignments

For $^{12}CH_3I$, the FIR assignments derived through the double-resonance investigation of ref. (5) have been reported in Table I. The strongest lines are originated by the ν_6 absorptions, while absorption on hot-bands or overtones contribute to the remaining FIR emission. Several lines are not assigned but the information on hot-bands is incomplete, while the assignments of ref. (5) were based on the radiofrequency double-resonance investigation of the quadrupole splitting. It should be noted that FIR emission from $^{13}CH_3I$ in natural abundance was observed in both the experiments by Dyubko et al[(17)] and Chang and McGee[(10)]. For two IR lines, only a tentative assignment, in square brackets in Table 1, is reported. For the first one (at 529.28μm) the assignment was not confirmed by the radiofrequency double-resonance. However the double-resonance requires a close coincidence of the CO_2 pump with the absorption line, while the FIR emission may occur also through a Raman process. For the second one (at 964.0μm) the assignment of ref. (5) does not correspond to the selection rules for the polarization of the line.

The assignment of $^{12}CD_3I$ FIR lines have been reported in Table 2, including those reported by Dyubko et al[(17)], Duxbury and coworkers[(31,37)] and a few ones obtained by the present author. Once again the hot-band absorptions are not well characterized and all the assignments involving those bands should be considered tentative ones.

The ν_5, ν_2 and $2\nu_3$ bands have been described through the molecular constants of Matsuura and Shimanouchi[35] including the Coriolis coupling between those bands. It turns out from Table 2 that the infrared and far infrared transitions belonging to the ν_5 and $2\nu_3$ bands are well fitted by the constants. Large disagreements results for the ν_2 transitions, while Duxbury and coworkers[31] have obtained better results using the data of Kawaguchi et al[26]. The assignment by Duxbury and coworkers[31] ($^rR(44,13)\nu_5 \leftarrow 0$) for the 555.2$\mu$m line was not reported in the table because it was unsatisfactory.

All these FIR results may be used in the future to derive a better description of the methyl iodide bands.

Acknowledgements.

The author wishes to thank Prof. Eizi Hirota for providing preliminary results on the Stark spectroscopy of CD_3I and Dr. D.J.E. Knight for useful comments on the manuscript.

References

(1) G. Amat, H.H. Nielsen and G. Tarrago, "Rotation--Vibration of Polyatomic Molecules" Dekker, New York 1971.

(2) D.R. Anderson and J. Overend, Spectrochimica Acta 28A 1231 (1972)

(3) D.R. Anderson and J. Overend, Spectrochimica Acta 28A 1637 (1972)

(4) R. Anttila, M. Koivusaari, J. Kauppinen and E. Kyrö, J. Mol.Spectrosc. 84 225 (1980)

(5) E. Arimondo, P. Glorieux and T. Oka, Phys. Rev. A17 1375 (1978)

(6) E. Arimondo and P. Glorieux, Phys. Rev. A19 1067 (1979)

(7) G. Bionducci, M. Inguscio, A. Moretti and F.Strumia, Inf. Phys. 19 297 (1979)

(8) D. Boucher, J. Burie, D. Dangoisse, J. Demaison and A. Dubrulle, Chem. Phys. 29 323 (1978)

(9) G.L. Caldow, G. Duxbury and L.A. Evans, J. Mol. Spectrosc. 69 239 (1978)

(10) T.Y. Chang and J.D. McGee, IEEE J. Quantum Electron. QE-9 62 (1976)

(11) Polash P.Das, V. Malathy Devi, and K. Narahari Rao, J. Mol.Spectrosc. 84 305 (1980)

(12) Polash P. Das, V. Malathy Devi and K. Narahari Rao, J. Mol. Spectrosc. 86 202 (1981)

(13) A. Dubrulle, J. Burie, D. Boucher, F. Herlemont and J. Demaison, J. Mol. Spectrosc. 88 394 (1981)

(14) J.L. Duncan and A. Allan, Spectrochimica Acta A25 901 (1969)

(15) G. Duxbury and H. Herman, Abstracts of papers at 3rd Int.Conf. on Submillimetre Waves and their applications, University of Surrey 1978, pag. 175; J. Phys. B: Atom. Molec. Phys. 11 935 (1978)

(16) S.F. Dyubko, V.A. Svich and L.D. Fesenko, Opt. Spektr. 37 208 (1974) Engl. transl. Opt. Spectr. 37 118 (1974)

(17) S.F. Dyubko, L.D. Fesenko, O.I. Baskakov and V.A. Svich, Zh.Prikl. Spektrosk. 23 317 (1975)

(18) S.F. Dyubko, L.D. Fesenko and O.I.Baskakov, Sov. J. Quantum Electron. 7 859 (1977)

(19) S.F. Dyubko, O.I. Baskakov, M.V. Moskiensko and L. D. Fesenko, Abstracts of papers at 3rd Int. Conf. on Submillimetre Waves and their applications, University of Surrey 1978, pag. 68

(20) S.F. Dyubko and L.D. Fesenko, ibidem, pag.70

(21) P.L. Fenton, F.F. Cleveland and A.G. Meister, J. Chem. Phys. 19 1561 (1951)

(22) C. Freed, L.C. Bradley and R.G. O'Donnel. IEEE J. Quantum Electron. QE-16 1195 (1980)

(23) G. Graner, Opt. Commun. 14 67 (1975)

(24) E.W. Jones, R.J.L. Popplewell and H.W.Thompson, Proc. Roy. Soc. A288 39 (1965)

(25) E.W. Jones and H.W. Thompson, Proc. Roy. Soc. A288 50 (1965)

(26) K. Kawaguchi, C. Yamada, T. Tanaka and E. Hirota, J. Mol. Spectrosc. 64 125 (1977)

(27) K.Kawaguchi and E. Hirota, unpublished results

(28) M. Koivusaari, J. Kauppinen, A.M. Kelhälä and R. Anttila, J. Mol. Spectrosc. 84 342 (1980)

(29) G. Kramer and C.O. Weiss, Appl. Phys. 10 187 (1976)

(30) M.L. Le Lerre, H. Kato, J. McCombie, J.C. Petersen and G. Duxbury, Abstracts of papers at 5th Int. Conf. on Infrared and Millimetre Waves, University of Wurzburg, 1980, pag. 142

(31) J.McCombie, J.C. Petersen and G. Duxbury, Vth

National Quantum Electronics Conference, Hull 1981, paper G2, in press by J. Wiley.

(32) O.F. Manila, Sov. J. Quantum Electron. 10 363 (1980)

(33) H. Matsuura and J. Overend, Spectrochim. Acta A27 2165 (1971)

(34) H. Matsuura, T. Nakagawa and J. Overend, J. Chem Phys. 59 1449 (1973)

(35) H. Matsuura and T. Shimanouchi, J. Mol. Spectrosc. 60 93 (1976)

(36) Y. Morino and J. Nakamura, Bull. Chem. Soc. Japan 38 443 (1965)

(37) J.C. Petersen, J. McCombie and G. Duxbury, Abstract of papers at 5th Int. Conf. on Infrared and Millimetre Waves, University of Wurzburg 1980, pag. 140;

(38) H.E. Radford, F.R. Peterson, D.A. Jennings and J.A. Mucha, IEEE J. Quantum Electron. QE-13 92 (1977)

(39) J. Sakai and M. Katayama, Chem. Phys. Lett. 35 395 (1975), Appl. Phys. Lett. 28 119 (1976)

(40) J. Sakai, Proceedings of the Meeting on Molecular Structure, Tokio, 1976; ibidem, Sapporo, 1977

(41) F. Strumia and M. Inguscio, in "Infrared and Millimetre Waves" vol. 5, ed. by K. Button, Academic Press 1982.

(42) C.H. Townes and A.L. Schawlow, "Microwave Spectroscopy" McGraw-Hill (1955)

(43) M. Wahlen and G. Tucker, Opt. Commun. 36 93 (1981)

FAR-INFRARED LASER LINES FROM OPTICALLY PUMPED CH_3OH

M. Inguscio and F. Strumia

Istituto di Fisica dell'Università
Piazza Torricelli, 2 I-56100 Pisa, Italy

and

J. O. Henningsen
Physics Lab. I, H. C. Ørsted Institute
University of Copenhagen, Copenhagen
Denmark

Far Infrared Emissions

Methyl alcohol illustrates in a remarkable way the potential of the optical pumping technique. It was recognized early as a source of strong FIR lasing by Chang et al. (1970). Since then, hundreds of lines have been generated by the various isotopic modifications of the molecule. At the present time, normal CH_3OH has generated more than 300 lines, ranging from 19 μm to 1223 μm, in CW or quasi-CW operation. About 200 of these lines have been pumped by conventional CO_2 lasers using the normal isotopic modification of CO_2 (Tables Ia and II). Information on wavelength, relative polarization, relative power, and pressure is generally available. Also, precise FIR frequency measurements are at our disposal for 98 FIR lines (Petersen et al., 1980 and unpublished). This is helpful for the assignment of the lines and provides a comb of precise reference FIR frequencies from 245 GHz to 8 THz, using a single laser source.

Experimental Conditions

Lasing action from CH_3OH has been obtained under widely differing experimental conditions, as summarized in Table II. Both open structures and dielectric or metal waveguide configurations have been used for the resonator. Operation with metal-dielectric hybrid waveguides has allowed the observation of Stark effect on the emission, and the Stark tuning into resonance of new pumping absorptions (Table III). In most cases, the simple hole configuration has been used both for the input and output coupling of the radiation. A wide variety of pump sources have been successfully used, including conventional, TEA, waveguide, sequence band, and isotopically substituted CO_2 lasers, and many new experimental arrangements have been tested using CH_3OH as active medium.

The large variety of configurations has allowed tc optimize the particular characteristics required for a given experiment, such as power performance for plasma diagnostics or spectral purity and frequency tuning for high resolution spectroscopy and frequency synthesis.

Spectroscopic Information

The success of CH_3OH as active FIR medium derives from many particular spectroscopic features of the molecule. A first reason is the excellent overlap that exists between the strongly absorbing C-O stretch band and the CO_2 laser spectrum. Coincidences are observed for almost all the 9 μm R and P lines as well as for a large number of 10 μm lines (Fig.1). Other reasons are the complexity of the rotational spectrum, caused by the internal degree of freedom referred to as torsion or internal rotation, and the fairly large permanent dipole moments both along and orthogonal to the quasi-symmetry axis, which make the electric dipole selection rules less restrictive than for symmetric top molecules. The torsion-rotation energy level structure has been discussed by Lees and Baker (1968) and Kwan and Dennison (1972). For characterizing the

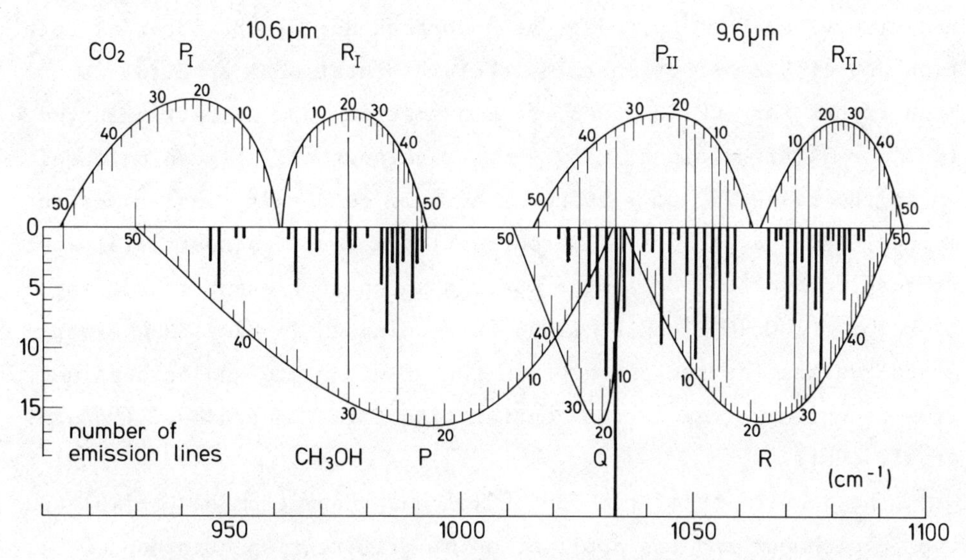

Fig. 1 Spectrum of the normal CO_2 laser superimposed on the C-O stretch of CH_3OH. Heavy lines indicate the number of FIR emissions observed for a given CO_2 line. Thin lines indicate that some of the emission lines have been assigned to the C-O stretch. (Updated from (15)).

levels, we use the notation $(n\tau K,J)$, where J and K are the symmetric top angular momentum quantum numbers, measuring the total angular momentum and its projection on the quasi symmetry axis respectively, while $n = 0,1,2,\cdots$ and $\tau = 1,2$ or 3 define the torsional state. The levels can be classified according to the irreducible representations of the point group C_3 as A, E_1 or E_2 for $\tau + K = 3N + 1$, $3N$, $3N + 2$ respectively, N being integral. E_1 and E_2 states are always doubly degenerate, whereas A states for low K are split into doublets by the asymmetry, the two components being labeled by a $\pm$ superscript. Within each internal rotation symmetry type, the selection rules for the transitions are: $\Delta J = 0,\pm 1$, $\Delta K = 0,\pm 1$, and Δn arbitrary, whereas transitions among states of different symmetry are strictly forbidden. Transitions between asymmetry split A states obey the rules $\pm \leftrightarrow \pm$ for $\Delta J = \pm 1$, $\pm \leftrightarrow \mp$ for $\Delta J = 0$, provided Δn is even, and vice versa for Δn odd. As a consequence of the internal rotation and of the selection rules, the vibrorotational spectrum is much richer for CH_3OH than for a symmetric top. In particular, due to $\Delta n \neq 0$ transitions, many short wavelengths lines can be obtained in the range 200 μm - 30 μm. The same complexity that makes the molecule such a prolific FIR source also makes it a nontrivial spectroscopic object. A review of what is known at present of the spectroscopy of CH_3OH can be found in Henningsen (1982c). New experimental values for the energies in the ground state can be obtained from recent high resolution Fourier transform spectroscopy (Moruzzi et al. 1981).

The complexity of the IR-FIR molecular spectrum has stimulated the development and the application of different spectroscopic techniques for solving the puzzle of the assignment of the lines. For instance IR-FIR-RF intracavity triple resonance measurements were reported by Airmondo et al. (1980) on transitions between levels with K-splitting. Also, Stark effect was investigated on different transitions, with a large variety of rotational quantum numbers and selection rules, as reviewed by Strumia and Inguscio (1982).

If several absorption lines overlap within the Doppler profile, classification of the observed FIR emissions according to their pump transition can be done unambiguously using the Transferred Lamb Dip (TLD) technique (Inguscio et al. 1979b). A typical situation is illustrated in Fig.2 for the 10R34 CO_2 pump line. Nine FIR emissions are obtained (Table Ia), and the optoacoustic spectrum reveals a broad absorption in which at least three unresolved transitions are overlapping. If the absorption line center is located within the pump tuning range, the TLD technique allows to associate a precise pump offset to each FIR emission and thus provides the required classification.

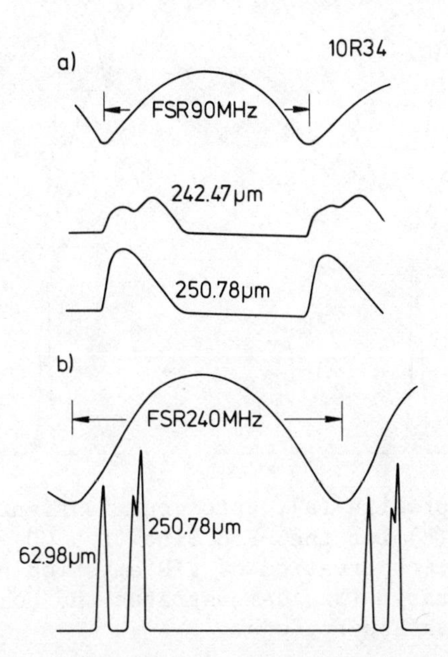

Fig.2 FIR emission from three different absorptions pumped by 10R34 of a conventional CO_2 laser (a) and a waveguide CO_2 laser (b). In (a) only the TLD for the 242.47 µm line is within the tuning range, whereas the TLD for the 250.78 µm line is outside. In (b) the 250.78 µm line is inside, and a third line, without TLD, is seen at an even larger offset. (Adapted from (50) and (61)).

It is also worth noting that weak absorptions can pump CH_3OH laser lines. This is another feature which makes the spectrum more rich and its interpretation more difficult. A typical situation is illustrated in Fig.3 for the 9P18 CO_2 laser pump. The large offset lines No. 109 and 112 can be pumped by Stark tuning, by a TEA source, or by a tunable waveguide laser. Recently a new line (No. 113) has been reported by Dyubko et al. (1981), who used a

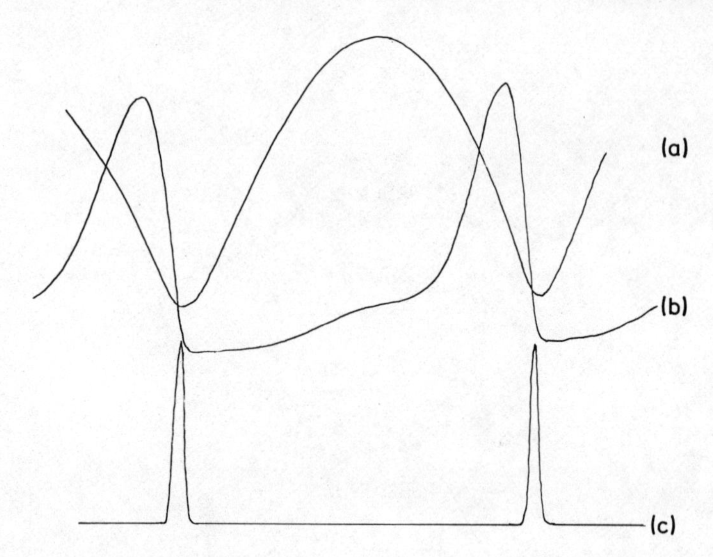

Fig.3 CO_2 tuning profile (a), optoacoustic signal (b) and 205.0 μm FIR output (c) for the 9P18 line of a CO_2 waveguide laser (55). Recent observation of FIR emission near the center of the tuning range (52) demonstrates the possibility of pumping very weak absorptions.

much less tunable pump source. This line is evidently pumped by the much weaker absorption, optoacoustically detected nearly at the center of the tuning range.

Assignments

From the strength of the transition matrix elements it follows that the typical emission pattern will consist of triads of lines as indicated in Fig.4. In the most frequently encountered case (a), the energy increases with K , and an approximate combination relation exists among the three lines. Case (b) may occur if the energy decreases with K , and again an approximate combination relation can be constructed by scaling one of the emission frequencies by $(J + 1)/J$. A large number of such triads have been identified, and together with calculations based on the works of Lees and Baker

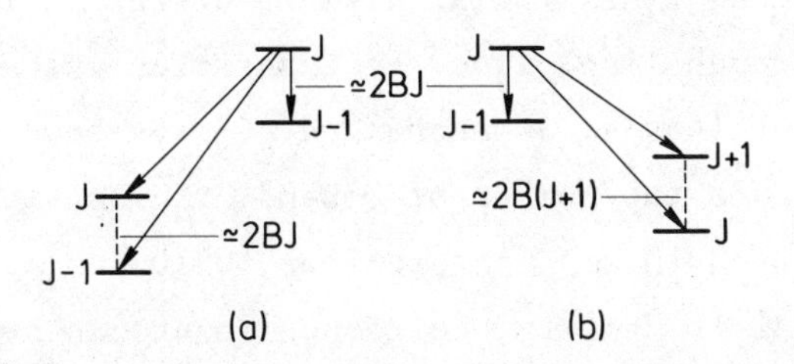

Fig.4 Construction of combination relations among FIR emission lines for energy increasing (a) and decreasing with K .

(1968) and Kwan and Dennison (1972), they serve as the main basis for the assignments listed in Table IV. The entire assignment procedure, including the role played by polarization measurements, Stark effect measurements, saturated absorption spectroscopy and diode laser spectroscopy has been discussed in details by Henningsen (1982c), where the relevant references can be found.

Included in Table IV are only those lines for which all torsional and rotational quantum numbers have been identified. Most of the assigned lines can be associated with the C-O stretch vibrational mode, but a different mode, labeled x , and identified as the CH_3 in-plane rock mode (Henningsen, 1982b) has also been pumped. The notation $(n\ \tau\ K)^u$ and $(n\ \tau\ K)^\ell$ refers to upper and lower components of states which are hybridizations of K states of the C-O stretch and K-1 states of x . Henningsen (1982a) lists 90 lines which are associated with the torsional ground state of the C-O stretch, and the measured frequencies are compared with calculations based on a numerical model (Henningsen 1981). Table IV in addition contains assignments of lines which are entirely within the torsionally excited manifold.

Reference to incomplete, tentative and dubious assignments, is given in Table V. In order to justify this classification a few remarks are needed. The crucial element of the assignment procedure is the Q-branch line, i.e. the transition which changes K by +1 or -1, but leaves J unchanged. Fig.5 shows for the vibrational ground state the number of Q-band origins with $n \leq 3$, $K \leq 20$, inside each 10 cm^{-1} interval up to 100 cm^{-1}. The density is seen to be fairly uniform, corresponding to an average spacing of 0.6 cm^{-1}. The observed lines frequently correspond to high

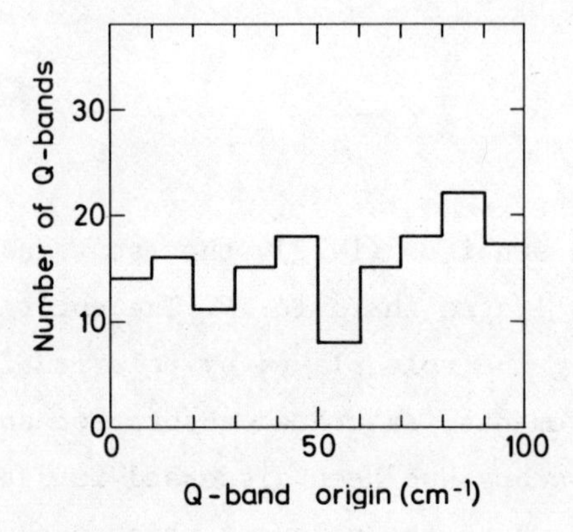

Fig.5 Number of Q-band origins with $n \leq 3$ and $K \leq 20$ inside each 10 cm^{-1} interval up to 100 cm^{-1} for the vibrational ground state.

J-values where the Q-band lines can easily be shifted several cm^{-1} away from the origins, and similar shifts are induced by realistic variations in the C-O stretch molecular parameters with torsional excitation. As a consequence of these effects, any measured frequency will usually lead to a number of possible assignments, even if the pump frequency seems to suggest that it belongs to the C-O stretch. That early assignments, performed by comparing only a single Q-band frequency with calculations for the vibrational ground state, nevertheless proved to be correct, can be attributed to the fact that only lines pumped by strong absorptions associated with n = 0 or 1 were considered. To establish a safe assignment for lines pumped by weaker absorptions, additional evidence is needed, such as Stark data, or the inclusion of the lines in a larger, self consistent scheme. Assignments are listed as tentative if such evidence is not presented, and as dubious if they are inconsistent with other assignments which are, in the opinion of the reviewers, well founded.

Spectroscopic Constants

Infrared spectra have been recorded by Barnes and Hallam (1970) and by Serralach et al. (1974), who both give the frequencies of all 12 normal modes. The C-O stretch has been studied with high resolution by diode laser spectroscopy by Sattler et al. (1979), and microwave transitions within the C-O stretch have been identified by Lees (1972). Molecular constants for the vibrational ground state as derived from the works of Lees and Baker (1968) and Kwan and Dennison (1972) are given in Table VI, which also contains the constants for the C-O stretch as determined from a combined analysis of diode laser spectra and optically pumped laser emission (Henningsen, 1981). The C-O stretch is perturbed by a vibrational interaction and the effect of this perturbation is quantified by Henningsen (1981).

Suggestion for Further Measurements

A majority of the CH_3OH laser IR-FIR transitions are still unassigned. To remedy this, precise pump offsets and precise FIR frequency measurements are required for as many lines as possible. In addition, extending the pump power to the 50-100 Watt range over a sizable portion of the CO_2 laser spectrum may produce many new lines, in particular if full advantage is taken of isotopic pump lasers and waveguide lasers. Many of the unassigned lines must arise from highly excited torsional states of the C-O stretch, or from CH_3 rock bands pumped by perpendicular transitions. Owing to the high accuracy of the data, a full analysis will considerably deepen our understanding of these molecular states.

Table Ia. CH_3OH lines a)

No	Pump Line	Wavelength (μm)	Frequency (GHz)	Rel. Pol.	Rel. Pwr. b)	Offset (MHz) c)	Press. (Pa)	Comments	Ref. n)
1	9R34	29			M			TEA,d)	41
2	9R32	33			M			TEA,d)	41
3	9R28	198.9±0.1						TEA	47
4		289 ±1			0.1		13	CW,e)	52
5	9R26	151.25369	1982.0506	‖	9		23	CW	5,46,53
6		159.67568	1877.5085	‖	7		23	CW	3,46,53
7		290 ±1		‖	10		13	CW,f)	52
8		346 ±1		⊥	3		13	CW,f)	52
9		461 ±1		⊥	1		13	CW,f)	52
10		781 ±2		‖	2		13	CW,f)	52
11	9R24	241.0±0.2		‖	W		20	CW	59
12		288.7±0.3		‖	W		20	CW	59
13	9R22	232.78845	1287.8322	‖	1	+15	16	CW	3,46,22
14	9R20	186.3±0.2		‖	M		32	CW	59
15	9R18	33.54±0.05		⊥			50	CW	71
16		36.69±0.05		‖			50	CW	71
17		49.50±0.05		‖			50	CW	71
18		61.61331	4865.7098	⊥	3		41	CW	46
19		67.49536	4441.6752	⊥	0.5		33	CW	46
20		186.04220	1611.4219	‖	10	+5	41	CW	46,5
21		251.43239	1192.3383	‖	4	+5	33	CW	46,3,22
22		280.93413	1067.1276	‖	2	+5	23	CW	46,3,22
23		48.90±0.15			M		7	STARK	36
24		63.78±0.15			M		6	STARK	36
25		25-40			W			TEA,d)	41

No	Pump Line	Wavelength (μm)	Frequency (GHz)	Rel. Pol.	Rel. Pwr. b)	Offset (MHz) c)	Press. (Pa)	Comments	Ref. n)
26		64.0±0.1						TEA	47
27	9R16	33.6±0.1					70	CW	71
28		188.9±0.3					70	CW	71
29		56.9±0.3		⊥			29	CW,g)	53,71
30		66.3±0.3		||			29	CW,g)	53,71
31		191.63±0.05		||			32,75	CW,g)	3,5,53,59
32		310 ±1		||	5		13	CW	52
33		419 ±1		||	10		13	CW	52
34	9R14	100.80647	2973.9406	||	9		28,81	CW	46,53,59
35		194.06319	1544.8187	⊥	3.6		27,44	CW	46,22,53,59
36		209.93023	1428.0576	||	3.8		27,44	CW	3,46,22,53,59
37		216.8±0.2		||	M		44	CW	59
38		319 ±1			1		13	CW	52
39	9R12	34.3±0.4						TEA	47
40		430 ±1		⊥	0.5			CW	52
41		449 ±1		||	0.2			CW	52
42	9R10	96.52239	3105.9368	||	20,S	+3	80,40	CW	5,46,22,58
43		164.78325	1819.3140	⊥	27	+3	73,35	CW	3,46,22
44		232.93906	1286.9995	||	20	+3	73	CW	3,46,22
45		242.2±0.3		||	W		71	CW	59
46		285 ±1		||	5		13	CW	52
47		216.8±0.2				large		TEA	47
48		190.65±0.10		||		>-110		CW,WG	61
49		224.53±0.05		⊥		>-110		CW,WG	61
50	9R8	77.40565	3873.0051	||	5	+28	50	CW	46,55,53
51		86.23938	3476.2825	||	1		50	CW	46,3,53
52		113.73188	2635.9580	⊥	1.5		40	CW	46
53		225.51589	1329.3629	||	0.5		40	CW	46,3
54		232.93±0.05		||				CW,h)	5

(continued)

Table Ia (Continued)

No	Pump Line	Wavelength (μm)	Frequency (GHz)	Rel. Pol.	Rel. Pwr. b)	Offset (MHz) c)	Press. (Pa)	Comments	Ref. n)
55		259 ±1		$\parallel$	5			CW	52
56		461 ±1		$\perp$	3			CW	52
57	9R6	235.8±0.3		$\parallel$	M		17	CW	59
58	9R4	235.1±0.3		$\parallel$	M		23	CW	59
59	9R2	94.7±0.5		$\parallel$			16	CW,g)	53
60		105.1±0.5		$\parallel$			16	CW,g)	53,44
61		151.9±0.7		$\perp$	S		16	CW,g)	53,45
62		176 ±1.5		$\perp$	S			CW	45,44
63		261.5±1		$\perp,\parallel$	S			CW,i)	45,44,53
64	9P6	46.23±0.08		$\parallel$			50	CW	71
65		117.8±0.1		$\perp$	W		39	CW	59
66		167.8±0.2		$\parallel$	S		52	CW	59,71
67		183.9±0.2		$\perp$	W		40	CW	59
68		419 ±1		$\parallel$	8		13	CW	52
69	9P8	77.6±0.1						TEA	47
70		150			S			TEA,d)	41
71		134.4±0.4					4	STARK	36
72		181.1±0.2		$\parallel$			16	CW	59
73	9P10	45.5±0.1			0.2	+77	27	CW,WG	46,55
74		46.2±0.1						TEA,CW	47,71
75		50.3±0.1		$\perp$				CW	71
76		214.35±0.06		$\parallel$	0.1	-20	4,31	CW	22
77		218.22±0.06		$\parallel$	0.1	-20	4,31	CW	22
78		289.70±0.06		$\perp$		-20	29	CW	56
79		232 ±1		$\parallel$	1		13	CW	52
80	9P12	46.4						TEA,d)	16
81		206.78508	1449.7780	$\parallel$	0.3	-5	4	CW,g)	46,22
82		211.31476	1418.7010	$\perp$	1	-5	5	CW,g)	46,22
83		290.6±0.1		$\perp$		-5	5	CW,g)	22

No	Pump Line	Wavelength (μm)	Frequency (GHz)	Rel. Pol.	Rel. Pwr. b)	Offset (MHz) c)	Press. (Pa)	Comments	Ref. n)
84		163.57355	1832.7686	∥	M	+85	7,25	CW,WG,TEA,STARK	42,2,36,59,73
85		257.8±0.3		⊥	W	+85	7,31	CW,WG,STARK	42,36,59
86		448.45537	668.5001	∥	M	+85	8,25	CW,WG	42,59,73
87		452 ±1						TEA	8
88		781 ±1		∥	6		13	CW	52
89	9P14	37						CW,d)	46
90		51.48±0.02		⊥		-30		CW	71
91		55.9±0.5				+500		TEA	16
92		183.1±0.6						STARK	36
93		117.95954	2541.4856	∥	10	-30	23	CW	46,22
94		164.50757	1822.3627	⊥	12	-30	13	CW	46,12
95		301.99432	992.7089	⊥	4	-30	13	CW	46,22
96		386.33926	775.9824	∥	2	-30	13	CW	46,22
97		416.52243	719.7511	∥	5	-30	15	CW	46,12
98	9P16	570.568643	525.4275	∥	50	+64	13	CW,WG	1,46,49
99		627.3±0.6		∥		+64		CW	5
100		1223.65981	244.9966	∥	0.3	+64	7	CW	46,7
101		35.86±0.04		⊥		-56	25	CW	71
102		41.91±0.01		⊥		-56	25	CW	71
103		164.60038	1821.3352	⊥	8,M	-56	11,43	CW,WG	1,46,49,59
104		223.5±0.3		∥	M		44	CW	1,46,59
105		369.11369	812.1954	∥	15,M		20,44	CW	1,46,59
106		306.3±0.3		∥	M		41	CW	59
107		44.18±0.03		⊥	2.5	+64	11,25	CW	46,71
108		<30						TEA,d)	47
109	9P18	205.0±0.1		∥	S	+140	11	CW,WG,TEA,STARK	2,55,36,59,64
110		203.2±0.2						TEA	47
111		225.2±0.2						TEA	47
112		304.8±0.2		⊥	W	+140	11	WG	64

(continued)

Table Ia. (Continued)

No	Pump Line	Wavelength (μm)	Frequency (GHz)	Rel. Pol.	Rel. Pwr. b)	Offset (MHz) c)	Press. (Pa)	Comments	Ref. n)
113		676		∥	1.5		13	CW	52
114	9P20	51.3±0.1						TEA	47
115	9P22	47.8±0.1		⊥				CW	56
116		57.3±0.1		∥				CW	56
117		213.46246	1404.4270	∥	0.9		15	CW	46,3
118		346.48751	865.2331	∥	1.8		15	CW	46,3
119	9P24	43.97±0.03		∥		0	50	CW	71
120		47.80±0.03		∥		0	50	CW	71
121		92.54392	3239.4616	∥	5	0	17	CW	46,22
122		133.11956	2252.0542	∥	12.5	0	20	CW	46,22
123		164.69747	1820.2615	⊥	7.5	0	13	CW	46,3
124		311.2±0.5		⊥				CW	3
125		470		∥	30		13	CW	52
126		602.48698	497.5916	∥	2	0	13	CW	46,3
127		614.28518	488.0347	∥	1.5	0	13	CW	46,22
128		694.18923	431.8598	∥	2	0	13	CW	46,22
129	9P26	516 ±1		∥	10		13	CW	52
130		973 ±2		∥	5		13	CW	52
131	9P28	314 ±3		∥	5,W		13,23	CW	52,59
132		417 ±1		∥	10		13	CW	52
133	9P30	25-40						TEA,d)	41
134		41.90±0.05						TEA	47
135		210 ±0.5		∥			13	CW	52
136		314 ±0.5		∥			13	CW	52
137	9P32	37.85421	7919.6602	⊥	15	-16	33	CW	46,4,50
138		42.15908	7110.9814	∥	10	-16	53	CW	46,4,50
139		240 ±0.5		∥			13	CW	52
140		274.3±0.5		∥			33	CW	42,56,52
141		372 ±1		⊥			13	CW	52

No	Pump Line	Wavelength (μm)	Frequency (GHz)	Rel. Pol.	Rel. Pwr. b)	Offset (MHz) c)	Press. (Pa)	Comments	Ref. n)
142		418 ±1		⊥			13	CW	52
143		141.7±0.2						TEA	47
144	9P34	39.92423	7509.0362	⊥	10	+24	17	CW	46,4,20
145		42.30±0.02		||	10	+24		CW	4,69,20
146		63.36954	4730.8604	||	33	+24	10	CW	46,4
147		70.51163	4251.6740	⊥	45	+24	13	CW	46,1,58
148		80.3±2						CW	4
149		180.67639	1659.2786			+24	8	CW	46
150		185.50040	1616.1284	⊥	10	+24	9	CW	46,1
151		186.31913	1609.0267	⊥		+24	7	CW	46
152		190.72590	1571.8497	⊥	7	+24	9	CW	46,1
153		237.6±0.05		||		+24		CW	1
154		253.55297	1182.3662	||	6	+24	9	CW	46,1
155		254.04149	1180.0925	||		+24	7	CW	46,1
156		263.68315	1136.9420	||		+24	7	CW	46,1
157		264.53590	1133.2770	||	4	+24	8	CW	46,1
158		292.5±0.05		⊥		+24		CW	1
159		303 ±0.6		||	300	+24	13	CW	52
160		362.6±0.4		⊥	400	+24	13,11	CW	52,59
161		622 ±1		⊥	500	+24	13	CW	52
162		699.42258	428.6285	⊥	27	+24	15	CW	46,1
163		205.6±0.2		⊥	S	+110	7	WG,STARK	49,46,28,64
164		208.3±0.2		⊥	S	+130	7	WG,STARK	49,46,26,64
165		255.5±0.3		||	M	+120	7	WG	64
166		204.5±0.6			M		4	STARK	36
167		206.5±0.6			M		4	STARK	36
168	9P36	118.83409	2522.7816	⊥	160	+24	17	CW	1,46
169		170.57638	1757.5263	||	58	+24	8	CW	1,46
170		202.4±0.1		||		+24		CW	1,46

(continued)

Table Ia. (Continued)

No	Pump Line	Wavelength (μm)	Frequency (GHz)	Rel. Pol.	Rel. Pwr. b)	Offset (MHz) c)	Press. (Pa)	Comments	Ref. n)
171		392.06872	764.6426	⊥	35	+24	19	CW	1,46
172		417.8±0.05		⊥		+24		CW	1
173		418.08268	717.0650	⊥	0.8	+24	9	CW	46
174		332 ±1		||	150		13	CW	52
175		110.71647	2707.7493	⊥	VS	-80	7	WG,CW,TEA	49,47,73
176		162.21804	1848.0838	||	S	-80	11	WG,CW	49,73
177		99.28±0.05		⊥	M	-89	7	WG,CW	49
178		135.71±0.05		||	S	-89	11	WG,CW	49
179		117.8±0.1						TEA	47
180	9P38	141.7±0.2			W			TEA	47
181		193.14159	1552.1901	⊥	2.2	0,+13	11	CW	46,1,50,56
182		198.66433	1509.0402	⊥	2.4	0,+13	11	CW	46,1,50,56
183		278.80483	1075.2771	||	2.4	0,+13	11	CW	46,1,50,56
184		292.14149	1026.1893	||	2	0,+13	11	CW	46,1,50,56
185		624.43013	480.1057	⊥	0.6	0,+13	7	CW	46
186	9P40	55.37004	5414.3441	⊥	1	-10	15	CW	46,22
187		60.17327	4982.1531	||	0.7	-10	9	CW	46,22
188		73.30643	4089.5796	||	0.1	-10	20	CW	46,22
189		85.60094	3502.2102	||	0.1	-10	20	CW	46,22
190		697 ±1.5		⊥	50		13	CW	52
191	9P42	255.04±0.7			M		4	STARK	36
192	9P44	51.53±0.15			W		3	STARK	36
193		112.94550	2654.3107	⊥	W	-31	20	WG	73
194		196.56407	1525.1641	⊥	W	-31	16	WG	73
195	9P46	52.59±0.15			W		3	STARK	36
196	10R48	164 ±2			M		40	CW	50
197		286.15500	1047.6576	||	S	-11,-28	42,26	CW,WG	12,50,73
198	10R46	53 ±1			0.5	+15	11	CW	50
199		65 ±1				+15		CW	50

No	Pump Line	Wavelength (μm)	Frequency (GHz)	Rel. Pol.	Rel. Pwr. b)	Offset (MHz) c)	Press. (Pa)	Comments	Ref. n)
200		274.24523	1093.1547	‖	S	>+45	20	CW,WG	12,46,50,73
201	10R44	120.90247	2479.6222	⊥	0.5	+9	13,33	CW,WG	12,46,73
202		162.5±0.2		‖	W		45	CW	59
203		231.3±0.2		⊥	M		45	CW	59
204		251.91181	1190.0691	‖	0.4	+9	13,33	CW,WG	12,46,73
205		284.0±0.3		‖	M		45	CW	59
206		381.5±0.4		⊥	W		45	CW	59
207	10R40	97.51854	3074.2100	⊥	14	0	37	CW	46,12,22
208		167.58699	1788.8766	‖	7		27	CW	46
209		244 ±1		‖	1		13	CW	52
210	10R38	163.03353	1838.8393	‖	30,M	+27	13,29	CW	46,50,56,6
211		213.9±0.2		‖	M		32	CW	59
212		251.11398	1193.7283	‖	4,M	+27	13,29	CW	46,50,56,6
213		251.4±0.3		⊥	W		25	CW	59
214		261.7±0.3		‖	M		32	CW	59,50,56,6
215		469.02331	639.1846	⊥	50,M	+27	13,25	CW	46
216	10R36	43.71±0.02		‖	2.7	+23	10,40	CW	42,46,66
217		53.86087	5566.0527	⊥	7	+23	10,40	CW	42,46,50
218		233 ±0.5		‖	1		13	CW	52
219	10R34	43.47±0.02		‖				CW	22
220		92.66285	3235.2536	‖	1			CW	22,46
221		163.01±0.02						CW	5
222		242.47269	1236.3968	‖	10	−30	17	CW	46,12,50
223		129.54972	2314.1113	‖	5	−46	13	CW	9,46,50,61
224		250.78129	1195.4339	‖	11	−46	13	CW	11,46,50,61
225		267.44316	1120.9577	⊥	7	−46	15	CW	46,50,61,22
226		48.77±0.02		⊥		−83	13	CW	61
227		62.98±0.01		‖		−83	13	CW	61

(continued)

Table Ia. (Continued)

No	Pump Line	Wavelength (μm)	Frequency (GHz)	Rel. Pol.	Rel. Pwr. b)	Offset (MHz) c)	Press. (Pa)	Comments	Ref. n)
228	10R32	100.16598	2992.9570	||	W	+84	21	CW,WG	71,73
229		145.25243	2063.9411	||	W	-26	37	CW	50,59,73
				⊥,||		< -45	27	WG,ℓ)	42
230		242.84714	1234.4904	||	0.1,M	<-45,-26	27,37	CW,WG	5,46,50,59,73
231		362.0±0.3		⊥	M		37	CW	59
232		390 ±1		||	7		13	CW	52
233	10R26	329 ±1		⊥	3		13	CW	52
234	10R22	262.7±0.3		||	W		19	CW	59
235		622 ±1		||	10		13	CW	52
236	10R20	145.05±0.1		||		+130		CW	64
237		209.03±0.1		||		+130		CW	64
238	10R16	62.96597	4761.1824	||	10	-14	16	CW	46,50
239		69.67956	4302.4449	⊥	6		20	CW	46,22
240		77.90489	3848.1855	||	10	-14	25	CW	46,12,50
241		208					27	TEA,d)	35
242		564.±1		||	20		13	CW	52
243		695.34994	431.1390	||	1		17	CW	46,12
244	10R10	191.61961	1564.5187	||	50	-15	23	CW	5,46,22
245		293.82167	1020.3211	||	5	-15	27	CW	11,46,22
246	10R8	151 ±0.5		||	0.3		13	CW	52
247		262 ±0.5		⊥	1		13	CW	52
248	10R4	179.72791	1668.0350	||	11		20	CW	46
249		191.2±0.1						CW	5
250		211.26289	1419.0493	||	16		35	CW	12,46
251		493.54142	607.4312	⊥	0.5	-18	20,35	CW,WG	12,46,73
252	10R2	178 ±3		||	VW			CW	45
253	10P10	469 ±1		⊥	0.6		13	CW	52
254	10P12	368.3±0.35		||	0.6,W		13,19	CW	52,59

No	Pump Line	Wavelength (μm) m)	Frequency (GHz)	Rel. Pol. b)	Rel. Pwr. c)	Offset (MHz)	Press. (Pa)	Comments	Ref. n)
255	10P16	84.91±0.05		‖				CW	56
256		99.86132	3002.0875	⊥	W	-20	36	CW,WG	56,73
257		123.64±0.05		‖				CW	56
258		480 ±1		‖			13	CW	52
259		566 ±1		‖			13	CW	52
260	10P18	296 ±1		‖	3		13	CW	52
261		523 ±1		‖	1		13	CW	52
262		663 ±1		⊥	0.5		13	CW	52
263	S-9P13	107.7		‖			27	CW	56,67
264		147.9		⊥			27	CW	56,67
265		393.2		‖				CW	19
266	S-9P15	61.2		⊥			20	CW	56,67
267		67.9		‖			27	CW	56,67
268		702.0		‖			20	CW	56,67
269	S-9P17	376.0		‖			16	CW	56,67
270		720.0		⊥			13	CW	56,67
271		782.2		‖			13	CW	56,67
272	S-9P18	85.3		‖				CW	56,68
273		480		‖				CW	56,68
274		902		‖				CW	56,68
275	S-9P21	57.5		‖			27	CW	56,67
276		61.5		‖			20	CW	56,67
277		80.7		⊥			54	CW	56,19
278	S-9P31	159.3		⊥			27	CW	56,19
279		230.2		‖			26	CW	56,67
280		243.3		⊥			20	CW	56,67
281		353.3		‖			20	CW	56,67

(continued)

Table Ia. (Continued)

No	Pump Line	Wavelength (μm) m)	Frequency (GHz)	Rel. Pol. b)	Rel. Pwr. c)	Offset (MHz)	Press. (Pa)	Comments	Ref. n)
282	S-10R33	126.4						CW	56,68
283		250						CW	56,68
284	13-9R26	56.23		∥			7	CW	63
285		52.48		⊥			7	CW	63
286	13-9R22	674		⊥			12	CW	63
287	13-9R18	126.6		⊥			8	CW,g)	63
288		177.0		∥			8	CW,g)	63
289		212.2		(∥)			8	CW,g)	63
290		448.3					8	CW,g)	63
291	13-9R14	85.3		⊥			7	CW	63
292	13-9R12	124.7		⊥			7	CW	63
293		145.0		∥			7	CW	63
294	13-9P14	19.52					13	CW	63
295		419.3					13	CW	63
296	13-9P16	119.7		∥			12	CW,g)	63
297		148.0		⊥			12	CW,g)	63
298	13-9P18	51.85					16	CW,g)	63
299		60.35		∥				CW,g)	63
300	13-9P20	351.1		∥			12	CW	63
301	13-9P26	132.9					13	CW	63
302		297.7		∥			13	CW	63
303	18-9P12	60.58		⊥			13	CW	63
304	18-9P14	102.3		∥			13	CW	63
305		193.5		⊥			13	CW	63
306		216.5		∥			13	CW	63
307	18-9P16	108.8		∥			13	CW	63
308		211.0		⊥			13	CW	63
309		224.7		∥			13	CW	63

No	Pump Line	Wavelength (μm) m)	Frequency (GHz)	Rel. Pol. b)	Rel. Pwr. c)	Offset (MHz)	Press. (Pa)	Comments	Ref. n)
310	18-9P22	95.25		∥			13	CW	63
311		149.8		⊥			13	CW	63
312		262.4		∥			13	CW	63
313	18-9P24	272.9		∥			12	CW	63
314	18-9P26	79.98					7	CW	63
315	18-9P28	125.9					9	CW	63
316		205.1		⊥			12	CW	63
317		312.1		∥			12	CW	63
318	18-9P30	117.4					12	CW	63
319		176.5		⊥			12	CW	63
320		348.3		∥			12	CW	63
321	18-9P34	73.2		∥			12	CW	63
322		131.6		∥			12	CW	63
323		167.1		⊥			12	CW	63
324	18-9P36	41.06		⊥			9	CW	63
325	18-10R16	75.06		∥			10	CW,g)	63
326		109.4		⊥			10	CW,g)	63
327		240.4		∥			10	CW,g)	63
328	18-10R24	233.4		∥			11	CW	63
329		271.5		(∥)			11	CW	63
330	18-10R30	247.4		∥			5	CW,g)	63
331		552.0		⊥			5	CW,g)	63

(continued)

Table Ia. (Continued)

FIR Laser Pumped with a Continuously Tunable TEA Laser

	Pump Waven. cm^{-1}		Offset (MHz)	Comments	Ref. n)
332	1050.30	210.7	±300	TEA	75
333	1052.40	210.2	±300	TEA	75
334	1053.65	210.2	±300	TEA	75
335	1055.10	210.7	±300	TEA	75
336	1056.35	210.2	±300	TEA	75

Table Ia - Footnotes

a) The table contains lines pumped by regular $^{12}C^{16}O_2$ lasers (no prefix), $^{12}C^{16}O_2$ sequence lasers (prefix S-), regular $^{13}C^{16}O_2$ lasers (prefix 13-), and regular $^{12}C^{18}O_2$ lasers (prefix 18-). Emission lines are quoted without uncertainty if the frequency has been measured. Frequencies are taken from Petersen et al. (1980) with an estimated accuracy of $\pm 5\times 10^{-7}$. For lines pumped by sequence - and isotopic lasers, the estimated accuracy is $\pm 5\times 10^{-3}$.

b) The relative power of a number of FIR laser lines pumped by different CO_2 lines can be found in Petersen et al. (1980) and Dyubko et al. (1981). All reported values are given in the table, but the values of the two works cannot be compared. Other papers occasionally report a different amplitude sequence for FIR emissions from a given pump line, depending on the experimental conditions. The power of FIR lines not included in Petersen et al. (1980) and Dyubko et al. (1981) is indicated either as S (strong), M (medium), W (weak).

c) The pump offset is given whenever measured on the FIR lines. Offsets measured in absorption only, and not associated with particular FIR lines are given in Table Ib.

d) Wavelength accuracy not reported.

e) The pump could also be 9R26

f) The pump could also be 9R24

g) Relative intensity sequence given by refs. 22, 53 and 63.

h) Could be line No. 44, pumped by 9R10.

i) || from Kon and Fukutani (1980); ⊥ from Landsberg (1980).

ℓ) Observed as partial ⊥ polarization by Inguscio et al. (1980a).

m) Wavelength accuracy 5×10^{-3} for lines pumped by sequence- and isotopic lasers.

n) The correspondence number-reference is the following:

1) Chang et al. (1970), 2) Fetterman et al. (1972), 3) Dyubko et al. (1973), 4) Hodges et al. (1973), 5) Wagner et al. (1973), 6) Domnin et al. (1974), 7) Tanaka et al. (1974), 8) Izatt et al. (1975), 9) Kon et al. (1975), 10) Petersen et al. (1975), 11) Radford (1975), 12) Tanaka et al. (1975), 13) Kramer and Weiss (1976), 14) Danielewicz and Coleman (1977), 15) Henningsen (1977), 16) Mathieu and Izatt (1977), 17) Radford et al. (1977), 18) Tobin and Jensen (1977), 19) Weiss et al. (1977), 20) Heppner et al. (1977), 21) Dyubko and Fesenko (1978), 22) Henningsen (1978), 23) Sattler et al. (1978), 24) Worchesky (1978), 25) Arimondo (1979), 26) Bionducci et al. (1979), 27) Feld (1979), 28) Inguscio et al. (1979a), 29) Inguscio et al. (1979b), 30) Inguscio et al. (1979c), 31) Koo and Claspy (1979), 32) Lees et al. (1979), 33) Sattler et al. (1979), 34) Young et al. (1979), 35) Bluyssen et al. (1980), 36) Gastaud et al. (1980), 37) Arimondo et al. (1980), 38) Forber and Feld (1980), 39) Henningsen (1980a), 40) Henningsen (1980b), 41) Hirose and Kon (1980), 42) Inguscio et al. (1980a), 43) Inguscio et al. (1980b), 44) Kon and Fukutani (1980), 45) Landsberg (1980), 46) Petersen et al. (1980), 47) Bernard et al. (1981) 48) Henningsen (1981), 49) Inguscio et al. (1981a), 50) Inguscio et al. (1981b), 51) Yoshida et al. (1981), 52) Dyubko et al. (1981), 53) Sokabe et al. (1981), 54) Lees et al. (1981), 55) Inguscio et al. (1981c), 56) Henningsen (1982c), 57) Strumia and Inguscio (1982), 58)Dahmani (1981), 59) Hartmann and Lindgreen (1982), 60) Henningsen (1982a), 61) Henningsen et al. (1982), 62) Henningsen (1982), 63) Petersen and Duxbury (1982), 64) Strumia et al. (unpublished) 65) Tsunawaki et al. (1981), 66) Strumia (unpublished), 67) Danielewicz and Weiss (unpublished), 68) Willenberg (ubpublished), 69) Weiss (unpublished), 70) Yano and Kon (unpublished), 71) Henningsen (unpublished), 72) Salomon (1979), 73) Vasconcellos et al. (1982), 74) Herlemont (unpublished), 75) Feld (1979).

Table Ib. Offsets Measured in Absorption

CO_2 Line	Offsets (MHz) (a)	Ref.
9R28	+30	65
9R26	−25±7	65
9R24	−19	65
9R22	+13±4	65
9R20	−13	65
9R18	−7	38
	+1±2	65
9R16	−9,−13	38
	−7±2	65
9R14	+18±5	65
9R12	−14	65
9R10	+4	38
	+7±2	65
9R8	+29	65
	+28	55
9P6	+22±5	65
9P8	−200	41
	< −120	55
	−13	65
9P10	+75	55
	+77,−20	59
9P14	−34	65
9P16	+64.74±.03	72
	+64,−56	49
9P18	+2±2	65,55
	+140±15	64
9P22	+7±6	65
9P24	−1	38
	−4±5	65
9P26	−60±5,+50±5	55
9P28	+12	65
9P30	−1	38
	+8±4	65
9P32	−15±3	65
	−11	38
9P34	+25±5	65
	+22	38
9P36	+24,±4	65
	+19,+6	38
9P38	−5±4	65
9P40	−5±4	65
10R46	−6±3	65
10R44	+14±4	65
10R40	−14±3	65
10R38	+26±2	65

CO_2 Line	Offsets (MHz) (a)	Ref.
10R36	+17±2	65
10R34	<-40	65
10R32	<-40	65
10R20	+130±10	64
10R16	-14±3	65
	-11±5	55
10R10	+11±3	65
10R8	+26	65
10R4	-11±2	65
10P10	-56±10	64
10P12	-50±15	64
10P16	-50±8	64
10P18	-40±8	64

Table Ib - Footnote

a) Unpublished intracavity absorption measurements have been reported by ref.74.

Table IIa. Pump Laser Characteristics

Ref. (a)	Conv.	WG	TEA	Power CW(W)	Power[c] LP(W)	Tuning ±MHz	Comments
1	x				20	50	
2			x			125	d
3	x			10		58	
4	x			25		62	e
5	x				150	53	
6	x			5		37	
7	x			1.3		14	
8			x				
9	x			3		37	
10	x					37	
11	x			25			
12	x				10	14	
13	x			30		30	
16			x			750	g
17	x						
18	x			20			
19	x			9		<40	i
20	x			22		42	
21	x			10		58	
22	x				25	40	
27			x			q	
28	x				40	45	
29	x				40	45	
30	x				40	45	
31	x			6			
35	n)						
36	x			30		28	
37	x				30	45	
39	x				50	40	
40	x				50	40	
41			x				r
42	x				40	45	
43	x				40	45	
45	x			10		37	
46	x			20		37	
47			x				
48	x				50	40	
49		x			20	120	
50	x				40	45	
51	x			25		39	
52	x			10		58	
53	x			45		55	
55		x			20	120	

Ref. (a)	Conv.	WG	TEA	Power CW(W)	Power LP(W)	Tuning ±MHz	Comments
56							s
57							s
58	x			18		25	
59	x				100	40	
61		x			20	120	
62	x				50	40	
63	x			15		30	u
64		x			20	170	
64		x			20	140	
73		x		20		70	
75			x				q

Table IIb. Fir Resonater Characteristics

Ref. (a)	Open	Metal WG	Diel. WG	Hyb. WG	Length (cm)	Cross-sect. (mm)b)	Comments
1	x				77		
2	x						d
3	x				120		
4	x				70		e
5	x				200		
6	x				105		
7		x			158	6-24	
8		x			70	10	
9	x				250		f
10		x			200	14	
11		x			90	11	
12		x			100	12	
13	x				80		
16		x			90	25	
17		x			200	14	
18				x	79	25×13	h
19			x		185	40	
20	x				80		ℓ
21	x				120		
22	x				100		
27		x			100	12	
28			x		100/150	40	
28				x	100/150	35×5,10	h,m
29				x	100/150	35×5	h,m
30				x	100/150	35×5,10	h,m
31	x				69	27×6	h,m
35		x	x		100	13-25	
36				x	120	40×5	o
37			x		150	20	p
39	x				140	85×5-20	h,m
40	x				140	85×5-20	h,m
41					80		r
42			x		150	40	
43				x	100/150	35×5,10	h,m
45	x				200		
46		x			200	14	
46	x				100		
47		x			91	25	
48	x				140	85×5-20	h,m
49			x		100	40	
49				x	150	35×5	h,m
50				x	100/150	35×5,10	h,m
51				x	130	40×10	h,m
52	x				120		

Ref. (a)	Open	Metal WG	Diel. WG	Hyb. WG	Length (cm)	Cross-sect. (mm)b)	Comments
53	x				92		
55			x		150	40	
56							s
57							s
58				x		24×22	h,m
59	x				70		t
61			x		100/150	40	
62	x				140	85×5-20	h,m
63	x				90		
63		x			70		
64			x		150	40	
73	x				100	50	
75		x			100	12	

Table II - Footnotes

a) The correspondence number - reference is the same as in Table I.
b) A single number denotes diameter of cylindrical waveguide. A product denotes width and height of rectangular waveguide.
c) Internally chopped CO_2 laser; ~150 μsec long pulses.
d) Pump power density ~700 KW/cm^2 . FIR resonator similar to that of ref.1.
e) Only half the CO_2 laser power effectively focussed into the FIR cavity.
f) Output coupling using a small intracavity mirror.
g) CO_2 bandwidth reduced using an intracavity etalon.
h) Stark effect on the power.
i) Hot intracavity cell used to pump with the sequence band lines.
ℓ) Used to perform FIR gain measurements.
m) Stark effect on the frequency.
n) The pump source was a conventional CO_2 laser Q switched with a rotating mirror and current pulsed. FIR pulses very short (~100 ns) were obtained with peak power > 1 W .
o) Stark tuning into resonance.
p) FIR waveguide put in a coaxial RF waveguide to apply RF power to the active medium and observe double resonance signals.
q) Continuously tunable high pressure TEA laser with a bandwidth of ~2 GHz and 200-400 kW peak power.
r) FIR resonator placed inside the CO_2 resonator.
s) Various CO_2 and FIR configurations are described.
t) Detection in dry nitrogen to avoid H_2O FIR absorption.
u) Isotopic CO_2 laser pump.

Table III. Stark Effect on CH_3OH Laser Lines

Pump Line	FIR Line (a)	Power (b)	Freq. (c)	Line Shift (d) MHz/(kV/cm) ΔM=±1	ΔM=0	Range kV (e)	Total Tuning MHz	Pump Line Tuning	Ref. (f)
9R18	48.9					2.2(σ)			36
	48.9					2.5(π)			36
	63.8					1.4–3.4			36
	67.5	↓	lin	33.1		0–1.6	53		51
9R10	96.5	↑	lin	11.4	12	0–3	34		26,50,57
	164.8	↑	b			0–0.5			43
	232.9	↓	b			0–0.5			57
9P8	134.4					10.8–15			36
9P12	163.6					3.0–6.6			36
	163.6		b						49
	257.8					3.7			36
	257.8	↑	lin	18.3		0–2.2	40		49
	448.5	↓	b						49
9P14	118.0	↓	lin	113		0–1.6	181		51
	183.1					8.3–12.2			36
9P16	164.6	–	–			0–0.8		–	50,49,29
	570.6	–	–			0–2		–	50,49,29
9P18	205.2					3.8–11.6			36
	205.2	↓	non-lin	35		0–0.8	29		64
9P24	133.1		non-lin	101	95	0–0.4	40		40
	133.1	↓	lin	99		0–1.6	155		51
9P32	37.9	↑	non-lin	59		0–2	90		50
	42.3	↓	b			0–0.1			50

Pump Line	FIR Line (a)	Power (b)	Freq. (c)	Line Shift (d) MHz/(kV/cm)		Range kV (e)	Total Tuning MHz	Pump Line Tuning	Ref. (f)
				ΔM=±1	ΔM=0				
9P34	63.4	↓	b			0-0.15			50
	70.5	↑	lin	44.2		0-1.4	62		18,42
	70.5	↑	lin	46.2	44.6	0-0.4	18		39
	204.5					6.7(π)			36
	204.5					2.8(σ)			36
	205.6	↑	non-lin	97.4	82.0	0-0.6	58	lin	64,30
	205.6		lin	(g)		0.6-1.9	127		64,30
	206.5					11.9-15.2			36
	208.3					1.3-2.7			36
	208.3	↑	lin	135		0-2.7	370	lin	30,62,64
9P36	99.3	↑	lin	29		0-1	29		49
	110.7	↑	lin	24.6		0-1.3	32		49
	110.7					4-8.6			36
	118.8	↑	lin(i)	26.2	28.8	0-6.5	172	lin	26,43,39 42,31,29
	135.7	↓	b						49
	162.6	↓	b						49
	170.6	↓	b						26,43
9P38	193.1	↑	lin	33.6	21.6	0-0.35	12	lin	40,50,62
	198.6	↑	b			0-0.6		non-lin	50
	198.6	↑	lin	24		0-0.35	10	lin	62
	278.8	↑	lin	38.4		0-0.3	12		62
	292.1	↑	lin	43.2		0-0.2	9		62
9P42	255.0					2.5-5			36
9P44	51.5					12.5-19			36
9P46	52.6					18			36

(continued)

Table III. (Continued)

Pump Line	FIR Line (a)	Power (b)	Freq. (c)	Line Shift (d) MHz/(kV/cm)		Range kV (e)	Total Tuning MHz	Pump Line Tuning	Ref. (f)
				ΔM=±1	ΔM=0				
10R48	164	↑	b			0-1			50
	286.2		b			0-1			50
10R46	53	↓	b			0-0.5			50
	65	↑	b			0-0.5			50
10R38	163.0	-	-					non-lin(h)	50
	251.1	-	-						50
	469.0	-	-						50
10R36	43.7	↓	b			0-0.7			50
	53.9	↓	b			0-0.7			50
10R34	129.5	↑	b			0-0.2			50
	242.5	↑	b			0-0.6			50
	250.8	↑	b			0-0.6			50
10R32	145.3	↓	b			0-0.8			50
	242.8	↓	b			0-0.1			50
10R16	63.0	↓	b			0-2			50
	77.9	↓	b			0-1			50

Table III - Footnotes

a) Wavelengths are listed according to table I and not necessarily identical to those given in the original references.

b) Effect on the power with ΔM=±1 pump selection rule. ↑ = power increase, ↓ = power decrease.

c) Linear or non-linear characterize the Stark shift of lines for which a splitting is observed. b indicates that only a Stark broadening is observed. Such lines are suitable for fast frequency modulation via Stark controlled pulling effect (57). - indicates that no Stark effect has been detected. Where nothing is entered the Stark effect is used to tune the pump transition into resonance, whereas no Stark effect is reported on the FIR emission.

d) For lines which exhibit a splitting into two components the peak separation is given. In case of non-linear shift at low fields, the numbers refer to the high field limit.

e) Actually applied, not to be considered as maximum. When two pump selection rules have been reported, the range refers to the $\Delta M=\pm 1$ case.

f) The correspondence number - reference is the same as for Table I.

g) One of the two components is subsplit into 4 components, with separation of 5.8 MHz/(kV/cm) . The overall splitting is 97.4 MHz/(kV/cm).

h) K splitting. $\Delta E = 370.8$ MHz (37).

i) High frequency (1 MHz) modulation observed by AC Stark effect (ref.42).

Table IV. CH_3OH Line Assignments

Pump Line [a]	Pump Assignment	Measured cm^{-1}	Emission μm	Emission Assignment $(n\tau K,J)\to(n'\tau'K',J')$	Calcul. [f] cm^{-1}	Comment	Ref. e)
9R14	R(039,29)	47.634874	209.9	039,30→039,29	47.632		22
		51.529607	194.1	→018,30	51.495		22
		99.199981	100.8	→018,29	99.155		60,59
		46.13	216.8	039,29→039,28	46.059	casc	59
9R10	R(0210,26)	42.929683	232.9	0210,27→0210,26	42.913		14,15,70
		60.685781	164.8	→039,27	60.786		14,15,70
		103.602901	96.5	→039,26	103.694		14,15,70
		41.28	242.2	0210,26→0210,25	41.333	casc	59
9R10	R(013^+,27)	44.54	224.5	013^+,28→013^+,27	44.504		61
		52.45	190.6	→022^+,27	52.505		61
9P8	R(017,15)	74.39	134.4	017,16→026,15	74.477		60
9P10	P(025,8)^x	46.65	214.3	025,7^x→025,6^ℓ	46.651	b	39
		45.83	218.2	→025,6^u	45.819		39
		34.52	289.7	→025,7^u	34.512		62
9P12	P(025,9)^x	12.80	781	025,8^x→025,7^x	12.814	b	c
		48.359388	206.8	→025,7^ℓ	48.361		39
		47.322770	211.3	→025,7^u	47.324		39
		34.41	290.6	→025,8^u	34.415		62
9P12	R(025,13)$^\ell$	22.298763	448.5	025,14^ℓ→025,13^ℓ	22.289		48
		38.79	257.8	→034,14	38.812		62
		61.134579	163.6	→034,13	61.132		48
9P16	R(010,10)	17.526375	570.6	010,11→010,10	17.527		15
		8.172206	1223.6	→031^+,10	8.162		15
		15.940	627.3	010,10→010,9	15.938	casc	15
		226.33	44.2	131^+,10°→010,10^o	225.78		71

Pump Line	Pump Assignment	Measured cm^{-1}	Emission μm	Emission Assignment $(n\tau K,J)\to(n'\tau'K',J')$	Calcul.[f)] cm^{-1}	Comment	Ref. e)
9P18	R(024,9)	48.78	205.0	024,10→033,9	48.747		48
		32.81	304.8	024,10→033,10	32.793	c	64
9P34	Q(125,9)	14.297508	699.4	125,9→125,8	14.254		15
		141.820579	70.5	→016,10		d	15
		157.804522	63.4	→016,9	157.806		15
		236.4	42.3	→034,9	236.123		15
		250.474487	39.9	→034,8	250.475		15
		53.671355	186.3	016,9→025,8^{ℓ}	53.671	casc	39
		52.431263	190.7	→025,8^{u}	52.431	casc	39
		39.363649	254.0	→025,9^{ℓ}	39.363	casc	39
		37.924304	263.7	→025,9^{u}	37.923	casc	39
		55.347575	180.7	016,10→025,9^{ℓ}	55.348	casc	39
		53.908241	185.5	→025,9^{u}	53.909	casc	39
		39.439490	253.5	→025,10^{ℓ}	39.440	casc	39
		37.802053	264.5	→025,10^{u}	37.803	casc	39
		34.19	292.5	034,8→013,7	34.205	casc	15
		124.5	80.3	234,8^{o}→125.9^{o}		v=0	
		42.09	237.6	134,9^{o}→125,9^{o}	42.139	v=0	15
		33.0	303	116,9→125,9	33.154	v=0	c
		27.58	362.6	134,8^{o}→125,9^{o}	27.670	v=0	c
		16.07	622	125,10^{o}→125,9	16.082	v=0	c
9P34	Q(016,6)	48.64	205.6	016,6→025,5^{ℓ}	48.600		48
		48.01	208.3	→025,5^{u}	48.001		48
		39.14	255.5	→025,6^{ℓ}	39.110		c
9P36	Q(018,16)	25.505733	392.0	018,16→018,15	25.506		14,15,70
		58.624765	170.5	→027,16	58.623		14,15,70
		84.150935	118.8	→027,15	84.150		14,15,70

(coninued)

Table IV. (Continued)

Pump Line	Pump Assignment	Measured cm^{-1}	Emission μm	Emission Assignment $(n\tau K,J)\rightarrow(n'\tau'K',J')$	Calcul.[f)] cm^{-1}	Comment	Ref. e)
		23.918714	418.1	$018,15\rightarrow018,14$	23.916	casc	56
		23.935	417.8	$027,15\rightarrow027,14$	23.934	casc	14,15,70
		49.41	202.4	$\rightarrow036,15$	49.422	casc	70
9P36	Q(0310,18)	61.645424	162.2	$0310,18\rightarrow019,18$	61.663		60,49
		90.320795	110.7	$\rightarrow019,17$	90.339		60,49
9P36	Q(0110,17)	73.69	135.7	$0110,17\rightarrow029,17$	73.696		49
		100.73	99.3	$\rightarrow029,16$	100.779		49
9P38	$^{P}Q(016,10)^{x}$	16.014602	624.4	$025,10^{x}\rightarrow025,9^{x}$		d	62
		50.336162	198.6	$\rightarrow025,9^{\ell}$	50.337		39
		51.775487	193.1	$\rightarrow025,9^{u}$	51.776		39
		34.229989	292.1	$\rightarrow025,10^{\ell}$	34.231		39
		35.867385	278.8	$\rightarrow025,10^{u}$	35.868		39
9P38	$Q(025,20)^{\ell}$	70.57	141.7	$025,20^{\ell}\rightarrow034,19$	70.608		60
9P42	P(024,5)	39.21	255.0	$024,4\rightarrow033,3$	39.197		60
10R38	$P(034^{+},26)$	39.818455	251.1	$034^{+},25\rightarrow034^{+},24$	39.822		14,15,70
		21.320902	469.0	$\rightarrow013^{-},25$	21.331		14,15,70
		61.337076	163.0	$\rightarrow013^{+},24$	61.345		14,15,70
		38.21	261.7	$034^{+},24\rightarrow034^{+},23$	38.267	casc	59
		46.75	213.9	$013^{+},24\rightarrow022^{+},23$	46.714	casc	59
		39.77	251.4	$013^{-},25\rightarrow013^{-},24$	39.828	casc	59
10R34	$^{P}P(016,26)^{x}$	39.875383	250.8	$025,25^{x}\rightarrow025,24^{x}$		d	60
		37.391122	267.4	$\rightarrow025,25^{\ell}$	37.390		60
		77.190443	129.5	$\rightarrow025,24^{\ell}$	77.192		60
S-9P15	R(123,8)	14.25	702	$123,9\rightarrow123,8$	14.254		67
		147.3	67.9	$\rightarrow014,10$	147.390		67
		163.3	61.2	$\rightarrow014,9$	163.336		67

Pump Line	Pump Assignment	Measured cm^{-1}	Emission µm	Emission Assignment (nτK,J)→(n'τ'K',J')	Calcul.[f] cm^{-1}	Comment	Ref. e)
S-9P17	R(032,7)	12.78	782	032,8→032,7	12.772		67
		13.89	720	→011,8	13.916		67
		26.60	376	→011,7	26.666		67
S-9P19	R(031$^+$,6)	11.09	902	031$^+$,7→031$^+$,6	11.072		24
		20.83	480	→010,6	20.893		24
S-9P31	Q(015,8)	28.31	353	015,8→024,8	28.263		48
		41.11	243	→024,7	41.022		48
13-9R22	Q(031$^+$,3)	14.82	674	031$^+$,3→010,2	14.991		60
13-9R18	Q(114,14)	22.31	448.3	114,14→114,13			63
		56.50	177.0	→123,14			63
		78.99	126.6	→123,13			63
18-9P14	R(039,28)	46.19	216.5	039,29→039,28	46.059		63
		51.68	193.5	→018,29	51.523		63
		97.75	102.3	→018,28	97.610		63
18-9P16	R(017,27)	44.50	224.7	017,28→017,27	44.527		63
		47.39	211.0	→026,28	47.320		63
		92.08	108.8	→026,27	92.091		63
18-9P22	R(019,23)	38.11	262.4	019,24→019,23	38.183		63
		66.76	149.8	→028,24	66.745		63
		104.99	95.2	→028,23	104.956		63
18-9P24	R(022$^+$,22)	36.64	272.9	022$^+$,23→022$^+$,22	36.710		60
18-9P28	R(017,19)	32.04	312.1	017,20→017,19	31.871		63
		48.76	205.1	→026,20	48.614		63
		80.71	125.9	→026,19	80.572		63

(continued)

Table IV. (Continued)

Pump Line	Pump Assignment	Measured Emission cm^{-1}	Emission μm	Emission Assignment $(n\tau K,J)\rightarrow(n'\tau'K',J')$	Calcul.[f)] cm^{-1}	Comment	Ref. e)
18-9P30	R(029,17)	28.71	448.3	029,18→029,17	28.672		63
		56.66	177.0	→028,18	56.750		63
		85.18	126.6	→038,17	85.419		63

Table IV - Footnotes

a) Precise values for the frequency of CO_2 laser lines including isotopically substituted CO_2, can be found in Freed et al. (1980). Values for the CO_2 sequence lines were reported by Siemsen and Whitford (1977).

b) Henningsen (1980a) gives the upper laser level as $n16^x$, whereas the more recent paper (Henningsen; 1982b) reports 025^x and identifies the x-state as the CH_3 in-plane rock vibration.

c) Not previously assigned.

d) Line included in a fit by Henningsen (1982a) to compute CH_3OH molecular constants.

e) The correspondence number-reference is the same as in Table I.

f) Calculated values based on ref.48.

Table V. Incomplete (i), dubious (d), or tentative (t) assignments for CH_3OH.

Pump Line	FIR Line (μm)	Assignment	Ref.
9R20	186.3	i	59
9R18	186.0	i	15
9R16	66.3	i	53
	191.6	i	15,53
9R8	232.9	t	53
	77.4	t	53
9R2	105.1	d	53,44
	176	d	53,44
	261.5	d	53,44
	151.9	d	53
	94.7	d	53
9P14	117.9	i	22
	164.5	i	22
	416.5	i	22
9P16	164.6	d	59
	223.5	d	15
	306.3	d	59
	369.0	d	15,59
9P22	346.5	t	25
	213.4	t	25
9P24	133.1	i	22
	164.7	i	22
	694.2	i	22
9P32	37.8	d,t	25,50
	42.1	d,t	25,50
10R48	286	t	42
	164	t	42
10R44	162.5	d	59
	284.0	d	59
	381.5	d	59
10R36	53.8	d	42
	43.7	d	42
10R32	144.9	d,t	59,42
	242.7	d,t	59,42
	362.0	d	59

(continued)

Table V (continued)

Pump Line	FIR Line (μm)	Assignment	Ref.
18-9P12	60.6	t	63
18-9P26	79.9	t	63
18-10R32	249	i,t	63
	148.9	i,t	63
	93.7	i,t	63
18-10R24	233.4	t	63
	271.5	t	63
18-10R16	240.4	i,t	63
	75.0	i,t	63
	109.4	i,t	63
13-9R26	56.2	t	63
	52.5	t	63

Table VI. Molecular Parameters for CH_3OH

	CH_3OH Ground State	CH_3OH C-O Stretch	Unit
I_b	34.003856	34.2828(26)	
I_c	35.306262	35.6380(26)	
I_{ab}	-0.1079		$kg\ m^2 \cdot 10^{-47}$
I_{a1}	1.2504	1.2523(8)	
I_{a2}	5.3331	5.3334(8)	
V_3	373.21	392.35(30)	
V_6	-0.52		
D_{KK}	0.38×10^{-4}		
k_1	-0.48×10^{-4}		
k_2	-18.41×10^{-4}		
k_3	-53.73×10^{-4}		
k_4	-85.50×10^{-4}		
k_5	137.07×10^{-4}		cm^{-1}
k_6	67.85×10^{-4}		
k_7	0		
F_v	-2.389×10^{-3}	$-6.546\ 10^{-3}$	
G_v	-1.168×10^{-4}	$-1.67\ 10^{-4}$	
L_v	-2.26×10^{-6}		
D_{JK}	9.54×10^{-6}		
D_{JJ}	1.6345×10^{-6}		
μ_a	2.952	3.055	$Cm \cdot 10^{-30}$
μ_b	4.80		

Moments of inertia are converted to amuÅ^2 by dividing the numbers of the table by 1.660531. Dipole moments are converted to Debye by dividing by 3.33564. Where nothing is entered for the C-O stretch, the ground state value may be used.

REFERENCES

Arimondo, E., 1979. IEEE J.Quantum Electron. QE-15, 1081.

Arimondo, E., Inguscio, M., Moretti, A., Pellegrino, M., and Strumia, F., 1980. Opt.Lett. 5, 87.

Bernard, P., Izatt, J.R., and Mathieu, P., 1981. Int.J. Infrared and Millimeter Waves 2, 65.

Barnes, A.J. and Hallam, H.E., 1970. Trans.Faraday Soc. 66, 1920.

Bluyssen, H.J.A., Van Etteger, A.F., Maan, J.C., and Wyder, P., 1980. IEEE J.Quantum Electron. QE-16, 1347.

Bionducci, G., Inguscio, M., Moretti, A., and Strumia, F., 1979. Infrared Phys. 19, 297.

Chang, T.Y., Bridges T.J., and Burkhardt, E.G., 1970. Appl.Phys. Lett. 17, 249.

Dahmani, B., 1981. 3ème cycle, thesis, Université Pierre et Marie Curie, Paris VI, 1981.

Danielewicz, E.J. and Coleman, P.D., 1977. IEEE J.Quantum Electron. QE-13, 485.

Danielewicz, E.J. and Weiss, C.O., (unpublished).

Domnin, Yu.S., Tatarenkov, V.M., and Shumyatskii, P.S., 1974. Kvant. Electron. 1, 603; Eng.Trans.Sov.J.Quantum Electron. 4, 401.

Dyubko, S.F. and Fesenko, L.D., 1978. 3rd Int.Conf.Submm. Waves, Guildford, Conf.Dig. p.70.

Dyubko, S.F., Svich, V.A., and Fesenko, L.D., 1973. Zh.Tekh.Fiz. 43, 1772; Eng.Transl.Sov.J.Tech.Phys. 18, 1121 (1974).

Dyubko, S.F., Fesenko, L.D., Shevyryov, A.S., and Yartsev, V.I., 1981. Kvant.Elektr. 8, 2048; Eng.Transl.Sov.J.Quantum Electron. 11, 1248.

Feld, M.S., 1979. 4th Int.Conf. IR and mm Waves, Miami, IEEE Cat. No. 79 CH 1384-7MTT, p.36.

Fetterman, H.R., Schlossberg, V., and Waldman, J., 1972. Opt.Commun. 6, 156.

Forber, R. and Feld, M.S., 1980. Int.J.Infrared and Millimeter Waves, 1, 527.

Freed, C., Bradley, L.C. and O'Donnel, R.G., 1980. IEEE J.Quantum Electron. 16, 1195-1206.

Gastaud, C., Sentz, A., Redon, M., and Fourrier, M., 1980. IEEE J. Quantum Electron. QE-16, 1285.

Hartmann, B. and Lindgreen, L., 1982. Int.J.Infrared and Millimeter Waves.

Henningsen, J.O., 1977. IEEE J.Quantum Electron. QE-13, 435.

Henningsen, J.O., 1978. IEEE J.Quantum Electron. QE-14, 958.

Henningsen, J.O., 1980a. J.Mol.Spectrosc. 83, 70.

Henningsen, J.O., 1980b. Conf.Dig. 5th Int.Conf. IR mm Waves, Würzburg, p.373.

Henningsen, J.O., 1981. J.Mol.Spectrosc. 85, 282.

Henningsen, J.O., Inguscio, M., Moretti, A., and Strumia, F., 1981. 6th Int.Conf. IR mm Waves, Miami,IEEE Cat. No. 81-CH, 1645-I MTT; J.Quantum Electron. QE-18, 1004.

Henningsen, J.O., 1982a. IEEE J.Quantum Electron. QE-18, 313.

Henningsen, J.O., 1982b. J.Mol.Spectrosc. 91, 430.

Henningsen, J.O., 1982c. In "Infrared and Millimeter Waves" vol.5, K.J. Button Ed., p.29,Academic Press.

Henningsen, J.O., (unpublished).

Heppner, J., Weiss, C.O., and Plainchamp, P., 1977. Opt.Commun. 23, 381.

Herlemont, F., (unpublished).

Hirose, H. and Kon, S., 1980 Jap.J.Appl.Phys. 19, 1131.

Hodges, D.T., Reel, R.D., and Barker, D.H., 1973. IEEE J.Quantum Electron. QE-9, 1159.

Inguscio, M., Minguzzi, P., Moretti, A., Strumia, F., and Tonelli, M., 1979a. Appl.Phys. 18, 261.

Inguscio, M., Moretti, A., and Strumia, F., 1979b. Opt.Commun. 30, 355.

Inguscio, M., Moretti, A., and Strumia, F., 1979c. 4th Int.Conf.

IR mm Waves, Miami, p.205, IEEE Cat.No. 79 CH 1384-7 MTT.

Inguscio, M., Moretti, A., and Strumia, F., 1980a. Opt.Commun. 32, 87.

Inguscio, M., Moretti, A., and Strumia, F., 1980b. IEEE J.Quantum Electron. QE-16, 955.

Inguscio, M., Ioli, N., Moretti, A., Moruzzi, G., Strumia, F., 1981. Opt.Commun. 37, 211.

Inguscio, M., Moretti, A., Moruzzi, G., and Strumia, F., 1981b. Int. J.Infrared and Millimeter Waves 2, 953.

Inguscio, M., Ioli, N., Moretti, A., and Strumia, F., 1981c. 6th Int.Conf.IR mm Waves, Miami, p.F4-3; IEEE Cat.No.81 CH 1645-I MTT.

Inguscio, M., Ioli, N., Moretti, A., and Strumia, F., 1982. To be published.

Izatt, J.R., Bean, B.L., and Caudle, G.F., 1975. Opt.Commun. 14, 385.

Kon, S., and Fukutani, M., 1980. 5th Int.Conf. IR mm Waves, Würzburg and Conf.Dig. p. 231.

Kon, S., Yano, T., Hagiwara, E., and Hirose, H., 1975. Jap.J.Appl. Phys. 14,1861.

Koo, K.P. and Claspy, P.C., 1979. Appl.Opt. 18, 1314.

Kramer, G. and Weiss, C.O., 1976. Appl.Phys. 10, 187.

Kwan, Y.Y. and Dennison D.M., 1972. J.Mol.Spectrosc. 43, 291.

Landsberg, B.M., 1980. IEEE J.Quantum Electron QE-16, 704.

Lees, R.M., 1972. J.Chem.Phys. 57, 2249.

Lees, R.M. and Baker, J.G., 1968. J.Chem.Phys. 48, 5299.

Lees, R.M., Young, C., Van der Linde, J., and Oliver, B.A., 1979. J.Mol.Spectrosc. 75, 161.

Lees, R.M., Walton, M.A., and Henningsen, J.O., 1981. J.Mol. Spectrosc. 88, 90.

Mathieu, P. and Izatt, J.R., 1977. IEEE J.Quantum Electron. QE-13, 465.

Moruzzi, G., Strumia, F., Bonetti, A., Carli, B., Mencaraglia, F.,

Carlotti, M., Di Lonardo, G., and Trombetti, A., 1981. 6th Int. Conf. on IR mm Waves, Miami IEEE Cat.No. 81 CH 1645-I MTT p.M3-3.

Petersen, F.R., Evenson, K.M., Jennings, D.A., Wells, J.S., Goto, K., and Jimenez, J.J., 1975. IEEE J.Quantum Electron. QE-11, 838.

Petersen, F.R., Evenson, K.M., Jennings, D.A., and Scalabrin, A., 1980. IEEE J.Quantum Electron. QE-16, 319.

Petersen, J.C. and Duxbury, G., 1982. Appl.Phys. B27, 19.

Radford, H.E., 1975. IEEE J.Quantum Electron. QE-11, 213.

Radford, H.E., Petersen, F.R., Jennings, D.A., and Mucha, J.A., 1977. IEEE J.Quantum Electron. QE-13, 92 and 881.

Salomon, C., 1979. 3rd Cycle Thesis, Paris Nord.

Sattler, J.P., Worchesky, T.L., and Riessler, W.A., 1978. Infrared Phys. 18, 521.

Sattler, J.P., Riessler, W.A., and Worchesky, T.L., 1979. Infrared Phys. 19, 217.

Serrallach, A., Meyer, R., and Günthard, Hs.H., 1974. J.Mol. Spectrosc. 52, 94.

Siemsen, K.J. and Whitford, B.G., 1977. Opt.Commun. 22, 11.

Sokabe, N., Sasabe, T., Kimura, T., Yasuda, Y., and Murai, A., 1981. Jap.J.Appl.Phys. 20, 2127.

Strumua, F., (unpublished).

Strumia, F. and Inguscio, M., 1982. In "Infrared and Millimeter Waves" vol.5, p.129, K.J. Button Ed., Academic Press.

Tanaka, A., Tanimoto, A., Murata, N., Yamanaka, M., and Yoshinaga, H., 1974. Jap.J.Appl.Phys. 13, 1491.

Tanaka, A., Yamanaka, M., and Yoshinaga, H., 1975. IEEE J.Quantum Electron.QE-11, 853.

Tobin, M.S. and Jensen, R.E., 1977. IEEE J.Quantum Electron. QE-13, 481.

Tsunawaki, Y., Yamanaka, M., Kabayashi, M., and Kujita, S., 1981. 6th Int.Conf. IR mm Waves, Miami, p.F4-4; IEEE Cat.No. 81 CH 1645-I; Appl.Opt. submitted.

Vasconcellos, E., Wyss, J., Evenson, K.E., and Petersen, F.R., Int. J. Infrared and Millimeter Waves (to be published).

Wagner, R.J., Zelano, A.J., and Ngai, L.H., 1973. Opt.Commun. 8, 46.

Walzer, K., Tacke, M., and Busse, G., 1979. Infrared Phys. 19, 175.

Weiss, C.O., (unpublished).

Weiss, C.O., Grinda, M., and Siemsen, K., 1977. IEEE J.Quantum Electron. QE-13, 892.

Willenberg, G.D., (unpublished).

Worchesky, T.J., 1978. Opt.Lett. 3, 232.

Yano, T. and Kon, S., (unpublished).

Yoshida, T., Kobayashi, M., Yishihara, T., Sakai, K., and Fujita, S., 1981. Opt.Commun. 40, 45.

Young, C., Lees, R.M., Van der Linde, J., and Oliver,B.A., 1979. J.Appl.Phys. 50, 3808.

FAR-INFRARED LASER LINES FROM OPTICALLY PUMPED $^{13}CH_3{}^{16}OH$

J.C. Petersen

National Research Council
K1A-OR6, Ottawa, Canada

J.O. Henningsen

Physics Laboratory I
H.C. Ørsted Institute
University of Copenhagen, Denmark

Introduction

The similarity between the infrared spectra of various isotopic modifications of CH_3OH suggest that all of them should be efficient Far Infrared laser molecules. However, while the deuterated methanols have been studied quite extensively, only two investigations have been published on $^{13}CH_3{}^{16}OH$ and none so far on any other modification involving isotopic substitution of carbon or oxygen.

Far Infrared Laser Lines

The initial work (Henningsen and Petersen, 1978) involved an open resonator, 1 m long with 10 cm mirrors, pumped by an electrically chopped CO_2 laser with a peak power of 25 W and a tuning range of ±40 MHz. The resonator was used as a scanning Fabry-Perot interferometer for determining the wavelengths to a relative accuracy of 3×10^{-4}. A subsequent investigation (Henningsen et al., 1979) employed a 14 mm i.d. copper waveguide, pumped by a 30 W true CW laser with a tuning range of ±25 MHz. Frequencies were measured to an accuracy of 1 MHz for 27 lines. Five of these were not reported earlier, and were identified as cascade lines. Their absence from the open resonator setup was attributed to its lower probability of being simultaneously resonant on the cascade transition and its parent transition. The observed lines are listed in Table I.

Assignments

Methanol is a weakly asymmetric top with hindered internal rotation. The permanent dipole moment has components both parallel to and orthogonal to the CH_3 axis, and the typical emission pattern leads to a triad of emission lines which satisfy an approximate combination relation. Five such triads were identified as indicated in Table II, 3 being associated with the torsional ground state, and 2 with the first torsionally excited state. In the 10R16 case, 5 associated cascade lines were identified, and the data for this pump line was used for deriving molecular constants for the C-O stretch state.

Spectroscopic Constants

Infrared spectra for matrix isolated $^{13}CH_3{}^{16}OH$ have been recorded by Barnes and Hallam (1970), who give the frequencies for 10 of the 12 normal modes. The only one which has been identified in the FIR laser data is the C-O stretch, which is located at 1018.5 cm^{-1} . This value agrees with the results of Serrallach et al. (1974) who observed the C-O stretch of $^{13}CH_3{}^{16}OH$ in gas phase spectra of normal methanol, owing to the natural abundance of C.

The frameworks for describing the rotation-internal rotation states of the methanols have been provided by Lees and Baker (1968) and Kwan and Dennison (1972). Constants for $^{13}CH_3{}^{16}OH$ are given in Table III. For the ground state, the moments of inertia and the internal rotation barrier are derived from Gerry et al. (1976) who quote unpublished results of R.M. Lees, while the interaction constants F_V and G_V are given by Lees et al. (1973). The remaining constants are of minor significance. They have not been determined for $^{13}CH_3{}^{16}OH$ and the values quoted are for normal methanol, which has been studied much more extensively. The constants listed for the C-O stretch were determined from the 10R16 data of Henningsen et al. (1979), assuming that none of the levels involved are significantly perturbed. Subsequent work on normal methanol has indicated that this condition is not necessarily satisfied (Henningsen, 1981), and the C-O stretch constants of Table III can therefore not be trusted to the same extent as the ground state constants.

Table I. $^{13}CH_3$ ^{16}OH Lines

No.	Pump Line	Wavelength (μm)	Frequency (GHz)	Rel. Pol.	Rel. Pwr.	Offset (MHz)a)	Pressure (Pa)b)
1	9P8	87.90	3410.6	$\perp$			40
2	9P10	86.1118	3481.4330	$\parallel$		+25	40
3		146.0974	2052.0041	$\perp$		+25	40
4		208.4121	1438.4603	$\parallel$		+25	40
5		236.5303	1267.4590	$\parallel$			
6	9P12	157.9285	1898.2799	$\parallel$	VS	-20	24
7		238.5227	1256.8718	$\parallel$	VS	-20	24
8		461.3848	649.7667	$\perp$	VS	-20	24
9	9P12	63.0964	4751.3409	$\parallel$		+25	24
10		237.5230	1262.1620	$\parallel$		+25	24
11		629.8443	475.9787	$\parallel$		+25	24
12	9P22	85.3173	3513.8534	$\parallel$	VS	-15	32
13		118.0131	2540.3310	$\perp$	VS	-15	32
14		307.78	974.0	$\parallel$		-15	32
15	9P22	103.4808	2897.0824	$\parallel$	VS	+15	32
16		149.2723	2008.3601	$\perp$	VS	+15	32
17		338.9638	884.4380	$\parallel$		+15	32
18	9P30	147.97	2026.0	$\perp$		-25	24
19	9P36	291.62	1028.0	$\parallel$		-20	17
20		325.17	922.0	$\parallel$		-20	17
21	9P40	168.84	1775.6	$\parallel$		+35	15
22		358.92	835.3	$\perp$		+35	15
23	10R28	85.79	3494.5	$\perp$		-5	40
24		121.20	2473.5	$\perp$		-5	40

No.	Pump Line	Wavelength (μm)	Frequency (GHz)	Rel. Pol.	Rel. Pwr.	Offset (MHz)a)	Pressure (Pa)b)
25	10R26	77.4894	3868.8189	$\parallel$		+40	31
26		103.5863	2894.1323	$\perp$		+40	31
27	10R22	34.79	8617.2	$\parallel$			61
28	10R18	105.1472	2851.1692	$\parallel$	VS	+25	27
29		110.4324	2714.7147	$\parallel$	VS	+25	27
30		171.7576	1745.4395	$\perp$	VS	+25	27
31	10R16	115.8232	2588.3617	$\parallel$	VS	+20	20
32		148.5904	2017.5761			+20	
33		152.0757	1971.3372			+20	
34		203.6358	1472.1993	$\perp$	VS	+20	20
35		268.5722	1116.2455	$\parallel$		+20	20
36		280.2183	1069.8534			+20	
37		280.2397	1069.7714			+20	
38		332.6034	901.3512			+20	
39	10P16	41.90	7155.0	$\perp$		-10	20
40	10P16	123.26	2432.2	$\parallel$		+40	39

a) Offsets indicate the pump laser condition during measurements and do not refer to the absorption line center.

b) Pressure measured with Pirani gauge calibrated for air.

Table II. $^{13}CH_3{}^{16}OH$ Line Assignments

Pump Line	Pump Assignment	Measured Emission cm^{-1}	Measured Emission μm	Emission Assignment $(n\tau K,J)\rightarrow(n'\tau'K',J')$ (a)	Calc. Emission cm^{-1}
9P10	$R(113^-,30)$	47.98186	208.4	$113^-,31\rightarrow113,30$	47.96
		68.44749	146.1	$\rightarrow122^+,31$	67.98
		116.12811	86.1	$\rightarrow122^-,30$	115.52
9P12	$R(034^-,26)$	41.92473	238.5	$034^-,27\rightarrow034^-,26$	41.93
		21.67388	461.4	$\rightarrow013^+,27$	21.68
		63.31980	157.9	$\rightarrow013^-,26$	63.29
9P22	R(019,18)	29.50168	339.0	$019,19\rightarrow019,18$	29.49
		66.99168	149.3	$\rightarrow028,19$	66.83
		96.63627	103.5	$\rightarrow028,18$	96.32
9P22	E(1310,20)	32.49	307.8	$1310,21\rightarrow1310,20$	32.49
		84.73632	118.0	$\rightarrow119,21$	85.29
		117.20953	85.3	$\rightarrow119,20$	117.77
10R16	P(027,25)	37.23394	268.6	$027,24\rightarrow027,23$	37.23
		49.10728	203.6	$\rightarrow036,24$	49.08
		86.33845	115.8	$\rightarrow036,23$	86.31
		30.06584	332.6	$036,24\rightarrow015,24$	30.15
		67.29910	148.6	$\rightarrow015,23$	67.38
		35.68647	280.2	$027,23\rightarrow027,22$	35.69
		35.68373	280.2	$036,23\rightarrow036,22$	35.69
		65.75673	152.1	$\rightarrow015,22$	65.83

a) Rotation-internal rotation states are characterized as $(n\tau K,J)$ where n and τ are internal rotation quantum numbers and K,J are the usual angular momentum quantum numbers. Superscript + and − refer to components of asymmetry split A-states.

Table III. Molecular Parameters

	$^{13}CH_3{}^{16}OH$ Ground State	$^{13}CH_3{}^{16}OH$ C-O Stretch	Units
I_b	34.8622	35.2553	
I_c	36.1608	36.6382	
I_{ab}	-0.123		$kg\ m^2 \cdot 10^{-47}$
I_{a1}	1.2528		
I_{a2}	5.3331	5.3874	
V_3	372.6	386.8	
V_6	-0.52		
D_{KK}	0.38×10^{-4}		
k_1	-0.48×10^{-4}		
k_2	-18.41×10^{-4}		
k_3	-53.73×10^{-4}		
k_4	-85.50×10^{-4}		cm^{-1}
k_5	132.07×10^{-4}		
k_6	67.85×10^{-4}		
k_7	0		
F_v	-2.323×10^{-3}		
G_v	-1.127×10^{-4}		
L_v	-2.26×10^{-6}		
D_{JK}	9.54×10^{-6}		
D_{JJ}	1.6345×10^{-6}		
μ_a	2.952		$Cm \cdot 10^{-30}$
μ_b	4.80		

Moments of inertia are converted to amuÅ^2 by dividing the numbers of the table by 1.660531. Dipole moments are converted to Debye by dividing by 3.33564.

References

Barnes, A.J. and Hallam, H.E., 1970. Trans.Farad.Soc. 66, 1920-1931.

Gerry, M.C.L., Lees, R.M., and Winnewisser, G., 1976. J.Mol. Spectrosc. 61, 231-242.

Henningsen, J.O. and Petersen, J.C., 1978. Infrared Physics 18, 475-479.

Henningsen, J.O., Petersen, J.C., Petersen, F.R., Jennings, D.A., and Evenson, K.M., 1979. J.Mol.Spectrosc. 77, 298-309.

Henningsen, J.O., 1981. J.Mol.Spectrosc. 85, 282-300.

Kwan, Y.Y. and Dennison, D.M., 1972. J.Mol.Spectrosc. 43, 291-319.

Lees, R.M. and Baker, J.G., 1968. J.Chem.Phys. 48, 5299-5318.

Lees, R.M., Lovas, F.J., Kirchhoff, W.H., and Johnson, D.R., 1973. J.Phys.Chem.Ref.Data 2, 205-214.

Serrallach, A., Meyer, R., and Günthard, Hs.H., 1974. J.Mol. Spectrosc. 52, 94-129.

OPTICALLY PUMPED FAR-INFRARED LASER LINES IN DEUTERATED METHYL ALCOHOL, CH_3OD, CH_2DOH, CHD_2OH, CD_3OH, CD_3OD

S. Kon

Department of Applied Physics, Nagoya University
Chkusa-ku, Nagoya 464, Japan

T. Kachi

Toyota Central Research & Development Lab. Inc.
Nagakute-cho, Aichi-ken 480-11, Japan

Y. Tsunawaki

Department of Chemistry, Osaka Industrial
University, Naka-Gaito, Daito, Osaka 574, Japan

M. Yamanaka

Electromagnetic Energy Engineering, Osaka University
Yamada-Oka, Suita, Osaka 565, Japan

Introduction

Normal methyl alcohol(CH_3OH) pumped by CO_2 laser is known as the best FIR laser molecule. The exchage of H by D in the molecule does not shift the CO_2 laser absorption appreciably; all of its deuterated species falls in the 900-1100 cm^{-1} region of CO_2 laser spectrum. Laser action in CH_3OD was first observed by Dyubko et al. in 1974, subsequently many new lines were reported by many authors. Their frequencies of several strong lines were measured by heterodyne method by Blaney et al. (1978). FIR laser action in fully deuterated methyl alcohol (CD_3OD) was first observed by Kon et al. (1975) and the accurate frequency measurement of the lines was reported by Vasconcellos et al. (1981). Another symmetrically deuterated methyl alcohol, CD_3OH, was used in optically pumped laser by Dyubko et al. (1975) and Danielewicz et al. (1978). Methyl alcohol molecules obtained by the exchange of H by D in the methyl group CH_3, CH_2DOH and CHD_2OH, have been observed to emit FIR laser lines by Ziegler et al. (1978). Subsequently, the frequency measurements of the molecules were made by Scalabrin et al. (1980).

Methyl alcohol has, inclusive of above 5 isotopes, a total of 7 deuterated species. Above-mentioned five species have been successfully found to give many laser lines when pumped by $^{12}CO_2$ lasers. Moreover, Davis et al. (1981) observed new lines in CH_3OD, CD_3OH and CD_3OD by the use of $^{13}CO_2$ in pump laser. It seems likely that all deuterated species, inclusive of CH_2DOD and CHD_2OD, of methyl alcohol will give strong FIR lines in the near future using $^{12}CO_2$, $^{13}CO_2$ or $C^{18}O_2$ lasers.

FIR laser lines

Initially, FIR emission has been obtained in an open resonator cavity. Subsequently waveguide resonators were used to reduce the threshold pump power and to obtain more powerful output. Therefore, optically pumped lasers have resulted in over 316 FIR transitions covering the spectral range from 35 μm to 1.29 mm. Table I to V list the lines of 64 in CH_3OD, 68 in CH_2DOH, 11 in CHD_2OH, 115 in CD_3OH and 59 in CD_3OD reported before 1982. The tables contain the pump line transition, the output wavelength, relative polarization, output power and the representative data for various conditions. Listed laser lines follow the set order of shorter wavelengths of CO_2 pump lines. FIR wavelengths measured with heterodyne methods indicate the values in a vacuum, while other wavelengths are the latest data which were obtained in the air. Emitted FIR laser is either parallel (||) or orthogonal (⊥) to pump radiation polarization but both results are written when different results were obtained by authors. FIR output power which is not indicated numerically is denoted by VS (very strong), S (strong), M (medium) and W (weak). In the column of FIR resonator, O and W mean open and waveguide resonator, respectively. The resonator of stark-field type is denoted by 'stark'. Numerical expression in pump power is not so accurate, because CO_2 pump intensity is not constant for each branch.

Laser line assignments

Deuterated methyl alcohol has 12 fundamental vibrational modes. The absorption band in deuterated methyl alcohol with respect to optical pumping with CO_2 laser is associated with the C-O vibrational stretch mode. Additional torsional motion or hindered internal rotation of OH (or OD) group for CH_3 (CHD_2, CH_2D or CD_3) group causes the spectrum of methyl alcohol to be very complex (Woods, 1970). Kwan et al. (1972) analyzed the torsion-rotation spectra of isotopic methyl alcohol molecules.

Assignments for FIR laser lines from optically pumped CH_3OH were reported by Henningsen (1977, 1981) and Danielewicz et al. (1977). Kachi and Kon (1982) made assignments of CH_3OD using molecular constants given by Lees et al. (1968). If necessary, assignments are required to satisfy the combination relation (Henningsen, 1981), the polarization rule (Chang, 1974, 1977) and the competitive or cascade couplings (Kachi and Kon, 1982) among these lines. The assignments for CH_3OD were shown in Table VI. Recently, for CH_3OH and CD_3OH, it was shown that Fermi resonance between the C-O stretch vibration and the higher harmonics of the torsional mode of the ground vibration was responsible to shift a pump transition (Weber et al. 1982).

Assignments of CD_3OH will be improved by introduction of Fermi resonance.

The rotational energy levels are described by four quantum numbers (n, τ, K)J, where J and K are the rotational quantum number and its component along the symmetry axis of the molecule. The letter n means the torsional quantum number and τ is the integral number which arises from the three fold of the torsional barrier and takes on three values 1, 2 and 3.

Table of reference number arranged
by author and active medium

(1)	Dyubko et al. (1974)	CH_3OD
(2)	Dyubko et al. (1975)	CD_3OH
(3)	Kon et al. (1975)	CD_3OD
(4)	Kon et al. (1975)	CH_3OD
(5)	Bean et al. (1977)	CH_3OD, CD_3OD
(6)	Radford et al. (1977)	CH_3OD
(7)	Blaney et al. (1978)	CH_3OD
(8)	Herman et al. (1978)	CH_3OD, CD_3OD
(9)	Ziegler et al. (1978)	CH_2DOH, CHD_2OH
(10)	Danielewicz et al. (1978)	CD_3OH
(11)	Grinda et al. (1978)	CD_3OH
(12)	Duxubury et al. (1978)	CD_3OD
(13)	Lund et al. (1979)	CH_3OD
(14)	Bluyssen et al. (1980)	CH_3OD
(15)	Ni et al. (1980)	CH_3OD
(16)	Landsberg (1980)	CH_3OD
(17)	Gastaud et al. (1980)	CH_3OD
(18)	Scalabrin et al. (1980)	CH_2DOH
(19)	Davis et al. (1981)	CH_3OD, CD_3OD, CD_3OH
(20)	Vasconcellos et al. (1981)	CD_3OD
(21)	Yoshida et al. (1981)	CD_3OD
(22)	Yoshida et al. (1982)	CH_3OD, CD_3OH
(23)	Vass et al. (1982)	CD_3OD
(24)	Kachi et al. (1982)	CH_3OD

Table I CH_3OD lines

CO_2 pump line (cm^{-1})		CH_3OD laser line		Rel. Pol.	FIR power (mW)	Press. (mTorr)	FIR resonator	Pump power (W)	Note	Reference
		λ(μm)	ν(MHz)							
9R(34)	1086.870	151.56			S	80	stark	30	E⊥CO_2[b]	(17)
9R(28)	1083.479	186		‖	W		O	6.0[a]		(16)
9R(26)	1082.296	106		‖	W		O	6.5[a]		(16)
		177			W	250	stark	17	E⊥CO_2[b]	(22)
		182		‖	W		O	6.0[a]		(16)
9R(22)	1079.852	169		‖	M		O stark	6.0[a]	E⊥CO_2[b] +15 MHz[h]	(16)(22)
9R(16)	1075.988	70.3		‖	~0.18	320	W stark	10	E⊥CO_2[b] -20 MHz[h]	(13)(22)
9R(14)	1074.646	215.37246	1391972.1	⊥,‖	0.1	10	O stark	1.5[a]	E⊥CO_2[b] -5 MHz[h]	(4)(7)(16)(22)
		233		‖	W	250	O stark	4[a]	E⊥CO_2[b] -15 MHz[h]	(15)(16)(22)
9R(8)	1070.462	46.7			~0.6	<150	W	>15		(5)
		57		⊥	~0.6	<150	O,W	>3		(4)(5)

(continued)

Table I CH_3OD lines (continued)

					5	825	W	~KW	Q	(14)
		294.81098	1016897.2	‖	~0.2	<150	O,W	>2		(4)(5)(7)
		305.72611	980591.6	‖	~0.2	<150	O,W	>3		(4)(5)(7)
$9R_I(6)$		69.5			~0.09	200	W	10		(13)
$9R_{II}(6)$		225			~0.6	200	W	10		(13)
9R(6)	1069.014	282			>10	270	W	~KW	Q	(14)
9R(4)	1067.539	212		‖	8		W,O	3.5[a]		(13)(16)
		330		⊥,‖	~0.3	<150	W,O	1.5[a]		(5)(16)
9R(2)	1066.037	106			M		O	1.5[a]		(16)
		114.96			S		stark	30	E⊥,‖CO_2[b]	(17)
9P(6)	1058.949	134.7		⊥	~0.3	<150	W	>15		(5)
					5000	250	W	~KW	Q	(14)
		182.1				200	O	<20		(15)
		229.1		‖	~0.7	<150	W	>15		(5)
					5000	250	W	~KW	Q	(14)

		417.1		⊥	~0.2	<150	W	>15		(5)
		510		‖	W	10	O	5		(24)
9P(10)	1055.625	134		⊥,‖	~2.4	200	O,W stark	5[a]	E⊥CO_2[b] +10 MHz[h]	(13)(16)(22)
		133			500	450	W	~KW	Q	(14)
9P(22)	1045.022	128.0		⊥	0.30		O stark	10	E⊥CO_2[b] +5 MHz[h]	(1)(22)
9P(24)	1043.163	133		‖	M		O	2.5[a]		(8)(16)
9P(26)	1041.279	100.8		⊥	~1.2	200	W	10		(13)
		101.6		‖	S	240	O	2.5[a]		(15)(16)
		107.67			W,M	40	stark	30	E⊥,‖CO_2[b]	(17)
		117.22707	2557365.4	‖		70	W stark	1.5[a]	E⊥CO_2[b] +10 MHz[h]	(1)(4)(6)(16)(22)
9P(30)	1037.434	81.9		⊥	S	250	O	<20		(15)
		89.6		‖	S	250	O	<20		(15)
		103.12463	2907088.9	⊥	~0.2	<150	O,W stark	>1[a]	E⊥CO_2[b] -15 MHz[h]	(1)(4)(5)(6)(22)

(continued)

Table I CH_3OD Lines (Continued)

		103			>1000	825	W	~KW	Q	(14)
		111.1			M	30	stark	30	E⊥CO_2[b]	(17)
		145.66171	2058141.8	∥	~0.2		0,W stark	>4	E⊥CO_2[b] -20 MHz[h]	(4)(5)(7)(22)
		168.1		⊥	0.10		0	10		(1)
		320.0		∥	~0.2	<150	W stark	>15	E⊥CO_2[b] +15 MHz[h]	(1)(5)(22)
		352.5		⊥	0.05		0	10		(1)
9P(32)	1035.474	80.0		⊥	0.80		0	2.5[a]		(1)(16)
		108		∥	M		0	3.5[a]		(16)
		108		∥	M		0	3.5[a]		(16)
		110.7		⊥	VW	220	0 stark	<20	E⊥CO_2[b] -20 MHz[h]	(15)(22)
		113.8		∥	W	220	0	<20		(15)
		141			W	150	stark	17	E⊥CO_2[b] -20 MHz[h]	(22)
		145.6		∥	VW	220	0	<20		(15)

		179.0		⊥	0.50		0 stark	2.5[a]	E⊥CO_2[b]	(1)(16)(22)
		279.4		⊥	0.10		0	2.5[a]		(1)(16)
		498.0		‖	0.05		0	10		(1)
10R(44)	989.647	110		‖	S	>10	0	1.5[a]		(4)(16)
		236		‖	M		0	2.0[a]		(16)
		241		‖	M		0	<20		(15)
10R(42)	988.647	141		⊥	W		0	2.5[a]		(16)
10R(34)	984.383	238		‖	VW	230	0		maybe CH_3OH	(15)
10R(22)	977.214	137		‖	M		0	4.0[a]		(16)
		137.17			S	200	stark	30	E⊥CO_2[b]	(17)
		224		‖	W		0	10.0[a]		(16)
10P(18)	945.980	280		‖	VW	250	0	<20		(15)
10P(26)	938.688	135.5			M	30	stark	30	E⊥CO_2[b]	(17)

(continued)

Table I CH_3OD Lines (Continued)

N_2O pump R(22)	954.651 or 957.056	135.9			W	30	stark	9.5	$E \perp N_2O$[b]	(17)
$^{13}CO_2$ pump										
9R(28)	1037.167	86.7			11.6		W	6.25		(19)
9R(24)	1034.838	917			0.3		W	11.25		(19)

R_I and R_{II} indicate different CO_2 pump frequencies.

a: Threshold power.

b: Stark field relative to CO_2 pump field.

In resonator column, O and W mean open and waveguide resoantor, respectively.

Q means Q switched pump.

k: Pump offset frequencies.

Table II CH_2DOH lines

CO_2 pump line (cm^{-1})		CH_2DOH laser line		Rel. Pol.	FIR power (mW)	Press. (mTorr)	FIR reso-nator	Pump power (W)	Note	Reference
		λ(μm)	ν(MH_z)							
$9R_I(24)$	1081.087	152.7		∥	0.20	100	0	31		(18)
$9R_{II}(24)$		219.09602	1368315.4	∥	0.13	100	0	30		(18)
		272.25164	1101159.4	∥	0.90	100	0	30[c]		(9)(18)
		682.6		⊥	0.03	100	0	30		(18)
9R(22)	1079.852	182.1		∥	0.063	70	0	30		(18)
		171.8		∥	0.008	40	0	30		(18)
		218.0		∥	0.025	45	0	30		(18)
9R(18)	1077.303	164			<0.5	80	0	25[c]	maybe CHD_2OH	(9)
9R(16)	1075.988	216.8		∥	0.05	110	0	29		(18)
9R(8)	1070.462	135.83350	2207058.3	⊥	0.038	90	0	26		(18)
		164.74645	1819720.3	∥	0.05	80	0	26		(18)
		422.15116	710154.3	⊥	0.015	50	0	26		(18)
9P(6)	1058.949	273.00372	1098125.9	∥	0.025	110	0	20		(18)

(continued)

Table II CH_2DOH Lines (Continued)

9P(10)	1055.625	183.62132	1632666.9	‖	1.0	110	0	24		(18)
		295.39666	1014881.0	‖	0.55	110	0	24^c		(9)(18)
9P(12)	1053.924	108.81776	2754995.7	‖	0.5	180	0	24^c		(9)(18)
		112.53224	2664058.3	‖	0.8	180	0	24	maybe J,K→J-1,k-1	(18)
		172.84619	1734446.4	⊥	0.2	100	0	24^c	maybe J,K→J,K-1	(9)(18)
		322.45221	929726.8	‖	0.5	135	0	24^c	maybe J,K→J-1,K	(9)(18)
9P(14)	1052.196	206.68741	1450463.1	‖	0.68	100	0	27^c		(9)(18)
		308.04046	973224.3	‖	0.5	110	0	27^c		(9)(18)
9P(16)	1050.441	102.02349	2938465.1	⊥	0.088	80	0	30		(18)
$9P_I(18)$	1048.661	87.1		⊥	0.20	145	0	32		(18)
		100.0		‖	0.25	160	0	32		(18)
		762.5		‖	0.063	130	0	32		(18)
$9P_{II}(18)$		167.54117	1789365.9	‖	0.63	70	0	27^c		(9)(18)

		396.0		‖	<0.5	80	0	25[c]		(9)(18)
9P(20)	1046.854	140.3		⊥	0.025	90	0	28		(18)
9P(26)	1041.279	468.23586	640259.5	‖	<1.0	80	0	31[c]		(9)(18)
		616.33505	486411.5	‖	1 ~ 2	80	0	31[c]		(9)(18)
9P(30)	1037.434	44		‖	0.075	180	0	30		(18)
$9P_I(32)$	1035.474	108.94124	2751872.9	⊥	0.050	130	0	28		(18)
		117.08507	2560467.0	⊥	0.063	130	0	28		(18)
$9P_{II}(32)$		167.35234	1791384.9	‖	0.50	150	0	29		(18)
		266.73524	1123932.7	‖	0.14	95	0	29		(18)
		451.47536	664028.4	⊥	0.075	85	0	29		(18)
9P(36)	1031.477	195.49556	1533499.9	‖	0.43	85	0	27		(18)
		336.24615	891586.3	‖	0.40	85	0	27		(18)
9P(38)	1029.442	42.5		⊥	0.025	210	0	21		(18)
		200		⊥	0.015	70	0	21		(18)
$9P_I(40)$	1027.382	87.9		⊥	0.11	180	0	19		(18)

(continued)

Table II CH_2DOH Lines (Continued)

$9P_{II}(40)$		387.55913	773539.9	‖	0.20	80	0	20		(18)
		523.09138	573116.8	‖	0.25	75	0	20		(18)
9P(46)	1021.057	226.29742	1324771.9	⊥	0.38	100	0	15		(18)
		452.4		⊥	0.40	80	0	15		(18)
10R(34)	984.383	150.81629	1987798.9	‖	2.0	160	0	34[c]		(9)(18)
		159.21794	1882906.3	‖	0.80	140	0	34		(18)
		295.63940	1014047.7	‖	0.6~2	120	0	34[c]		(9)(18)
10R(32)	983.252	135.17175	2217863.3	‖	0.13	130	0	29		(18)
		135.17256	2217849.9	‖	0.18	130	0	29		(18)
		149.61284	2003788.3	‖	0.063	150	0	29		(18)
		340.35664	880818.6	‖	1.0	165	0	29		(18)
10R(16)	973.289	212.5		‖		80	0			(18)
		363		‖	<0.5	80	0	25[c]	maybe CHD_2OH	(9)(18)
10P(18)	945.980	238			0.5~1	80	0	25[c]	maybe CHD_2OH	(9)

10P(26)	938.688	150.57167	1991028.3	⊥	0.31	110	0	31		(18)
		188.41100	1591162.2	‖	0.70	110	0	31		(18)
10P(28)	936.804	189.3		‖	0.018	120	0	31		(18)
		196.1		‖	0.025	120	0	31		(18)
10P(30)	934.894	90.4		‖	0.25	210	0	34		(18)
		162.7		⊥	0.003	70	0	34		(18)
10P(34)	931.001	124.43170	2409293.3	‖	1 ~ 2	160	0	24[c]		(9)(18)
		248.12204	1208246.0	⊥	0.75	115	0	24		(18)
		249.72035	1200512.7	‖	<0.5	115	0	24[c]		(9)(18)
10P(36)	929.017	149.38792	2006805.2	‖	0.020	100	0	24		(18)
		224.22562	1337012.5	‖	0.050	100	0	24		(18)
		427.2		⊥	0.004	150	0	24		(18)
10P(46)	918.718	374.08611	801399.6	‖	0.5~1	90	0	6		(9)(18)

P_I and P_{II} indicate different CO_2 pump frequencies.

c: In reference (9), pump power of the strongest line is 25 W.

Table III CHD_2OH lines

CO_2 pump line (cm^{-1})		CHD_2OH laser line		Rel. Pol.	FIR power (mW)	Press. (mTorr)	FIR reso-nator	Pump power (W)	Note	Reference
		λ(μm)	ν(MHz)							
9R(18)	1077.303	165			0.5~1	80	0	25		(9)
9P(6)	1058.949	483			<0.5	80	0	25		(9)
9P(20)	1046.854	346			<0.5	80	0	25		(9)
9P(30)	1037.434	518			0.5~1	80	0	25		(9)
10R(38)	986.567	168			0.5~1	80	0	25		(9)
		426			0.5~1	80	0	25		(9)
10R(20)	975.930	260			1~2	80	0	25		(9)
10R(16)	973.289	179			0.5~1	80	0	25		(9)
		363			1~2	80	0	25		(9)
10P(18)	945.980	238			1~2	80	0	25		(9)
		355			0.5~1	80	0	25		(9)

Pump power is that of the strongest laser line.

CO_2 pump line (cm^{-1})		CD_3OH laser line		Rel. Pol.	FIR power (mW)	Press. (mTorr)	FIR resonator	Pump power (W)	Note	Reference
		λ(μm)	ν(MHz)							
9R(44)	1092.007	407		∥	0.2		0	10[d]		(2)
9R(34)	1086.870	52.9		⊥	VS	225	0 stark	10[e]	E⊥CO_2[g] -5 MHz[h]	(11)(22)
		60.8		∥	VS	225	0	10[e]		(11)
		232		∥	0.4		0	10[d]		(2)
		409		⊥	0.06		0	10[d]		(2)
9R(32)	1085.765	336		∥	0.02		0	10[d]		(2)
		554		∥	0.6		0	10[d]		(2)
9R(28)	1083.479	49.8		⊥	VS	150	0	10[e]		(11)
		158		∥	S	150	0 stark	10[e]	E⊥CO_2[g] -10 MHz[h]	(11)(22)
		181			M	120	stark	17	E⊥CO_2[g],[h] 0 MHz	(22)
		352		∥	0.1		0	10[d]		(2)
9R(26)	1082.296	277		∥	0.15		0	10[d]		(2)

(continued)

Table IV CD_3OH Lines (Continued)

		370		∥	2.0		0	10^d		(2)
		745		∥	0.08		0	10^d		(2)
9R(22)	1079.852	483		∥	0.5		0	10^d		(2)
		583		∥	0.9	90	0	$10^{d,e}$		(2)(11)
9R(20)	1078.591	968		∥	0.15		0	10^d		(2)
9R(18)	1077.303	297		⊥	0.15		0	10^d		(2)
9R(16)	1075.988	321		⊥	0.04		0	10^d		(2)
		472		∥	0.04		0	10^d		(2)
9R(14)	1074.646	119			M	140	stark	16	E⊥CO_2[g] +15 MHz[h]	(22)
		179		⊥	S	150	0	10^e		(11)
		236		∥	M	150	0	10^e		(11)
		346		∥	M	150	0	10^e		(11)
		352		∥	1.0		0	10^d		(2)
		553		∥	4.5		0	10^d		(2)
9R(8)	1070.462	184		⊥	0.2		0	10^d		(2)

9R(6)	1069.014	299		∥	0.02		0	10[d]		(2)
9R(?)		353		∥	0.2		0	10[d]		(2)
9P(6)	1058.949	222		∥	0.1		0	10[d]		(2)
		680		∥	0.02		0	10[d]		(2)
9P(8)	1057.300	223		⊥	W	98	0	10[e]		(11)
		508		∥	0.1		0	10[d]		(2)
		711		∥	0.15		0	10[d]		(2)
9P(10)	1055.625	695		∥	1.5		0	10[d]		(2)
9P(12)	1053.924	1100		∥	0.2		0	10[d]		(2)
9P(14)	1052.196	268		∥	0.1		0	10[d]		(2)
9P(16)	1050.441	386		⊥	0.5		0	10[d]		(2)
		480		⊥	0.03		0	10[d]		(2)
9P(18)	1048.661	455		∥	0.2		0	10[d]		(2)
9P(20)	1046.854	258			M	130	stark	18	$E \perp CO_2$[g] -15 MHz[h]	(22)
		266		∥	0.25		0	10[d]		(2)

(continued)

Table IV CD_3OH Lines (Continued)

		422		∥	0.5		0	10^d		(2)
9P(22)	1045.022	551		⊥	0.3		0	10^d		(2)
9P(24)	1043.163	702		∥	0.4		0	10^d		(2)
		1146		∥	0.1		0	10^d		(2)
9P(28)	1039.369	370		∥	0.2		0	10^d		(2)
		435		⊥	VW	135	0	10^d		(11)
9P(30)	1037.434	385		⊥	0.01		0	10^d		(2)
		774		∥	0.01		0	10^d		(2)
9P(32)	1035.474	351		∥	VW	150	0	10^d		(2)(11)
		410		∥	0.2		0	10^d		(2)
9P(40)	1027.382	201		∥	M	150	0	10^d		(11)
		285		∥	0.3		0	10^d		(2)
10R(36)	985.488	253.2		⊥	S	200	0 stark	10^f	E⊥,∥ CO_2[g] -5 MHz[h]	(10)(22)
		419.0		⊥,∥	W	200	0 stark	$10^{d,f}$	E⊥ CO_2[g] -5 MHz[h]	(2)(10)(22)

		646		⊥	1.3		0	10^d		(2)
		703		⊥	1.2		0	10^d		(2)
10R(34)	984.383	37.6		‖	M	200	0	10^f		(10)
		102.6		⊥	W	150	0	10^f		(10)
		112.3		⊥	0.001	150	0	10^f		(10)
		128.7		‖	0.001	150	0 stark	10^f	E⊥CO_2[g] +20 MHz[h]	(10)(22)
		182.4		⊥	W	150	0	10^f		(10)
		191.9		‖	M	150	0	10^f		(10)
		265		‖	0.6		0	10^d		(2)
		266.2		⊥	M	150	0	10^f		(10)
		297		‖	0.5		0	10^d		(2)
		498.0		‖	M	150	0	10^f		(10)
		685		‖	0.08		0	10^d		(2)
10R(32)	983.252	83.6			W	270	stark	8	E⊥CO_2[g] -35 MHz[h]	(22)

(continued)

Table IV CD_3OH Lines (Continued)

		421		⊥	0.5		0	10[d]		(2)
10R(30)	982.096	336		⊥	0.1		0	10[d]		(2)
		351		⊥	0.1		0	10[d]		(2)
10R(28)	980.913	310		⊥	0.6		0	10[d]		(2)
		398		‖	0.5		0	10[d]		(2)
10R(24)	978.472	70.6			M	300	stark	19	$E\perp CO_2$[g] +25 MHz[h]	(22)
		278		‖	0.15		0	10[d]		(2)
10R(20)	975.930	55.4			W	230	stark	19	$E\perp CO_2$[g] +30 MHz[h]	(22)
		1290		‖	0.1		0	10[d]		(2)
10R(18)	974.622	41.8		⊥	VS	400	0	10[f]		(10)
		43.9		‖	VS	400	0	10[f]		(10)
		219.9		‖	W	150	0	10[f]		(10)
		495		‖	0.5		0	10[d]		(2)
		857		‖	6.2		0	10[d]		(2)
		862.0		‖	M	150	0	10[f]		(10)

10R(16)	973.289	81.2		∥	M	220	0 stark	10^{f}	$E\perp CO_2$[g] +15 MHz[h]	(10)(22)
		86.4		⊥	W	220	0	10^{f}		(10)
		599		∥	0.1		0	10^{d}		(2)
10R(14)	971.930	68.1			W	250	stark	21	$E\perp CO_2$[g] +10 MHz[h]	(22)
		267		∥	0.09		0	10^{d}		(2)
10R(12)	970.547	412		∥	0.1		0	10^{d}		(2)
10R(8)	967.707	41.5		⊥	M	225	0	10^{f}		(10)
		71.0		∥	M	150	0 stark	10^{f}	$E\perp CO_2$[g] -15 MHz[h]	(10)(22)
		553		∥	2.2		0	10^{d}		(2)
		648		∥	0.9		0	10^{d}		(2)
10P(18)	945.980	144.0		∥	M	200	0 stark	10^{f}	$E\perp CO_2$[g] +10 MHz[h]	(10)(22)
		287.4		∥	4.4	200	0 stark	$10^{d,f}$	$E\perp CO_2$[g] +10 MHz[h]	(2)(10)(22)
		290.0		⊥	M	200	0	10^{f}		(10)

(continued)

Table IV CD_3OH Lines (Continued)

		760		‖	0.07		0	10^{d}		(2)
10P(20)	944.194	309		‖	0.06		0	10^{d}		(2)
		722		⊥	0.05		0	10^{d}		(2)
10P(22)	942.383	34.8		‖	M	350	0	10^{f}		(10)
		40.1		⊥	M	350	0	10^{f}		(10)
		258.7		‖	W	250	0 stark	$10^{d,f}$	E⊥CO_2[g] +10 MHz[h]	(2)(10)(22)
10P(24)	940.548	238.3		⊥	W	200	0	10^{f}		(10)
		286.6		‖	W	200	0 stark	$10^{d,f}$	E⊥CO_2[g] +15 MHz[h]	(2)(10)(22)
10P(26)	938.688	483		‖	0.1		0	10^{d}		(2)
10P(28)	936.804	276		‖	0.8		0	10^{d}		(2)
10P(32)	932.960	76.1		‖	M	250	0	10^{f}		(10)
10P(42)	922.914	517		‖	0.15		0	10^{d}		(2)
$^{13}CO_2$ pump 9R(34)	1040.447	234.7			1.0		W	10.5		(19)

9P(16)	900.369	309.8			1.3		W	10		(19)
9P(18)	898.649	143.4			2.6		W	13.9		(19)
9P(20)	986.909	530.4			0.9		W	11.2		(19)
9P(36)	882.287	221.0			0.44		W	6		(19)

d: Peak power. e: Average power. f: 10R branch average power.

g: Stark field relative to CO_2 pump field.

h: Pump offset frequencies.

Table V CD_3OD lines

CO_2 pump line (cm^{-1})		CD_3OD laser line		Rel. Pol.	FIR power (mW)	Press. (mTorr)	FIR resonator	Pump power (W)	Note	Reference
		λ(μm)	ν(MH_z)							
9R(38)	1089.001	64.4		∥	0.002	75	0,stark	>5	0 MHz[h]	(20)(21)
9R(28)	1083.479	327.8		⊥	0.005	45	0,stark	20	0 MHz[h]	(20)(21)
9R(26)	1082.296	316			M	120	stark	17	-10 MHz[h]	(21)
9R(16)	1075.988	316			M	200	stark	18	-20 MHz[h]	(21)
		454			M	200	stark	18	-20 MHz[h]	(21)
9R(8)	1070.462	141.3		⊥	0.01	50	0	18		(20)
		270.73258	1107337.9	⊥	0.02	75	0,stark	>12	0 MHz[h]	(20)(21)
9R(4)	1067.539	124.5		⊥	0.002	40	0			(20)
		152.5		⊥	0.003	40	0			(20)
9P(14)	1052.196	353			W	150	stark	20		(21)
9P(16)	1050.441	129			M	180	stark	23	0 MHz[h]	(21)
9P(26)	1041.279	137			S	260	stark	21	+15 MHz[h]	(21)
9P(28)	1039.369	62.7			M	250	stark	24	-20 MHz[h]	(21)

9P(34)	1033.488	96.6			M	250	stark	20	-10 MHz[h]	(21)
		113			M	250	stark	20	-10 MHz[h]	(21)
		114			M	250	stark	20	-10 MHz[h]	(21)
10R(40)	987.620	210.5		‖	0.01	45	0	26		(20)
		255.3		⊥	0.03	45	0	26		(20)
10R(36)	985.488	255		⊥	S	>10	0	>3	maybe CD_3OH line	(3)(8)
10R(34)	984.383	181.5			0.16		W	27		(19)
10R(30)	982.096	150			0.1		0	30		(8)
$10R_I(30)$		192.5		⊥	0.03	105	0	38		(20)
$10R_{II}(30)$		232.4		‖	0.01	90	0	38		(20)
10R(28)	980.913	35			4.0		0	30		(8)
$10R_I(28)$		122.30421	2451203.1	⊥	0.2	130	0,stark	25	0 MHz[h]	(20)(21)
$10R_{II}(28)$		73.8		‖	0.005	100	0	25		(20)
		80.5		‖	0.005	100	0	25		(20)
10R(26)	979.705	183			M	150	stark	15	+35 MHz[h]	(21)

(continued)

Table V CD_3OD Lines (Continued)

		297			M	150	stark	15	+35 MHz[h]	(21)
$10R_I(26)$		119.13700	2518067.7	⊥	0.03	180	0,stark	32		(8)(20)(21)
		124.79788	2402224.0	⊥	0.008	130	0	32		(20)
$10R_{II}(26)$		97.5		⊥	0.03	115	0,stark	20		(20)(21)
10R(24)	978.472	184.76565	1622555.2	⊥	VS	150	0,W, stark	>3	+30 MHz[h]	(3)(5)(8) (20)(21)
		298.73596	1003536.6	∥	VS	>10	0,W, stark	>3	+30 MHz[h]	(3)(5)(8) (20)(21)
		486.5		⊥	0.01	155	0			(20)
		495		∥	0.8		0	30		(8)
10R(22)	977.214	165.60426	1810294.3	⊥	0.08	85	0,stark	32		(8)(20)(21)
10R(20)	975.930	80.5		⊥	0.2	135	0			(20)
10R(18)	974.622	41		⊥	60	40	0,W	33	maybe CD_3OH line	(3)(5)(8) (23)
		869			0.1	80	W	~4	-20 MHz[h]	(12)
$10R_I(16)$	973.289	52.4		∥	0.02	75	0	30		(20)
		82.2		∥	0.006	55	0	30		(20)

		354.17610	846450.3	∥	0.2	85	0,W, stark	30	-25 MHz[h]	(8)(19)(20)(21)
$10R_{II}(16)$		87.3		∥	0.006	65	0	30		(20)
$10R_{I}(12)$	970.547	104.3		∥	0.2	160	0	30		(20)
		107.53770	2787789.4	∥	0.3	180	0	37		(20)
		108.7		∥	0.2	125	0	37		(20)
$10R_{II}(12)$		410.71241	729932.8	∥	VS	115	0,W, stark	>3	-20 MHz[h]	(3)(5)(8)(20)(21)
10R(10)	969.140	86.5		∥	0.2	105	0	42		(20)
		227.66072	1316838.7	⊥	VS	>10	0	>3		(3)(8)(20)
		314.84059	952203.9	∥	VS	>10	0,stark	>3		(3)(8)(20)(21)
10R(6)	966.250	130.5		∥	0.008	85	0	30		(20)
10R(4)	964.769	344.77818	869522.7	⊥,∥	0.6	>10	0	>3		(3)(8)(20)
10P(10)	952.881	78			2.0		0	30		(8)
		149			W	180	stark	10		(21)

(continued)

Table V CD_3OD Lines (Continued)

$^{13}CO_2$ pump										
9P(12)	1007.798	182.0			5.2		W	6		(19)
9P(14)	1006.053	167.5			2.6		W	8.5		(19)
9P(16)	1004.280	310.0			2.6		W	13		(19)
9P(18)	1002.478	374.6			0.2		W	10		(19)
9P(20)	1000.647	533.0			0.12		W	7.5		(19)

R_I and R_{II} indicate different CO_2 pump frequencies.

In reference (21), stark field is vertical to CO_2 pump field.

h: Pump offset frequencies

Table VI Assignments of CH_3OD laser lines

CO_2 pump line (cm^{-1})	Observed wavelength (μm)	Rel. Pol.	Calculated wevelength (μm)	Assignment $(r,\tau,k)J \rightarrow (r',\tau',k')J'$
9R(4) 1067.539	330	∥	334.4	(0,2,3)20→(0,2,3)19
	212	∥	211.0	(0,2,3)20→(0,3,2)19
9P(6) 1058.949	510	∥	515.5	(0,3,5)13→(0,3,5)12
	417	⊥	416.7	(0,3,5)13→(0,1,4)13
	229	∥	230.4	(0,3,5)13→(0,1,4)12
	182.1		182.5	(0,2,10)13→(0,3,9)13
	134.7	∥	134.8	(0,2,10)13→(0,3,9)12
9P(30) 1037.434	352.5	⊥	352.1	(0,2,6)19→(0,2,6)18
	320.0	∥	318.5	(0,2,6)19→(0,3,5)19
	168.1	⊥	167.2	(0,2,6)19→(0,3,5)18
9P(32) 1035.474	279.4	⊥	279.3	(0,1,4)24→(0,1,4)23
	498.0	∥	497.5	(0,1,4)24→(0,2,3)24
	179.0	⊥	178.6	(0,1,4)24→(0,2,3)23
10R(44) 989.647	236	∥	239.8	(0,1,9)28→(0,1,9)27
	110	∥	109.8	(0,1,9)28→(0,1,9)27

References

Bean, B.L. and Perkowitz, S., 1977: "Complete Frequency Coverage for Submillimeter Laser Spectroscopy with Optically Pumped CH_3OH, CH_3OD, CD_3OD and CH_2CF_2," Optics Letts. 1, 202-204.

Blaney T.G., Knight D.J.E. and Lloyd E.K.M., 1978: "Frequency Measurements of Some Optically-Pumped Laser Lines in CH_3OD," Opt. Commun. 25, 176-178.

Bluyssen, H.J.A., van Etteger, A.F., Maan J.C. and Wyder P., 1980 "Very Short Far-Infrared Pulses from Optically Pumped CH_3OH/D, CH_3F and HCOOH Lasers Using an EQ-Switched CO_2 Laser as a Pump Source," IEEE J. Quantum Electron. QE-16. 1347-1351.

Chang T.Y., 1974: "Optically Pumped Submillimeter-Wave Sources," IEEE Trans. Microwave Theory Tech. MTT-22, 983-988.

Chang T.Y., 1977: in "Topics in Applied Physics"(Y.R. Shen, ed.), Vol. 16, 241-243, Springer-Verlag, Berlin, New York.

Danielewicz E.J. and Coleman P.D., 1977: "Assignments of the High Power Optically Pumped CW Laser Lines of CH_3OH," IEEE J. Quantum Electron. QE-13, 485-490.

Danielewicz E.J. and Weiss C.O., 1978: "New CW Far-Infrared Laser Lines from CO_2 Laser-Pumped CD_3OH," IEEE J. Quantum Electron. QE-14, 458-459.

Davis B.W., Vass A., Pidgeon C.R. and Allan G.R. 1981: "New FIR Laser Lines from Optically Pumped Far-Infrared Laser with Isotopic $C^{16}O_2$ Pumping," Opt. Commun. 37, 303-305.

Duxbury G. and Herman H., 1978: "Optically Pumped Millimeter Lasers," J. Phys. B: Atom. Mol. Phys. 11, 935-949.

Dyubko S.F., Svich V.A. and Fesenko L.D., 1974: "Submillimeter CH3OH and CH3OD Lasers with Optical Pumping," Sov. Phys. Tech. Phys. 18, 1121.

Dyubko S.F., Svich V.A. and Fesenko L.D., 1975: "Experimental Research of Radiation Spectra of Submillimeter Laser on Molecular CD_3OH," Radiofizika, 18, 1434-1437.

Gastaud G., Sentz A., Redon M. and Fourrier M., 1980: "New CW FIR Laser Action by Stark Tuning from Optically Pumped CH_3OH and CH_3OD," IEEE J. Quantum Electron. QE-16, 1285-1287.

Grinda M. and Weiss C.D., 1978: "New Far-Infrared Laser Lines from CD_3OH," Opt. Commun. 26, 91.

Henningsen J.O., 1977: "Assignment of Laser Lines in Optically Pumped CH_3OH," IEEE J. Quantum Electron. QE-13, 435-441.

Henningsen J.O., 1981: "Spectroscopy of Molecules by Far Infrared Laser Emission," in "Infrared and Millimeter Waves" (K.J.Button, ed.), Vol. 5, Academic Press, New York.

Heppner J. and Ghosh Roy D.N., 1981: "FIR Gain Measurements in CW Laser Pumped CH_3OD," Int. J. Infrared and Millimeter Waves. 2, 479-491.

Herman H. and Prewer B.E., 1978: "New FIR Laser Lines from Optically Pumped Methanol Analogues," Appl. Phys. 19, 241-242.

Kachi T. and Kon S., 1982: "Experimental Test of CH_3OH Laser Line Assignments with Competitive and Cascade Couplings," Infrared Phys. 22, 337-341.

Kachi T., Fukutani M. and Kon S., 1982: "Assignments of Optically Pumped CH_3OD Laser Lines," Int. J. Infrared and Millimeter Waves. 3, 401-408.

Kon S., Hagiwara E., Yano T. and Hirose H., 1975: "Far-Infrared Laser Action in Optically Pumped CD_3OD," Jpn. J. Appl. Phys. 14, 731-732.

Kon S., Hagiwara E., Yano T. and Hirose H., 1975: "Far-Infrared Laser Action in Optically Pumped CH_3OD," Jpn. J. Appl. Phys. 14, 1861-1862.

Kwan Y.Y. and Dennison D.M., 1972: "Analysis of the Torsion-Rotation Spectra of the Isotopic Methanol Molecules," J. Mol. Spectrosc. 43, 291-319.

Landsberg B.M., 1980: "New Optically Pumped CW Submillimeteer Emission Lines from OCS, CH_3OH and CH_3OD," IEEE J. Quantum Electron. QE-16, 704-706.

Lees R.M. and Baker J.G., 1968: "Torsion-Vibration-Rotation Interaction in Methanol. I. Millimeter Waves Spectrum," J. Chem. Phys. 48, 5299-5320.

Lund M.W. and Davis J.A., 1979: "New CW Far-Infrared Laser Lines from CO_2 Laser-Pumped CH_3OD," IEEE J. Quantum Electron. QE-15, 537-538.

Ni Y.C. and Heppner J., 1980: "New CW Laser Lines from CO_2-Laser Pumped CH_3OD," Opt. Commun. 32, 459-460.

Radford H.E., Petersen F.R., Jennings D.A. and Mucha J.A., 1977:

"Heterodyne Measurements of Submillimeter Laser Spectrometer Frequencies," IEEE J. Quantum Electron. QE-13, 92-94.

Scalabrin A., Petersen F.R., Evenson K.M. and Jennings D.A., 1980: "Optically Pumped CW CH_2DOH FIR Laser: New Lines and Frequency Measurements," Int. J. Infrared and Millimeter Waves 1, 117-125.

Vasconcellos E.C.C., Scalabrin A., Petersen F.R. and Evenson K.M., 1981: "New FIR Laser Lines and Frequency Measurements in CD_3OD," Int. J. Infrared and Millimeter Waves. 2, 533-539.

Vass A., Wood R.A., Davis B.W. and Pidgeon C.R., 1982: "Relaxation Oscillations in cw Optically Pumped CD_3OD and $^{15}NH_3$ Lasers," Appl. Phys. B27, 187-190.

Weber W.H. and Maker P.D., 1982: "Analysis of Doppler-Limited Spectra of the C-O Stretch Fundamental of CD_3OH," J. Mol. Spectrosc. 93, 131-153.

Woods D.R., 1970: Ph. D. Thesis, University of Michigan.

Yoshida T., Kobayashi M., Yoshihara T., Sakai K. and Fujita S., 1981: "Stark Effect in Submillimeter Laser Lines from Optically Pumped CH_3OH and CD_3OD," Opt. Commun. 40, 45-48.

Yoshida T., Yoshihara T., Sakai K. and Fujita S., 1982: "The Stark Effect on the Optically Pumped CH_3OD and CD_3OH Laser," Infrared Phys. 22, 293-298.

Ziegler G. and Dürr U., 1978: "Submillimeter Laser Action of CW Optically Pumped CD_2Cl_2, CH_2DOH and CHD_2OH," IEEE J. Quantum Electro. QE-14, 708.

OPTICALLY PUMPED CH_3CN, CD_3CN AND CH_3NC LASERS

M. Inguscio

Istituto di Fisica dell'Universita di Pisa, Pisa Italy
and
Gruppo Nazionale di Struttura della Materia del CNR

I - CH_3CN

Optically pumped lines

Laser action from optically pumped acetonitrile was first obtained by Chang and McGee (1971). New emissions were then reported by Chang and McGee (1976) and Radford (1975). At the present the CH_3CN molecule has generated 32 laser lines in the wavelength range from 281 μm to 1814 μm, as listed in Table I-1. Precise FIR frequency measurements were reported by Radford et al. (1977) (lines n. 21,27) and by Dyubko and Fesenko (1978) (lines n. 4,5,17,27).

Experimental details

Chang and McGee used as pump source a CO_2 laser 1.5 m long (free spectral range = 100 MHz), internally chopped. It delivered 150 μsec pulses of up to 200W peak power at 120 pps. The FIR resonator had 10 cm diameter internal gold mirrors separated by 93 cm. One of the mirrors was concave with 2.3 m radius of curvature. The other mirror was flat with 2.16 mm diameter coupling hole. The FIR wavelengths were measured by a scanning Fabry-Perot, corrected for diffraction effects with an accuracy of about ±0.05 μm, as can be confirmed on lines for which frequencies are also measured. The pump offsets, listed in Table I-2, were only roughly estimated from the mechanical tuning of the CO_2 laser and should be only considered with an uncertainty of at least ±10 MHz as indicative for best pumping action.

Radford (1975) used a CW CO_2 laser, with a maximum output power of 25W, and a waveguide configuration for the FIR resonator: 90 cm long, 11 mm ID copper tube. The mirrors were gold coated pyrex, one flat, the other spherical (R = 10 m). The pump input coupling was assured by a 1 mm diameter hold in the flat mirror and the FIR output by a 2 mm diameter hole in the spherical mirror. The accuracy of the wavelength measurements is poor (2%), also affected by a systematic shift (~1%) toward higher wavelengths because of the waveguide configuration. As a consequence FIR energies both in Tables I-1 and I-2 are listed using values by Chang and McGee (1971, 1976) or from direct frequency measurements, when available.

A waveguide FIR configuration was used by Radford et al. (1977) for the frequency measurements. These were carried out by mixing on a Shottky diode with harmonics of a phase-locked millimeter wave klystron. The accuracy of the frequency measurement was ±0.5 MHz.

In the frequency measurements by Dyubko and Fesenko (1978), the pumping power was in the range 10-20 W, with FIR output of the order of 1 mW. From other works of the same group (Dyubko et al. 1975), an accuracy of $2x10^{-6}$ can be inferred for the frequency measurements.

The self stability and resettability of FIR laser frequencies are confirmed by the agreement between the two independent measurements reported on Table I-1 for line n. 27.

The efficiency of CH_3CN as laser active medium can be inferred by the comparison between different lines listed in Table I-1 and by the maximum output/input and pump threshold values reported in Table I-2, together with indicative values of operation pressures. Table I-2 should not be considered as a general comparison between open resonator and waveguide structures. In fact it is more likely that Radford's waveguide configuration was designed to achieve lower pump thresholds and not to increase output power. As an example, 1 mw of output at 372 μm (by pumping with 30 W 10P20) was obtained by Hodges et al. (1977) using an optimized FIR waveguide resonator.

Spectroscopic information

CH_3CN methyl cyanide is a symmetric top molecule belonging to the point group C_{3v}. It has four totally symmetric parallel vibrations and four degenerate orthogonal vibrations. Extensive conventional experimental investigations have been performed on the infrared spectrum with low and high resolving power. Among all we shall refer to the most comprehensive (250 to 6500 cm^{-1}) by Nakagawa and Shimanouchi (1961). Spectra in the 10 μm region are rather complicated because of the overlapping and interaction of different fundamental and hot band systems. A comprehensive analysis of the rotovibrational spectrum around CO_2 laser lines was reported by Arimondo and Inguscio (1980). The importance of Fermi and Coriolis interactions had been pointed out by Duncan et al. (1971) and Kondo and Person (1974). They first studied the Fermi resonance of ν_6 with $\nu_7 + \nu_8$, the second investigated the Coriolis interactions between ν_4 and ν_7. A laser Stark spectroscopy investigation has been performed on the ν_4 band by Römheld (1978) and a reanalysis of the IR data had been reported by Ducan et al. (1978).

Coincidences with CO_2 laser lines were detected by Walzer et al. (1979) using the optoacoustic technique and with N_2O laser lines by Danielewicz (1979). Recently CH_3CN was used by Arimondo and Dinelli (1982) to obtain Q switching of a CO_2 laser and estimate absorption and saturation parameters for various transitions.

A set of useful constants for CH_3CN vibration in the 10 μm region is reported in Table I-3.

Due to the large value of the permanent dipole moment, the rotational structure of CH_3CN has been widely investigated using microwave techniques on ground and excited vibrational states. The rotational constants of interest for the present paper are listed on Table I-4. Data for the ground and ν_8 states are taken from the recent review by Boucher et al. (1980), in which also all the measured and predicted rotational transistions up to 300 GHz are included. Data for the excited ν_4 state are taken from Moskienko and Dyubko (1978) who measured rotational lines up to 700 GHz with an uncertainty of ±0.2 MHz using a video-Radio-spectrometer (Moskienko and Dyubko, 1977).

Assignment of the FIR transitions

Only for the seven lines first reported partial assignments had been proposed by the authors (Chang and McGee, 1971). Most of the 32 CH_3CN FIR lines were assigned by Arimondo and Inguscio (1980), as a result of the analysis of the rotovibration spectrum using simultaneously IR and FIR frequencies and relative polarizations.

Three lines were assigned to the fundamental parallel band $0 \rightarrow \nu_4$. Standard formula had been used to compute the IR and FIR energies, using rotation constants up to the quartic order.

More recently the sextic order constants, determined also for the excited state by Moskienko and Dyubko (1978) have been available (Table I-4) and made possible an even more precise determination of the IR energies. Also, the frequency of one FIR emission (line n. 4) has been precisely measured (Dyubko and Fesenko, 1978). That makes possible the precise $K = 0$ reassignment reported in Table I-1 for this which is among the strongest CH_3CN laser lines (Table I-2). It should be noted that the lack of precise FIR frequency measurements can lead to wrong K assignments, even in presence of precise spectroscopic data for the molecule (Arimondo and Inguscio, 1980).

Four lines were assigned to the $\nu_8 \rightarrow \nu_4 + \nu_8$ band. In this case the lack of precise molecular constants makes some assignment not definitive, in particular the vibrational constants are taken from low resolution spectra (Parker et al., 1957) and (A'-A") -- (B'-B") and $(D_K'-D_K'')$ values are fixed to those of the $0 \rightarrow \nu_4$ band. Moreover, the correction due to the degeneration of the $\nu_4 + \nu_8$ vibration should be taken into account in case more and more precise data were available. It should be noted that a very precise FIR frequency measurement can be actually useful only when other microwave and medium-accuracy infrared data are available. This is the case of the precise frequency measurement of line n. 5 which, differently from the case of line n. 4, at present cannot be used to significantly improve the spectroscopic knowledge of the molecule.

Nine lines, including cascade transitions, were assigned to the orthogonal $0 \rightarrow \nu_7$ band. This band has never been investigated under high resolution and essentially the investigation by Kondo and Person (1974) was used. It is worth noting that the band origin had to be shifted by 0.2 cm^{-1} in Arimondo and Inguscio (1980) to assign to the $0 \rightarrow \nu_7$ band the infrared transitions producing strong FIR lines, and in particular those creating collisional cascade transitions. In spite of the precise frequency measurements (lines 21, 27), no improvements in the spectroscopic constants were obtained because of the lack of other precise data, also considered the Coriolis coupling with ν_4 and the Fermi interaction with $3\nu_8$ to be taken in account, as discussed by Arimondo and Inguscio (1980).

Nine of the remaining lines were tentatively assigned to the $\nu_8 \rightarrow \nu_8 + \nu_7$ hot band. The vibrational state with $\nu_7 = \nu_8 = 1$ is composed by levels with E symmetry corresponding to the vibrational angular momenta $\ell_7 = \ell_8 = 1$ and $\ell_7 = \ell_8 = -1$ and by levels with $A_1 + A_2$ symmetry corresponding to $\ell_7 = 1$, $\ell_8 = -1$ and $\ell_7 = -1$, $\ell_8 = 1$ vibrational angular momenta. A discussion of the various interactions with ν_3 and ν_6 bands and of the roto-vibrational constants can be found in Arimondo and Inguscio (1980). The larger uncertainties in the constants and the lack of medium accuracy measurements do not allow an unambiguous assignment.

II - CD_3CN

Optically pumped lines

Laser action from optically pumped acetonitrile-d_3 was only reported by Dyubko and Fesenko (1978). They obtained three laser lines and measured the frequencies using the same apparatus described in the previous section for CH_3CN. No other information but that in Table II-1 is available.

Spectroscopy and assignments

The infrared spectrum of CD_3CN is similar to that of CH_3CN in many respects. Many overlapping fundamental and hot bands are active at 10 μm. The spectrum from 2 to 15 μm was early investigated by Fletcher and Shoup (1963). According to the recent detailed analysis by Duncan et al. (1978), accepted unperturbed vibration frequencies of fundamentals for CD_3CN are the following: $\nu_1 = 2128.8$, $\nu_2 = 2273.5$, $\nu_3 = 1110$, $\nu_4 = 827.25$, $\nu_5 = 2256.56$, $\nu_6 = 1046.45$, $\nu_7 = 847.11$, $\nu_8 = 334.80$ cm^{-1}. Among these, only the perpendicular ν_6 band can give coincidences with CO_2 laser lines. This band exhibits a Coriolis type interaction with the ν_2 band. The effect of this interaction between the normal coordinates involving the antisymmetric and symmetric CD_3 deformation modes results in a degradation inthe subband Q branches of the R wing of ν_6. The degradation is more pronounced to the red as ν_3 is approached and then reversed to the blue once the resonance is passed through. This A_1-E resonance has been investigated by Masri et al. (1973) and Kondo and Person (1974). Following the procedure introduced by Di Lauro and Mills (1966), a Coriolis interaction constant $\xi_6 = -0.375$ or -0.408 was given in the two works respectively.

An overlap with the CO_2 laser spectrum is also obtained with the $\nu_6 + \nu_8 - \nu_8$, $\nu_6 + 2\nu_8 - 2\nu_8$ and $\nu_6 + \nu_4 - \nu_4$ hot bands for which Masri et al. (1973) stated $\nu_0 + A_t - 2(A\xi_t) - B_t = 1049.29$, 1047.80 and 1050.86 cm^{-1} respectively.

The microwave spectrum in the ground state was recently investigated with high accuracy by Demaison et al. (1979). The rotational constants up to the quartic order are reported in Table II-2. As for the excited states, measurements were performed by Matsumura et al. (1962). They analyzed the $J = 2 \rightarrow 1$ transition in the ν_8, $2\nu_8$, $3\nu_8$, $4\nu_8$, $5\nu_8$, ν_7, ν_6, ν_4 and ν_3.

Unfortunately the ν_6 spectrum is only tentatively assigned ($\alpha_6 \simeq 36$ MHz). Moreover it must be pointed out that, since CD_3CN is a symmetric top molecule, the investigated $J = 2 \rightarrow 1$ transitions in the excited states of the E-type vibrations were split into triplets. Moreover, when in presence of a Coriolis interaction, the effective rotational constants of the states were much affected.

In spite of the poor spectroscopic information on CD_3CN, an assignment of the observed lines can be proposed.

Lines n. 1 and 2 can be assigned to the fundamental orthogonal ν_6 band. The CO_2 9P(30) line (1037.434 cm^{-1}) is coincident with the ν_6 $^PQ(J,2)$ transition which is measured and calculated by Kondo and Person (1974) at 1037.59 cm^{-1} and measured by Masri et al. (1973) at 1037.32 cm^{-1}. The latter compute the position, using a best fit set of constants, at 1034.4 cm^{-1}. This value is intermediate between the two experimental results and in fairly good agreement with the CO_2 laser line. The $J = 37$ is obtained from the FIR transition energy. A precise computation of the frequency cannot be done since the lack of precise constants in the ν_6 excited state. Anyway, if the D_J and D_{JK} values from the ground state are assumed, neglecting the Coriolis frequency shifts which are of the order of a few megahertz, a $B\nu_6$ = 7847 MHz can be computed from the FIR line. That is somewhat lower than the value for the ground state, in disagreement with the suggestion by Matsmura et al. (1963) for a negative ν_6 and with the general assumption of $\alpha_6 = 0$ in fitting the infrared band. It is worth noting that the J dependence of the orthogonal ν_6 band is given by $-\alpha_6 J(J + 1)$, with a shift up to 0.4 cm^{-1} for J = 37 and α_6 = 10 MHz.

The CO_2 9P(8) line (1057.300 cm^{-1}) is coincident with the ν_6 $^rQ(J,1)$ transition measured at 1057.56 cm^{-1} by Kondo and Person (1974) and at 1057.40 cm^{-1} by Masri et al. (1973). Again, J = 42 is obtained from the FIR transition and a $B\nu_6$ slightly lower than the g.s. value should be assumed (7853 MHz) to exactly compute the rotational frequency. Also in this case the relatively large disagreement with the measured IR transition can be ascribed to the bad knowledge of the molecular parameters, and in particular of ν_6, which can lead to relatively large shifting in presence of high J values.

Line n. 3 can be tentatively assigned to the hot band $\nu_6 + 2\nu_8 - 2\nu_8$. In fact from the measurements by Masri et al. (1973) it can be extrapolated for $^PQ(J,12)$ a value of 964.06 cm^{-1}, close to the value 964.769 cm^{-1} of the CO_2 9R(4). This assignment is only tentative, because of the too high rotational values assumed and involved in the extrapolation of the frequency.

III - CH_3NC

Optically pumped lines

Two works have been reported on optically pumped emission in CH_3NC. Landsberg et al. (1981) detected many strong optoacoustic absorption signals throughout the CO_2 laser 10 μm P branch and most of the 10 μm R branch. Of these only six resulted in FIR laser action. Part of the remaining absorptions generated the three longer wavelength FIR laser lines reported by Gilbert and Butcher (1981). Therefore we have in hand nine coincidences, giving rise to nine FIR laser lines ordered in Table III-1 by pump line.

Experimental details

In both the reported experiments the same pump CO_2 laser configuration described by Landsberg (1980) was used: 2 m long, 12 mm ID discharge tube; 150 line/mm grating; 80% reflection, 10 m radius of curvature output coupling mirror. The cw output power was higher than 10 W, as it can be deduced from the threshold reported in Table III-1 and in Table I of Landsberg (1980).

As for the FIR configuration, Landsberg et al. (1981) used an open structure, L = 2 m; 10 cm diameter and 2.5 m radius of curvature for both mirrors; 2 mm of diameter for the output coupling hole. This FIR resonator was diffraction limited for $\lambda > 550$ μm. Landsberg et al. (1981) give also the polarization of the FIR radiation relative to that of the pumping laser, a very useful information for the assignment. Optimum operation pressure was 200 mTorr. The output power was determined only roughly and is given as medium, weak and very weak. According to the private communication reported by Graner et al. (1982), M should correspond to the range 0.01-0.1 mW, W and VW being respectively one and two orders of magnitude smaller.

Gilbert and Butcher (1981) used a waveguide FIR resonator constituted by a cylindrical copper tube 96 cm long and 26 mm internal diameter. Two flat mirrors were used, with 2 mm diameter hole for input and output coupling. In addition to a number of other lines reported with the Fabry-Perot open resonator described by Landsberg et al. (1981), the waveguide configuration allowed the observation of long wavelength lines. Optimum pressures were of 25-75 mTorr for all the lines. Because of the copper waveguide configuration, the information on the FIR polarization is lost. The estimated wavelength accuracy was better than 1 percent.

The best confidence limit we estimate in general for the lines reported in Table III-1 is 1 percent.

Spectroscopic information

Methyl isocyanide is a symmetric top molecule with eight fundamental modes:

- four A_1 modes, giving parallel fundamentals ν_1, ν_2, ν_3 and ν_4 at 2965.8, 2165.9, 1427 and 944.9039 cm^{-1} respectively.
- four E modes, giving perpendicular fundamentals ν_5, ν_6, ν_7 and ν_8 at 3014.3, 1463.6, 1129.7 and 262.7 respectively.

Conventional infrared spectra, with a resolution of 0.15 cm^{-1}, were obtained by Thomas et al. (1972). References to previous conventional investigations can be found in that paper.

The ν_4 band is the only fundamental one located in the CO_2/N_2O laser spectral range.

It was analyzed to great precision by laser Stark spectroscopy by Römheld (1978). The laser-Stark spectrum was similar to that of the ν_4 band of the CH_3CN, but in general the signals were stronger. A list of about 50 methyl isocyanide Stark transitions tuned into resonance with 10.5 μm CO_2 lines from P(12) to P(26) was reported.

The microwave rotational spectra were investigated by Bauer and Bogey (1970) and Bauer and Godon (1975) in the ground and ν_4 states respectively. Further more precise results by Bauer for the ground state are reported by Römheld (1978) as private communication. The values were included in a fit to give improved rotational constants for the ground state.

In the laser-Stark investigation also the relatively large dipole moment in the ground and ν_4 excited state could be precisely determined.

The CH_3NC molecular constants, relative to the $0 \rightarrow \nu_4$ band and used in the present paper, are reported in Table III-2. Very recently Bauer and Godon (1982) have further improved the constants for the ground state B_o = 10052.8808 (20); D_J = 0.004680 (14); D_{JK} = 0.226706 (31) (all in MHz).

Absorption in the CO_2 laser region can be originated by the hot bands at 946.4 ($\nu_4 + \nu_8 - \nu_8$) and 948.1 cm^{-1} ($\nu_4 + 2\nu_8 - 2\nu_8$), as measured by Thomas et al. (1972). For the CH_3NC the ν_8 band is about 100 cm^{-1} lower than for the CH_3CN isomer and microwave spectroscopy measurements could be performed at room temperature by Bauer et al. (1971) for ν_8 = 1 and ν_8 = 2. New measurements allowed Godon and Bauer (1977) to perform a more precise analysis and to give the values for the molecular constants reported in Table III-2. Actually ν_8 and $2\nu_8$ are degenerate excited states and a more complete set of constants is required for the fit of the results, as discussed in details by Godon and Bauer (1977) and as necessary in case of more precise FIR frequencies values.

Assignment of the FIR lines

By fitting the FIR wavelengths and the CO_2 pumping frequencies simultaneously, Landsberg et al. (1981) and Gilbert and Butcher (1981) could assign the nine FIR transitions, 5 of them completely and 4 only partially. We have checked and partially revised and completed these assignments (Table III-1). Some of the K assignments are still tentative, due to the poor accuracy of the FIR wavelength measurements.

Five pump transitions can be ascribed to the $0 \rightarrow \nu_4$ band. The precise experimental IR offset given by Römheld (1978) allows us to unambiguously assign the emission n. 4. The comparison between the computed and experimental FIR values suggests that at least for this line the declared wavelength measurement uncertainty (1%) is overestimated. The precise spectroscopic data reported in Table III-2 allow the unambiguous K assignment for line n. 2.

Previously K = 0 or 1 had been proposed, but with K = 1 we compute a IR offset of about -240 MHz. As for the remaining three assignments to the $0 \rightarrow \nu_4$ band, the K values reported in Table III-1 should be considered only as tentative. In fact the IR energies for these high rotational quantum numbers have been computed using constants values (Table III-2) obtained from low J laser-Stark investigations (highest value J = 9). That explains also the relatively high Δ IR values. The FIR values cannot give any help because they are not sufficiently accurate.

One pump transition (line n. 1) has been assigned to the $\nu_8 \rightarrow \nu_8 + \nu_4$ parallel hot band. The IR value has been computed using the constants of Table III-2. A large source of uncertainty is in the not precise knowledge of $A-A_o$ value for the ν_8 band which we have assumed the same as for the ν_4 band. Moreover we have assumed $B_{\nu_4+\nu_8} = B_{\nu_8} + B_{\nu_4} - B_o$ and for the other rotational constants a medium value between the values for ν_4 and ν_8. The high rotational numbers values and the uncertainty of the constants cause ΔIR to be relatively high and K assignment tentative.

The remaining three transitions (lines 3,6,9) can be assigned to the $2\nu_8 \rightarrow 2\nu_8 + \nu_4$ band. The procedure for computing the IR values has been the same as for the $\nu_8 + \nu_4$ band. Again, the high IR values are caused by the lack of spectroscopic information (also considered the high K values tentatively proposed) and are anyway, comparable to the uncertainty of the center band value. The spread in the ΔIR offsets signs, differently from the CH_3CN case, does not suggest any shifting of the band origin.

Suggestion for further investigations

As for CH_3CN, the fundamental $0 \rightarrow \nu_4$ band is well known. In this case the precise measurement of pump offsets, not available, and the detection of further coincidences with infrared laser lines could be useful. In fact it could allow a simultaneous fit of all the existing data, hence improving the accuracy of the rotational constants (mainly the sextic order ones) and checking the consistency of measurements carried out in different spectral regions and in different laboratories. There is lack of medium and high accuracy measurements for all the remaining bands active in the 10 μm region. Infrared and microwave data could be added to the already existing precise FIR frequency measurements possibly obtaining a good set of rotovibrational constants for these bands and their various interactions (*).

In the case of CH_3NC improved values for the FIR frequencies are necessary, as well as measurements of offsets (CH_3NC-CO_2 laser) for all coincidences (leading or not FIR action). That should allow a definitive assignment. Since the measurements refer to high rotational numbers, an improved set of molecular constants could be obtained for the Laser-Stark investigated $0 \rightarrow \nu_4$ band. As for the hot bands, the suggested investigations should lead to a determination of $A-A_o$ and possibly of the band centers.

As for CD_3CN, there is lack of information on the FIR action, such as polarization, pump offsets, relative power. The very precise FIR lines frequencies are not sufficient to improve the knowledge of the molecular structure because of the too many parameters to be taken in account (rotation constants, Coriolis interactions, ℓ-type doubling) and of the absence of measurements on other transitions. Microwave spectroscopy of the excited states, in particular ν_6, could be very useful. Also, a higher

(*) see also the note added in proof.

accuracy reinvestigation of the IR spectra around 10 μm, using the observed coincidences with CO_2 laser lines for calibration, is suggested.

Finally, due to the high μ values, FIR Stark measurements should be interesting on all the three molecules. In particular a measurement on lines n.4 of Table I-1 and n.2 of Table III-1, in case of total absence of electric field effect should confirm the assignment.

Acknowledgements

The author wishes to thank E. Arimondo and F. Strumia for many stimulating discussions and A. Bauer for communicating unpublished results.

Note added in proof (CH_3CN)

When the present paper was ready for publication a work on the laser-Stark spectroscopy of CH_3CN in the 10μm region was published by Rackley et al.(1982). Results were obtained using $^{12}CO_2$, $^{13}CO_2$ and N_2O laser lines and the data, combined with microwave measurements of Bauer and Godon (1975) were fitted to a model which includes l-type doubling in ν_7, ν_4-ν_7 Coriolis coupling, and ν_7-$3\nu_8^1$ Coriolis and Fermi couplings. The vibration-rotation parameters, determined for the ν_4, ν_7 and ν_8 bands were partially included in Table I-3. The improvement in the spectroscopic information is particularly dramatic for the ν_7 band to which several strong FIR emissions have been assigned (Table I-1). It is worth noting that to assign the FIR emissions Arimondo and Inguscio(1979) had to assume a shift of the ν_7 band origin of 0.2 cm^{-1} with respect to the previous experimental determination (Kondo and Person, 1974). The recent accurate experiment by Rackley et al. (1982) confirms the assumption made. That also confirms the powerfulness of the IR-FIR data by themselves in determining molecular parameters. It has to be noticed that the precise frequency measurements available for FIR lines pumped via the ν_4 and ν_7 bands (TableI-1) were not included in the fit by Rackley et al. (1982). The inclusion of these measurements could have led to a further improvement in the accuracy and possibly to the determination of sextic order constants. That could have repaied for the increased difficulty related to the increased "size" of a fit computer program including high rotational quantum numbers.

Table I-1
FIR Lines in CH_3CN

	Ref.	CO_2 line	ν_{IR} (cm^{-1})	ν_{FIR} (GHz)	Pol.	I(*)	Assignment	$\Delta\nu_{IR}$ (°)	$\Delta\nu_{FIR}$ (°)
1	a	10P46	918.72	165.23(1)	//	?	$\nu_4+\nu_8-\nu_8$ $^qP(10,0)$	-0.12	0.01
2	a	10P32	932.96	420.04(6)	//	?	ν_4 $^qR(22,10)$	-0.02	-0.04
3	a	10P24	940.55	710.17(17)	//	?	ν_4 $^qR(38,14)$	-0.07	0.0
4	a,e,f	10P20	944.19	804.1348	//	?	ν_4 $^qR(43,0)$	+0.0	0.0
5	a,e	10P18	945.98	696.4109	//	?	$\nu_4+\nu_8-\nu_8$ $^qR(37,6)$	-0.14	-0.0
6	a	10P16	947.74	787.5(2)	//	?	$\nu_4+\nu_8-\nu_8$ $^qR(42,9)$	+0.16	-0.4
7	a	10P10	952.88	987.6(3)	//	?	$\nu_4+\nu_8-\nu_8$ $^qR(53,9)$	+0.08	-0.07
8	c	9P50	1016.72	1063.2(4)	//	s	ν_7 ^rP (59,2)	-0.01	0.04
9	c	"	cascade	1045.0(4)	//	s	"	"	0.0
10	c	9P46	1021.06	775.8(2)	//	vw	?		

	Ref.	CO_2 line	ν_{IR} (cm^{-1})	ν_{FIR} (GHz)	Pol.	$I^{(*)}$	Assignment	$\Delta\nu_{IR}$(°)	$\Delta\nu_{FIR}$(°)
11	c	9P40	1027.38	275.83(2)	$\perp$	m	$(\nu_7+\nu_8)^A-\nu_8$ $^PQ(15,4)$	0.07	+0.71
12	c	9P34	1033.49	1066.2(4)	$\perp$	w	?		
13	b	9P30	1037.34	459.33(7)	//	m	ν_7 $^rP(26,2)$	+0.08	0.03
14	c	9P26	1041.28	702.02(16)	$\perp$	w	?		
15	c	9P22	1045.02	771.9(2)	//	vw	$(\nu_7+\nu_8)^E-\nu_8$ $^PP(43,1)^{(+)}$	1.36	+1.57
16	c	9P16	1050.44	865.6(2)	//	s	$(\nu_7+\nu_8)^A-\nu_8$ $^rP(48,5)$	1.3	-0.9
17	c,e	"	"	350.8048	//	m	$(\nu_7+\nu_8)^A-\nu_8$ $^rR(18,2)$	-0.21	-0.55
18	d	"	"	400(?)	?	?	$(\nu_7+\nu_8)^E-\nu_8$ $^rR(21,1)$	0.37	+5
19	c	9P10	1055.62	257.37(2)	//	m	?		
20	c	9P8	1057.30	294.98(3)	//	m	$(\nu_7+\nu_8)^A-\nu_8$ $^rP(17,3)$	0.17	+0.05
21	b	9P6	1058.95	606.0747(10)	$\perp$	vs	ν_7 $^rQ(33,3)$	-0.003	-0.029

(continued)

Table I-1 (Continued)

	Ref.	CO_2 line	$\nu_{IR}(cm^{-1})$	$\nu_{FIR}(GHz)$	Pol.	I(*)	Assignment	$\Delta\nu_{IR}(°)$	$\Delta\nu_{FIR}(°)$
22	b	9P6	cascade	587.64(11)	$\perp$	m	$\nu_7 \; {}^{r}Q(33,3)$		0.07
23	c	9R8	1070.46	534.00(9)	//	m	$(\nu_7+\nu_8)^E-\nu_8 \; {}^{r}P(30,9)$	0.6	-0.56
24	c	"	"	404.24(7)	$\perp$	m	$\nu_7 \; {}^{r}Q(22,5)$	+0.02	-0.16
25	b	9R12	1073.28	774.0(2)	$\perp$	s	$(\nu_7+\nu_8)^A-\nu_8 \; {}^{r}Q(42,4)$	-0.02	-0.9
26	c	9R14	1074.65	295.39(3)	$\perp$	w	?		
27	b	9R16	1075.99	661.2134(10)	$\perp$	s	$\nu_7 \; {}^{r}Q(36,6)$	-0.01	-0.60
	e			661.2118(13)					
28	c	9R16	cascade	642.99(14)	$\perp$	s	"	"	-0.57
29	c	"	cascade	624.55(13)	$\perp$	m	"	"	-0.69
30	c	"	lower state	679.57(15)	$\perp$	m	"	"	-0.09
31	c	9R20	1078.59	221.78(2)	$\perp$	w	?		
32	c	9R34	1086.87	425.52(6)	//	m	$(\nu_7+\nu_8)^E-\nu_8 \; {}^{P}R(22,1)^{(+)}$	0.33	-1.52

(*) vs, s, m, w, vw max out/in in a logaritmic scale

(∘) $\Delta\nu = \nu_{calc.} - \nu_{meas}$

(+) for this line the ν_6, $(\nu_7+\nu_8)^E$ mixing is very large

a) Chang and McGee (1971); b) Radford (1975); c) Chang and McGee (1976);

d) Knight (1979); e) Dyubko and Fesenko (1978); f) Moskienko and Dyubko (1978)

Table I-2 Power and pressure performances of CH_3CN laser under different experimental configurations. The lines are listed in the same order of Table I - 1, but the FIR transitions are given in microns. Cascade lines are in parentheses.

	CO_2 PUMP	OFFSET (MHz)	FIR(μm)	MaxOUT/IN mW/W	THRESH W	Press mTorr	Ref.
1	10P46	0	1814.37	0.34/48	12	70	a
2	10P32	-40	713.72	0.68/44	11	70	a
3	10P24	+20	422.14	0.017/110	34	70	a
4	10P20	-30	°372.814	3.5/84	1.7	155	a
				.01	2	10	b
5	10P18	-15	°430.55	0.19/46	8	60	a
6	10P16	-15	380.71	0.007/88	29	40	a
7	10P10	-45	303.54	0.003/68	36	80	a
8	9P50	+50	281.98	0.67/38	2	95	c
9	9P50	+50	(286.88)	0.29/38		95	c
10	9P46	-15	386.41	0.003/42	17	43	c
11	9P40	+40	1086.89	0.15/46	12	24	c
12	9P34	-20	281.18	0.018/44	12	95	c
13	9P30	-15	652.68	0.32/77	29	20	c
				0.0006	5	10	b
14	9P26	+15	427.04	0.024/61	15	43	c
15	9P22	+30	388.39	0.005/69	43	24	c
16	9P16	-10	346.32	0.45/70	7	95	c
17	9P16	-10	°854.585	0.13/70	18	35	c
18	9P16		750				d
19	9P10	+45	1164.83	.050/45	24	20	c
20	9P8	-40	1016.33	.045/33	5	15	c

	CO_2 PUMP	OFFSET (MHz)	FIR(μm)	MaxOUT/IN mW/W	THRESH W	Press mTorr	Ref.
21	9P6	-10	°494.646	4.0/51	4	25	c
				.006	2	10	b
22	9P6	-10	(510.16)	0.19/51		25	c
23	9R8	+40	561.41	0.18/42	19	41	c
24	9R8	+35	741.62	0.23/48	7	24	c
25	9R12	+15	387.31	0.38/44	11	40	c
				0.0005	6	10	b
26	9R14	-20	1014.89	0.018/43	24	20	c
27	9R16	0	°453.398	0.75/43	6	43	c
				0.004	2	10	b
28	9R16	0	(466.25)	0.56/43		35	c
29	9R16	0	(480.01)	0.20:43		17	c
30	9R16	0	441.15+	0.18/43		43	c
31	9R20	-15	1351.78	0.02/49	47	20	c
32	9R34	-15	704.53	0.12/41	11	28	c

a Chang and McGee(1971)
b Radford (1975)
c Chang and McGee(1976)
d Knight(1979)
\+ ground state
° from the precise frequency measurements.

Table I-3

Vibrational Constants of CH_3CN (in cm^{-1}) [a]

Parallel bands	$\nu_\circ$	$A'-A''-(B'-B'')$	$D'_k - D''_k$
$\nu_4 \leftarrow 0$	920.28847(1)	5.2729630(11)-5.28[f] -(0.3054253(13)-0.306842252(3))	(8.6228(24)-8.4[g])10^{-5}
$\nu_4+\nu_8 \leftarrow \nu_8$	924.9(?)[b]	c	c

Perpendicular bands	$\nu_\circ+A'(1-2\zeta_i)-B'$	$A'(1-\zeta_i)-B'$	$A'-A''-(B'-B'')$	ζ_i
$\nu_6 \leftarrow 0$	1456.14(21)[d]	6.517(55)[d]	-0.043(2)[d]	-0.303(10)[d]
$\nu_7 \leftarrow 0$	1042.3819(2)	2.77337	0.0298	0.4199
	1042.34(1)[b]			
$\nu_8 \leftarrow 0$	366.3(?)[d]	0.369	0 (assumed)	0.872(6)[f]
$(\nu_7+\nu_8)^E \leftarrow 0$	1428.79(43)[d]	11.81(12)[d]	0.058(15)[d]	-1.27(2)[d]
$(\nu_7+\nu_8)^A \leftarrow \nu_8$	1050.3(?)[h]	2.82(?)[h]	0.04(?)[h]	-

$\zeta_{47} = 0.2015(10)$ $\quad W_{678} = 15.58(15)$[d]

Dipole moments (Debye)

$\mu_o = 3.925191(48)$ $\mu_4 = 3.935513(55)$ $\mu_7 = 3.929493(92)$

a) When not differently indicated, data are taken or computed from Rackley et al.(1982). See also the note added in proof.

b) Arimondo and Inguscio (1980)

c) fixed to the $\nu_4 \leftarrow 0$ value

d) Duncan et al.(1971)

e) calculated from $A_o (1-\zeta_8) - B_o$

f) Bauer and Maes (1969)

g) Kondo and Person (1974)

h) Thompson and Williams (1952)

Table I-4 Useful rotational constants for CH_3CN bands active in the 10 μm region

	B(MHz)	D_j(MHz)	D_{jk}(KHz)	D_k(MHz)	H_j(Hz)	H_{jk}(Hz)	H_{kj}(Hz)
g.s.	9198.899299(80)	3.8048(15)	177.417(5)	2.840(a)	-0.0140(56)	1.071(19)	6.006(52)
ν_4	9152.6361(18)	3.7939(24)	184.325(46)		-0.0039(10)	1.328(16)	-11.60(32)
ν_8	9226.6402(11)	3.886(15)	177.972(61)				
ν_7	9191.268(23)	3.394(33)	184.1 (12)	2.665(6)			
$\nu_4+\nu_8$	9180.4(c)	(b)	(b)				
$\nu_7+\nu_8$	9221.4(c)	(b)	(b)				

(a) calculated; (b) fixed to g.s. value; (c) calculated (Arimondo and Inguscio, 1980)
(d) see note added in proof.

Table II-1 FIR Laser Emission from CD_3CN

	CO_2 Pump	FIR (MHz)	FIR (μm)	Ref.	Assignment°
1	9P30	580708.2+	516.253	a	$\nu_6\ {}^{P}Q(37,2)$
2	9P8	658778.6	455.073	a	$\nu_6\ {}^{r}Q(42,1)$
3	9R4	565774.2	529.880	a	$\nu_6+2\nu_8-2\nu_8\ {}^{P}Q(36,12)$

a - Dyubko and Fesenko (1978)

\+ most intense

° Pump transition, proposed in the present work.

Table II-2 CD_3CN Rotational constants in the ground state (from Demaison et al.(1979))

B_o(MHz)	D_J(KHz)	D_{JK}(KHz)
7857.98222(13)	2.7451(11)	110.661(27)
H_J(Hz)	H_{JK}(Hz)	H_{KJ}(Hz)
-0.041(26)	0.586(72)	2.107(76)

Table III-1 FIR Laser Emission from CH_3NC

	CO_2 PUMP	FIR[+] (μm)	REL.POL.	THRESHOLD (W)	I	ASSIGNMENT°		ΔIR°° MHz	FIR	REF.
1	10P42	481	//	3.5	W	$\nu_4+\nu_8-\nu_8$	$^qP(32,13)$	1600	483	a
2	10P32	938		6	M	ν_4	$^qP(17,0)$	-48	937	b
3	10P30	823		9.5	VW	$\nu_4+2\nu_8-2\nu_8$	$^qP(19,7)$	-1418	828	b
4	10P14	2140		4	S	ν_4	$^qR(6,1)$	14.2	2143	b
5	10R4	454	//	0.4	M	ν_4	$^qR(32,2)$	-118	455	a
6	10R12	402	//	1.5	M	$\nu_4+2\nu_8-2\nu_8$	$^qR(36,14)$	-1616	404	a
7	10R18	284	//	1.3	S	ν_4	$^qR(52,4)$	-286	284	a
8	10R22	250	//	1.5	M	ν_4	$^qR(59,9)$	+600	251	a
9	10R24	277	//	4	W	$\nu_4+2\nu_8-2\nu_8$	$^qR(55,21)$	+1864	278	a

a- Landsberg et al. 1981; b- Gilbert and Butcher 1981; °° $\nu_{CH_3NC}-\nu_{CO_2}$ + Because of the relatively poor accuracy (≃1%), the FIR energies are given in microns. ° IR transitions, some K assignment is tentative: see text.

Table III-2 Molecular Constants for the CH_3NC Levels Involved in FIR Action. (all in MHz except μ in Debye).

	V = 0	$V_4 = 1$	$V_8 = 1$	$V_8 = 2$
ν_o		28327507.7(10)	28372358(1500?)	28423323(1500?)
$A-A_o$		-254.00(26)		
B	10052.8855(31)	9992.9474(74)	10092.3169	10131.2928
D_J	$4690(21)x10^{-3}$	$4.693(60)x10^{-3}$	$4.875x10^{-3}$	$5.087x10^{-3}$
D_{JK}	$227.16(14)x10^{-3}$	$225.37(20)x10^{-3}$	$226.9x10^{-3}$	$227.3x10^{-3}$
$D_K-D_K^o$		$63(19)x10^{-3}$		
H_{HJ}	$7.3(20)x10^{-6}$			
$H_K-H_K^o$		$1.38(36)x10^{-3}$		
μ	3.88674(20)	3.91701(18)		
$\mu-\mu_o$		0.030272(93)		

References

Arimondo, E. and Inguscio, M., 1980: "Assignments of Laser Lines in Optically Pumped CH_3CN", Int.J. IR mm waves 1, 437-458.

Arimondo, E. and Dinelli, B.M. 1982: to be published.

Bauer, A. and Maes, S.: (1969) J. Phys. 30, 169.

Bauer, A. and Bogey, M. 1970: "Spectre de rotation de la carbylamine dans l'état fondamental de vibration", C.R.Acad.Sc. Paris B271,892-893.

Bauer,A., Bogey, M. and Maes, S., 1971: "Spectre de rotation de la carbylamine dan les états excités de vibration ν_8=1 et ν_8=2", J. de Phys. 32, 763-772.

Bauer, A. and Godon, M., 1975: "Microwave Spectra in the ν_4 Vibrational State of Methyl Cyanide and Methyl Isocyanide and Their ^{15}N Derivatives", Can. J. Phys. 53, 1154-1156.

Bauer, A. and Godon, M.,1982: private communication, unpublished.

Boucher,D., Burie,J., Bauer,A., Dubrulle,A., Demaison,J., 1980: "Microwave Spectra of Molecules of Astrophysical Interest. XIX Methyl Cyanide", J. Phys. Chem. Ref. Data 9, 659-719.

Chang,T.Y. and McGee, J.D., 1971: "Millimeter and Submillimeter Wave Laser Action in Symmetric Top Molecules Optically Pumped via Parallel Absorption Bands", Appl. Phys. Lett. 19, 103-105.

Chang, T.Y. and McGee, J.D., 1976; "Millimeter and Submillimeter-Wave Laser Action in Symmetric Top Molecules Optically Pumped Via Perpendicular Absorption Bands", IEEE J. Quant. Electron. QE-12, 62-65.

Danielewicz, E.D., 1979 private communication.

Demaison, J., Dubrulle, A. Boucher, D. and Burie, J., 1979: "Microwave Spectra, Centrifugal Distortion Constants, and r_z Structure of Acetonitrile and its Isotopic Species," J. Mol. Spectros. 76, 1-16.

DiLauro, C. and Mills, I., 1966: "Coriolis Interactions about X-Y axes in Symmetric Tops," J. Mol. Spectros. 21, 386-413.

Duncan, J.L., Ellis, D. and Wright, I.J., 1971: "Analysis of the ν_3, ν_6, $\nu_7 + \nu_8$ Fermi and Coriolis Interacting Band Systems in methyl cyanide," Mol. Phys. 20, 673-685.

Duncan, J.L., McKean, D.C., Tullini, F., Nivellini, G.D. and Perez Pena, J., 1978: "Methyl Cyanide: Spectroscopic Studies of Isotopically Substituted Species, and the Harmonic Potential Function," J. Mol. Spectrosc. 69, 123-140.

Dyubko, S.F., Fesenko, L.D., Baskakov, O.I. and Svich, W.A., 1975: "Use of CD_3I, CH_3I, CD_3Cl molecules as active substances for submillimeter lasers with optical pumping," Zh. Prikl. Spektr. 23, 317; Engl. Transl.: J. Appl. Spectr. 23, 1114-1116.

Dyubko, S.F., Fesenko, L.D., 1978: "Frequencies of Optically Pumped Submillimeter Lasers," Proc. 3rd Int. Conf. SUBMM waves and their applications, p. 70 (Guilford, University of Surrey).

Fletcher, W.H. and Shoup, C.S., 1963: "The Infrared Spectrum of Methyl-d_3 Cyanide," J. Mol. Spectros. 10, 300-308.

Gilbert, B. and Butcher, R.J., 1981: "New Optically Pumped Millimeter Wave CW Laser Lines from Methyl Isocyanide," IEEE J. Quantum Electron. QE-17, 827-828.

Godon, M. and Bauer, A., 1977: "Microwave spectra in the ν_8 and $2\nu_8$ states of methyl isocyanide and its ^{15}N derivative," J. Mol. Structure 38, 9-16.

Graner, G., Deroche, J.C. and Betrencourt-Stirnemann, C., 1982: "Far-Infrared Laser Lines Obtained by Optical Pumping of CD_3Br Molecule," Rev. IR mm waves this issue.

Hodges, D.T., Foote, F.B. and Reel, R.D., 1977: "High-Power Operation and Scaling Behavior of CW Optically Pumped FIR Waveguide Lasers," IEEE J. Quantum Electron. QE-13, 491.

Knight, D.J.E.,1979: "Ordered List of cw FIR lines", Nat.Phys. Lab., Teddington Middlesex, UK.,NPL rep. n. QU-45, 4th issue; reported as H.E.Radford, private communication.

Kondo,S. and Person, W.B., 1974: "Infrared Spectrum of Acetonitrile: Analysis of Coriolis Resonance", J.Mol.Spectrosc. 52, 287-300.

Landsberg, B.M., 1980: "Optically Pumped CW Submillimeter Emission Lines from Methyl Mercaptan CH_3SH", IEEE J. Quantum Electron. QE-16, 684-685.

Landsberg, B.M., Shafik, M.S. and Butcher, R.J., 1981: "CW Optically Pumped Far-Infrared Emissions from Acetaldeyde,Vinyl Chloride, and Methyl Isocyanide", IEEE J. Quantum Electron. QE-17, 828-829.

Masri, F.N., Duncan, J.L. and Speirs, G.K., 1973: "Vibration-Rotation Spectra of CD_3CN", J.Mol.Spectros. 47, 163-168.

Matsumura, C., Hirota, E., Oka,T. and Morino,Y.,1962: "Microwave spectrum of Acetonitrile-d_3, CD_3CN", J.Mol.Spectros. 9, 366-380.

Moskienko,M.V. and Dyubko, S.F.,1977: "Submillimeter rotational Spectrum of Methyl Cyanide," Journal of Physics of Ucraina 22, 235-237 (in Russian).

Moskienko, M.V. and Dyubko, S.F.,1978: "Identification of generation lines of submillimeter methyl bromide and acetonitrile molecule laser", Radiophysics 21, 951-960 (in Russian).

Nakagawa, I. and Shimanouchi, T., 1961: "Rotation-vibration spectra and rotational, Coriolis coupling, centrifugal distortion and potential constants of methyl cyanide," Spectrochim. Acta, 18, 513-539.

Parker, F.W., Nielsen,S.H. and Fletcher, W.H., 1975: "The Infrared Absorption Spectrum of Methyl Cyanide Vapor", J. Mol Spectrosc. 1, 107-123.

Rackey,S.A., Butcher, R.J., Ramheld,M., Freund,S.M. and Oka,T., 1982: J. Mol. Spectros. 92, 203.

Radford, H.E., 1975: "New CW Lines from a Submillimeter Waveguide Laser", IEEE J. Quant. Electron. QE-11, 213-214.

Radford, H.E., Petersen,F.R., Jennings, D.S. and Mucha, J.A., 1977: "Heterodyne Measurements of Submillimeter Laser Spectrometer Frequencies", IEEE J. Quant. Electron. QE-13, 92-94; corrected in J.Quant. Electron. QE-13, 881.

Romheld,M.,1978: "Laser-Stark-Spectrum and Einige Neuere Laser Spektroskopische Effekte, beobachtet an den Molekulen $^{12}CH_3F$, $^{13}CH_3F$, CH_3CN and CH_3NC," Ph.D. Thesis, University of Ulm.

Thomas, R.K., Leisegang, E.C. and Thompson, H., 1972: "Vibration-rotation bands of methyl isocyanide and its d_3-derivate", Proc. R. Soc. Lond. A. 330, 15-28.

Thompson, H.W. and Williams, R.L., 1952: Trans. Faraday Soc. 48, 502.

Walzer, K., Tacke, M. and Busse, G., 1979: "Optoacoustic spectra of some Far-Infrared Laser active Molecules", Infr. Phys. 19, 175.

THE OPTICALLY PUMPED DIFLUOROMETHANE

FAR-INFRARED LASER

Edward J. Danielewicz

Laser Power Optics Corporation
11211-V Sorrento Valley Road
San Diego, CA 92121

I- Introduction

Difluoromethane, CH_2F_2, is a relatively new CW optically-pumped Far Infrared (FIR) laser. A set of selection criteria, which were based on key molecular properties, was used to successfully predict the discovery of efficient FIR laser operation from this molecule by Danielewicz and Weiss in 1978. Subsequently, Galantowicz et al., 1979 showed that the operating characteristics of the laser together with its molecular dynamical properties were consistent with a collisionally deactivated laser system. This was the first report of a nondiffusion-limited optically pumped far infrared laser. The fortunate combination of very strong absorption of the CO_2 pump radiation together with the high operating pressures of the CH_2F_2 molecule, has produced conversion efficiencies up to 32 percent of the maximum theoretical limit.

This was a factor of two improvements in conversion efficiency over the best previously reported optically pumped CW FIR lasers. Scalabrin and Evenson, 1979 discovered an additional 25 lines and extended the wavelength coverage over the range from 95.6 μm to 1,448.1 μm. A parametric study to maximize the performance by Danielewicz et al., 1979 confirmed the high efficiency capability of the CH_2F_2 laser and clarified some of the limits on scaling the laser to higher output powers.

Many of the strongest emission lines were assigned by Danielewicz 1979 who used the assignments to show that individual transition parameters play a major role in determining the performance of optically pumped lasers. The need for line assignments was prompted by gain measurements by Heppner and Weiss, 1980 who reported single pass gain for the λ = 117 μm line of 34 percent/m which was the highest gain of any CW FIR line at that time. The assignments were needed to check the theoretical models which they were developing to quantitatively predict the gain properties of CW FIR lasers.

The assignments also made possible the theoretical modeling of the pressure behavior of the gain by Galantowicz et al., 1979 who showed that pump depletion was a significant factor in the high pressure limit of the non-diffusion limited CH_2F_2 laser system. All of the laser lines are strong enough that their frequencies could be measured accurately by Peterson et al., 1980 who measured a total of 48 lines. Optimization of the performance on the λ = 117.7 μm and λ = 184.3 μm transitions was performed by Julien and Lourtioz, 1981 who used a uniform transmission output coupler to achieve 65 mW and 150 mW respectively at the two laser wavelengths. The high powers and low beam divergence achieved with these two laser lines combined with excellent long term stability make them of great interest for high density plasma diagnostics.

Most recently, the use by Petersen and Duxbury, 1982 of an isotopically labeled CO_2 pump laser has added an additional 39 new lines which bring the total number of emission lines of CH_2F_2 to 88. Thus the large variety of wavelengths and high output powers make the CH_2F_2 molecule one of the most useful CW FIR lasers.

II. Spectroscopic Details

The structure of the difluoromethane molecule has been established from early microwave studies by Lide, 1952. Hirota et al.,

1970 later refined the data and extended it to the deuterated variety. Most recently, Hirota 1978a has determined the complete equilibrium structure and anharmonic potential function. Difluoromethane is a near-prolate asymmetric-top rotor (Ray's $\kappa = -0.93$) with symmetry point group C_{2v}. It has nine nondegenerate fundamental modes of vibration of which eight IR-active modes were identified by Stewart and Nielsen, 1949. Medium resolution infrared studies by Suzuki and Shimanouki, 1973 have clarified the earlier assignments.

The permanent dipole moment lies solely along the b-principal axis, which is the axis of intermediate moment of inertia. Using laser Stark spectroscopy, Kawaguchi and Tanaka, 1977, have measured the values $\mu'' = 1.9785 \pm .0021$D. and $\mu' = 2.0100 \pm .0014$D. for the ground and ν_9 excited vibrational states, respectively.

The absorption spectrum in the vicinity of the CO_2 laser emission bands is complicated. Figure 1 shows a portion of the high-resolution spectrum measured by Mead, 1979. The P and R branches of the $^{12}C^{16}O_2$ laser overlap the A-type ν_9 mode which has a parallel type band envelope in the symmetric top limit. Analysis of the laser Stark spectroscopy data by Kawaguchi and Tanada (1977) yielded an accurate value for the band origin of $\nu_9 = 1090.1266 \pm .0001$ cm^{-1}.

The IR-absorption of CH_2F_2 is very strong. The ν_9 absorption band strength S = 1196 cm^{-2} atm^{-1} at 300° K given by Pugh and Rao, 1976, converts to an infrared transition dipole moment of $/\mu_v/ =$ 0.33D. Gamss and Ronn, 1975,confirmed the existence of coincidences between CO_2 laser lines and CH_2F_2 absorption lines from laser-induced fluorescense measurements. They also measured the relaxation rate ($\gamma_{v-t} = 43.9$ msec^{-1} Torr^{-1}) from the ν_9 mode. This relaxation rate is fast compared to many other optically pumped laser molecules, and it is an important factor in the CW operation of the CH_2F_2 laser.

Although the main absorption is associated with the ν_9 mode, some of the absorption at CO_2 laser frequencies is associated with hot band transitions. The frequency of the ν_4 mode is very low at 528.5 cm^{-1} and some transitions from the ν_4 to the $\nu_4 + \nu_9$

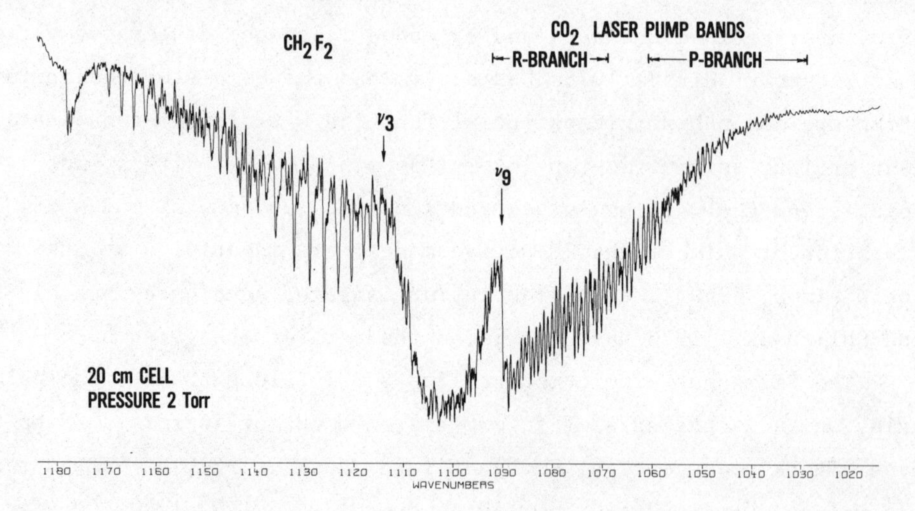

Figure 1. High resolution absorption spectrum of diflouromethane in the vicinity of the CO_2 laser emission bands, (After Mead, 1979).

combination band lead to FIR emission lines (Galantowicz et al., 1979). In addition to the ν_9 mode, the CF_2 symmetric stretch vibration has its fundamental frequency at ν_3 = 1113.2 cm^{-1}. This is a Type B band with a perpendicular band envelope in the symmetric top limit. There is a strong Coriolis coupling between the ν_9 and ν_3 modes which complicates the process of assigning the lines.

The molecular constants in seven of the nine fundamental vibrational states have been obtained from the microwave spectrum of $^{12}CH_2F_2$ by Hirota, 1978b. A set of constants for the ground and ν_9 excited state have also been obtained by Kawaguchi and Tanaka, 1977, from the laser Stark spectroscopy results. Table I lists the results for the ground plus the ν_9, ν_3 and $\nu_4 + \nu_9$ states using the notation of Kivelson and Wilson, 1952. The Coriolis interaction between ν_3 and ν_9 is strong as can be seen from the Coriolis coupling constant $\phi_{s,s'}$ = 0.704 obtained by Hirota 1978b, so that any analysis of either vibrational state must account for the Coriolis

TABLE I - Molecular Constants of $^{12}CH_2F_2$ in the ν_9, ν_3 and $\nu_4 + \nu_9$ States (MHz)[a]

CONSTANT	GROUND STATE	ν_9 [b]	ν_3 [b]	$\nu_4 + \nu_9$
$\tilde{A}$	49,142.818 (18)	48,704.26 (44)	48,295.44 (51)	49,040.455(94)
$\tilde{B}$	10,604.7050(44)	10,524.25 (22)	10,544.10 (19)	[10,509.837]
$\tilde{C}$	9,249.8437(33)	--	--	8,983.087(26)
C	--	9,147.3 (96)	9,261.2 (95)	--
τ_{aaaa}	-2.30061 (77)	[-2.3006]	[-2.3006]	[-2.318]
τ_{bbbb}	-0.06151 (18)	[-0.0615]	[-0.0615]	[-0.0579]
τ_{cccc}	-0.02723 (13)	[-0.0272]	[-0.0272]	0.1481 (18)
T_1	0.1575 (23)	[0.1575]	[0.1575]	[0.176]
T_2	-0.03756 (19)	[-0.0376]	[-0.0376]	[-0.0362]

a. Values in parentheses are standard deviations, and those in brackets are fixed, after Hirota, 1978.

b. The constants have been obtained from A simultaneous analysis of both modes including Coriolis Coupling. The resulting common constants D=10,189.(336) and F=-116.(17)were also with $\nu_3-\nu_9$ = 628.3 (.78) x 10^3 MHz interaction in the effective Hamiltonian

For these states a first-order centrifugal distortion treatment is not satisfactory, and an effective Hamiltonian comprised of two vibrational states must be used. The values in Table I for the ν_9 and ν_3 modes were obtained from the analysis including Coriolis Coupling with the values of the centrifugal distortion constants fixed to the ground state values.

III. Experimental Results

The far infrared laser lines from $^{12}CH_2F_2$ pumped by the $^{12}C^{16}O_2$ laser are summarized in Table II. The vacuum frequencies of 48 of the lines have been measured relative to stabilized CO_2 lasers by Petersen et al., 1980. They estimated the fractional frequency uncertainty to be about 5 parts in 10^7 which is an estimate of the reproductibility of the FIR laser frequencies in an actual experiment. No corrections have been applied for pressure, power, Stark, or other types of frequency shifts. The listed wavelengths have been calculated from the measured frequencies. The fourth column indicates the polarization of the FIR lines relative to the CO_2 pumplines. The pressures in column five correspond to the values where maximum FIR output has been observed for each laser line.

For the frequency measurements, the FIR radiation was generated in a 1 mm long Fabry-Perot cavity with a variable output coupler as described by Scalabrin and Evenson, 1979. This cavity design is convenient and effective over the entire far-infrared range, but it is not optimized for maximum power output. An absolute power measurement yielded only 4.0 mW on the 184.3 μm line, which was the strongest line obtained with this type of resonator. This power level was assigned a value of 170 in relative intensity units, which are given for many of the lines in Table II where no actual power measurement has been reported. The FIR powers are the highest values which have been reported for each line.

The CO_2 pump powers are the values measured at the pump frequency offset, which gave maximum FIR output power. No pump frequency offsets have been listed because there is wide variation among the values reported in the literature. Using Doppler-free optoacoustic spectroscopy, Tsunawaki et al., 1981, have measured the offsets of various absorption lines of CH_2F_2 and compared these values to offsets measured by other methods. Their results show that there is not necessarily a direct correspondence between absorption offsets measured by optoacoustic methods and FIR lasing. Some FIR laser lines are pumped at offsets which give no strong absorption signal. Also, as noted in Table II, there are several pumplines which produce FIR lines at more than one offset.

The salient operating features of the CH_2F_2 laser can be illustrated best by a discussion of the results obtained with various resonator and output coupling schemes. The initial study of CH_2F_2 by Danielewicz and Weiss, 1978, used a Chang-type open resonator to test the prediction of efficient FIR performance from the molecule. Conversion efficiencies up to 20 percent of the theoretical limit were observed with this non-optimized system due to the favorable intrinsic properties of CH_2F_2. The main reason for the high conversion efficiency, which was a factor of two higher than the highest previously reported value, was the more efficient absorption of the CO_2 pump laser. This is a direct consequence of the higher optimum operating pressures compared to other FIR laser molecules, because of the rapid V-T relaxation from the excited vibrational states back to the ground state. The wide range of operating pressure is well illustrated by the data of Julien and Lourtioz, 1981, in Figure 2, which shows the output power of the 184.3 μm line as a function of pressure for several values of infrared pump power. Galantowicz et al., 1979b, showed that the higher optimum pressure and slow decrease of output power with increasing pressure are characteristic of collisionally-deactivated systems, which do

TABLE II. Observed CW Far Infrared Laser Lines from $^{12}CH_2F_2$ Pumped by the $^{12}C^{16}O_2$ Laser

$^{12}C^{16}O_2$ PUMP LINE(d)	WAVE-LENGTH a) (μm)	FREQUENCY a) (cm^{-1})	REL POL	OPTIMUM PRESSURE (m Torr)	FIR POWER (mW)	CO_2 POWER (W)	REFS.
9R(46)	588.0	17.006 005	—	60	4.5[b]	7	a,b
9R(44)	642.6	15.561 783	∥	50	12.0[b]	9	a,b
	1448.1	6.905 619	—	50	0.3[b]	9	a,b
9R(42)	230.1	43.458 258	⊥	70	30[b]	14	a,b
	541.0	18.484 755	∥	65	6,0[b]	14	a,b
9R(40)	1694.0[c]	5.903[c]	⊥	110	<1.0	13	c
	1089.0[c]	9.183[c]	⊥	110	<1.0	13	c
9R(36)'	298.2	33.533 310	∥	100	1.5[b]	20	a,b
9R(36)"	382.0	26.178 312	∥	100	4.0[b]	22	a,b
9R(34)	214.6	46.602 859	⊥	200	60.0	16.5	a,b,e,g
	287.7	34.762 394	∥	400	9.0	16.5	a,b,e,g
9R(32)	184.3	54.257 623	⊥	140	150.0	24	a,b,e,f,g
	235.7	42.435 071	∥	170	4.5	17.5	a,b,e,g
9R(28)	511.4	19.552 438	⊥	70	4.0[b]	26	a,b
	567.5	17.620 164	—	70	1.0[b]	26	a,b

$^{12}C^{16}O_2$ PUMP LINE d)	WAVE-LENGTH a) (μm)	FREQUENCY a) (cm^{-1})	REL POL	OPTIMUM PRESSURE (m Torr)	FIR POWER (mW)	CO_2 POWER (W)	REFS.
9R(22)'	122.5	81.655 439	⊥	400	2.0	12.0	a,b,e,g
	166.7	59.996 406	∥	200	30.0[b]	27	a,b
9R(22)"	193.9	51.571 792	∥	70	17.0[b]	29	a,b
	270.0	37.036.287	⊥	65	6.0[b]	29	a,b
9R(20)	117.7	84.941 929	⊥	440	65.0	23	a,b,e,f,g
	166.6	60.012 827	∥	400	2.5	20.5	a,b,e,g
9R(14)	326.4	30.635 092	⊥	80	1.0[b]	30	a,b
9R(12)	95.6	104.656 087	⊥	280	2.0[b]	25	a,b
	194.4	51.427 733	⊥	90	1.0[b]	25	a,b
	418.3	23.907 982	∥	80	0.5[b]	25	a,b
9R(6)	202.5	49.391 266	∥	70	24.0[b]	23	a,b,g
	236.59	42.266 939	∥	100	120 [b]	23	a,b,e,g
	236.60	42.265 288	∥	100	120 [b]	23	a,b,e,g
	435.0	22.991 074	⊥	65	27.0[b]	23	a,b,g
	503.1	19.878 475	⊥	40	20.0[b]	23	a,b

(continued)

TABLE II. Observed CW Far Infrared Laser Lines from $^{12}CH_2F_2$ Pumped by the $^{12}C^{16}O_2$ Laser (Continued)

$^{12}C^{16}O_2$ PUMP LINE d)	WAVE-LENGTH a) (μm)	FREQUENCY [a] (cm^{-1})	REL POL	OPTIMUM PRESSURE (m Torr)	FIR POWER (mW)	CO_2 POWER (W)	REFS.
9P(4)	289.5	34.542 319	‖	75	2.5	16	a,b
	724.9	13.794 620	⊥	75	20.0[b]	16	a,b
9P(6)	394.7	23.335 638	‖	75	40.0[b]	25	a,b
	464.4	21.532 594	‖	70	8.0[b]	25	a,b
9P(8)	122.5	81.655 645	‖	130	0.5[b]	24	a,b
	355.1	28.159 009	‖	120	1.5[b]	24	a,b
9P(10)'	158.5	63.086 120	‖	225	15.0	16	a,b,e,g
	272.3	36.718 958	⊥	140	0.5	19	a,b,e
9P(10)"	382.6	26.134 280	‖	100	45.0[b]	25	a,b
	657.2	15.215 162	⊥	70	20.0[b]	25	a,b
9P(16)	105.5	94.770 326	‖	290	15.0[b]	29	a,b
9P(18)	227.7	43.925 737	⊥	120	0.5[b]	30	a,b
9P(20)	159.0	62.908 831	‖	85	0.5[b]	30	a,b
	293.9	34.025 006	‖	80	0.5[b]	30	a,b
9P(22)	134.0	74.631 512	⊥	230	40[b]	28	a,b
	191.8	52.124 591	‖	200	54[b]	28	a,b

$^{12}C^{16}O_2$ PUMP LINE d)	WAVE-LENGTH a) (μm)	FREQUENCY (cm^{-1})	REL POL	OPTIMUM PRESSURE (m Torr)	FIR POWER (mW)	CO_2 POWER (W)	REFS.
9P(24)	109.3	91.494 831	‖	400	4.5	22.0	a,b,e
	135.3	73.926.593	‖	450	3.0	22.0	a,b,e,g
	256.0	39.058 387	⊥	450	2.0	22.0	a,b,e,g
9P(38)	261.7	38.207 437	‖	190	0.3	19.0	a,b,e

a) From Peterson et al., 1980

b) From Scalabrin and Evenson, 1979. These are relative intensities only where the value 170 corresponds to 4.0 mW of output power.

c) Previously unpublished.

d) ' and " indicate different CO_2 Laser frequency offsets.

e) Danielewicz et al., 1979.

f) Julien and Lourtioz, 1981.

g) Galantowicz et al., 1979.

not show the "vibrational bottleneck" of diffusion limited systems.

A parametric study was performed by Danielewicz et al., 1979, to maximize the performance of the laser. The experimental setup was the same one used by Hodges et al., 1976, to achieve high-power operation with CH_3OH. The FIR cavity consisted of a 1.1 m length of 44 mm diameter hollow dielectric waveguide with plane external mirrors. The output coupler was a hybrid-hole-coupler on a dielectrically coated Si substrate. This type of resonator favors operation of the shorter ($\lambda > 500$ μm) wavelength FIR lines.

Optimization of the performance on the 184.3 μm line was done by systematically testing several different sized output coupling apertures in the FIR resonator. A CO_2 power of 17.5 W yielded the highest output power of 80 mW using a 16 mm diameter coupling hole. The conversion efficiency calculated for this case was 33 percent of the theoretical limit. High output powers of 60mW on the 214.6 μm and 35 mW on the 117.7 μm line were also obtained with this system.

Scaling of the CO_2 pump showed no decrease in the conversion efficiency with increasing power up to the maximum available power from the laser. This indicated that further increases in output power could be expected from increasing the pump power and optimizing the FIR cavity length. In fact, the FIR output power scales approximately linearly with pump power as can be seen from Figure 2. With the longer hollow dielectric waveguide of 1.8 m length and 25 mm dia., Julien and Lourtioz, 1981, obtained the maximum power of 150mW with 24 W of pump power. Their value of 27 percent for conversion efficiency confirmed the earlier results for this transition.

The main advantage of their FIR resonator is the innovative output coupling technique, which allows the output power to be optimized for each transition without changing mirrors. At the same time, the couplers have uniform transmission over the full aperture which minimizes output beam divergence. Two kinds of output couplers have been used. The first one is composed of a double silicon plate

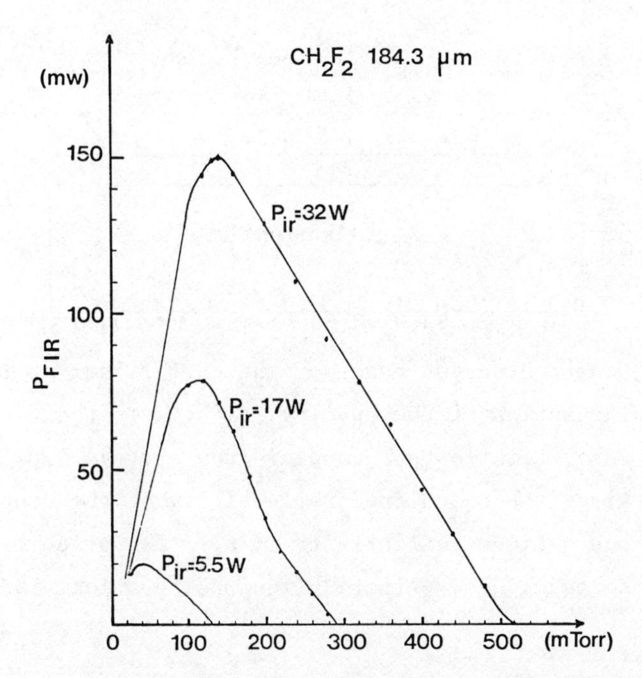

Figure 2. Experimentally measured curves of the FIR output power as a function of operating pressure for the 184.3 μm. Line of CH_2F_2 for three different values of CO_2 pump power coupled into the FIR cavity.

Fabry-Perot interferometer and of a grating, which couples the pump beam back into the resonator and couples out the FIR beam. This variable coupler allows the optimum coupling reflectivity to be determined for a given FIR transition as shown in Figure 3. The second coupler is composed only of a single silicon plate and the

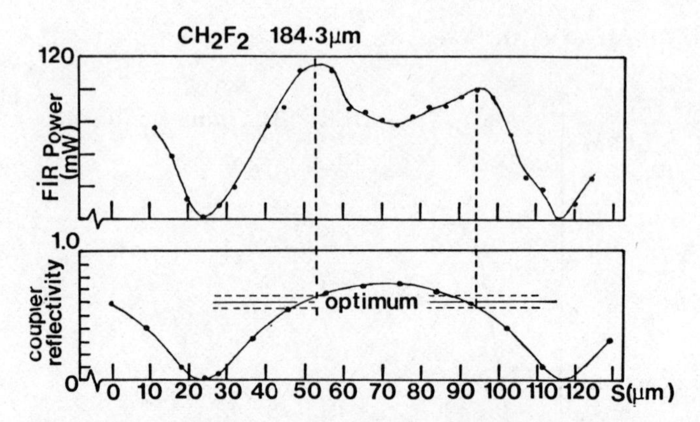

Figure 3. Measured output power of the CH_2F_2 laser at 184.3 μm as a function of the spacing, s, between the plates of the Fabry-Perot output coupler (top curve). The bottom curve gives the calculated reflectivity of the coupler and the optimum reflectivity of 61 ± 5% is indicated by the dotted lines, (After Julien and Lourtioz, 1981).

grating. The reflectivity of the silicon etalon can be adjusted between 0 percent and 70 percent to match a given transition's optimum reflectivity to achieve maximum output for any given FIR line. For the 184.3 μm line, the optimum reflectivity was found to be 61 ±5 percent. This high value of output coupling is related to the high gain of this laser transition. A value of 0.70 m^{-1} was determined for the unsaturated gain at 184.3 μm and 0.48 ± .15 percent at 117.7 μm. These values are in relatively good agreement with the values of Heppner et al., 1980 although they measured the gain in an amplifier cell.

In the long wavelength region, little work has been done to optimize the performance of any FIR lines. Gamble and Danielewicz, 1981, used a rectangular metallic guide with internal hole-coupled mirrors to investigate transitions with wavelengths longer than 1000 μm. Using this resonator, two new FIR lines were found with the 9R(40) pump line as listed in Table II. Kawaguchi and Tanaka, 1977 showed that the 9R(40) CO_2 line coincides with the $3_{2,2} \rightarrow 3_{2,1}$ Q-branch absorption line of CH_2F_2. The two new laser lines, which were observed, played an important role in confirming the method of assigning many of the FIR laser lines as will be shown in the next section.

Recently, the number of lines observed has been considerably increased with the use of the $^{12}C^{18}O_2$ isotopic pump laser by Peterson and Duxbury, 1982. They used a 1 m open resonator for finding 39 new FIR lines as listed in Table III. The wavelengths of the lines have been measured with an accuracy of within 5 x 10^{-3} by using the FIR resonator itself as a scanning Fabry-Perot interferometer. The lines indicated with a star are lines with an estimated output power of between 1 and 10 mW. This brings the total of known FIR lines from CH_2F_2 to 88. They also reported that no emission lines were found from CH_2F_2 using a $^{13}C^{16}O_2$ isotopic pump laser, since the emission bands do not overlap any absorption bands of CH_2F_2.

It seems likely that the number of emission lines from CH_2F_2 will be increased in the future by using waveguide and sequence band CO_2 pump lasers. In addition the output power of all the lines will continue to be increased as the resonators are optimized for each transition. Scaling of the CO_2 pump power will also continue to produce higher output powers since saturation has not yet been reached.

IV. Laser Line Assignments

Assignments for the laser lines are important for theoretical modeling of laser performance and for refining the molecular con - stants. Fourteen of the emission lines of CH_2F_2 pumped by the $^{12}C^{16}O_2$ laser have been assigned by analogy with the CH_3OH molecule, which has similar FIR laser transitions. Figure 4 shows the

TABLE III. Observed CW Far Infrared Laser Lines from $^{12}CH_2F_2$ Pumped by the $^{12}C^{18}O_2$ Laser [(b)].

CO_2 Pump Line[(a)]	CH_2F_2 Laser Line $\lambda(\mu m)$	Relative Polarization	CH_2F_2 Laser Press (m Torr)	CO_2 Pump Power (watts)
9R(42)	142.9	∥	70	3.0
	236.4	⊥	70	3.0
9R(40)	283.2*	⊥	75	5.5
	718.7*	∥	75	5.5
9R(38)	261.6	∥	70	6.0
9R(36)	298.4*	∥	70	6.5
9R(34)	247.3	--	75	7.0
9R(30)	243.1*	∥	75	7.0
9R(26)	193.2*	∥	75	6.5
	268.0*	⊥	75	6.5
9R(22)	163.4*	∥	70	7.5
	309.8*	⊥	70	7.5
9R(18)	284.8	∥	75	7.0
	439.4	∥	75	7.0
9R(14)'	358.3*	∥	80	8.0
9R(14)''	190.3	--	75	7.0
9R(8)	360.1	--	70	5.5
9R(6)	535.2	∥	70	2.5
9P(4)	273.4*	⊥	50	5.0
	404.1*	∥	50	5.0
9P(6)	237.0*	∥	50	9.0
9P(8)	116.0	⊥	100	7.5
	166.8	∥	100	7.5
9P(12)	134.9	⊥	70	7.0
	208.4	∥	70	7.0

CO_2 Pump Line(a)	CH_2F_2 Laser Line λ(μm)	Relative Polarization	CH_2F_2 Laser Press (m Torr)	CO_2 Pump Power (watts)
9P(14)	281.6*	11	100	6.5
	677.4	--	100	6.5
9P(16)	586.8*	11	80	6.0
	1090.7	⊥	80	6.0
9P(24)	289.6*	11	100	7.5
	154.6	⊥	100	7.5
9P(26)	282.9	⊥	100	7.5
9P(28)	280.5	⊥	100	7.0
	153.1	11	100	7.0
9P(30)	126.4*	11	90	7.0
9P(38)	402.2	11	80	6.0
9P(40)	223.6	11	100	6.0

(a) ' and " indicate different CO_2 laser offsets.

(b) Table III is taken from Petersen and Duxbury, 1982.

assignments of two of the strongest lines of CH_2F_2 and compares these to transitions of CH_3F and CH_3OH. The CH_2F_2 molecule is a near-prolate asymmetric top with a large permanent dipole moment (μ_b) solely along the b-axis of intermediate moment of inertia. CH_3OH is also a near prolate asymmetric top but has a permanent moment (μ_a) along the a-axis as well as the b-axis. CH_3F is symmetric top with permanent moment solely along the a-axis of the molecule. As can be seen from Figure 4, the components of the

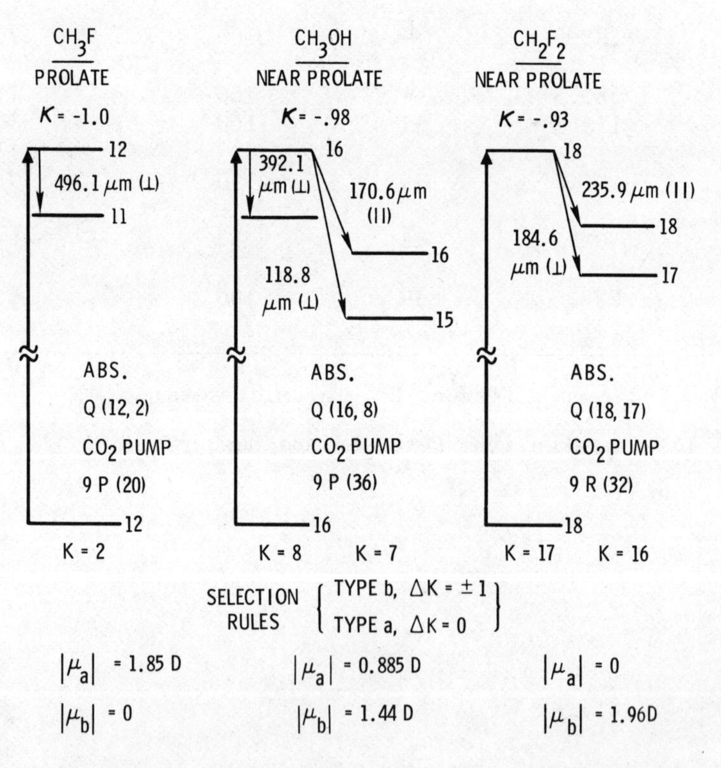

Figure 4. Comparison of the line assignments of various strong FIR laser transitions of CH_3F, CH_3OH, and CH_2F_2, (After Danielewicz, 1979).

permanent dipole moment which are non-zero, strongly affect the number and type of FIR laser transitions. The selection rules which govern the types of allowed transitions are also given in the figure for type-a and type-b transitions. So, for example, CH_3F which has only a μ_a component can only have type-a FIR transitions where the quantum number K remains unchanged. CH_2F_2 on the other hand, must have rotational transitions where K changes by ± 1 and CH_3OH can have both types of rotational transitions.

The rotational energy levels for an asymmetric top like CH_2F_2 may be denoted by the three numbers J, K_p and K_o. As defined by Townes and Schawlow, 1955, J is the total rotational quantum number, and K_p and K_o are the quantum numbers associated with the projection of J on the unique axis for the limiting prolate and oblate symmetric tops, respectively. Since CH_2F_2 is so close to a prolate symmetric top, the rotational energy levels can be calculated by the approximate formula given by Herzberg, 1951.

$$E\,(J,K) \cong 1/2\,(\tilde{B} + \tilde{C})\,J(J+1) + [\tilde{A} - 1/2\,(\tilde{B} + \tilde{C})]K^2 \qquad (1)$$

The K in equation (1) corresponds to K_p for the asymmetric molecule, and this is the K referred to in Figure 4. The constants $\tilde{A}$, $\tilde{B}$ and $\tilde{C}$ are given for the various vibrational modes of CH_2F_2 in Table I. This equation neglects asymmetry effects, which can be large at small values of K and centrifugal distortion which is large for large values of J and K.

However, it has been accurate enough to tentatively assign fourteen of the lines of $^{12}CH_2F_2$, pumped by the $^{12}C^{16}O_2$ laser as listed in Table IV.

Applying the selection rules to equation (1) gives the two equations for the frequencies of the Q-branch and R-branch FIR emission lines for each pumpline of CH_2F_2 given by

$$\nu_Q = [\tilde{A} - 1/2(\tilde{B} + \tilde{C})]\,(2K + 1) \qquad (2)$$

$$\nu_R = \nu_Q + (\tilde{B} + \tilde{C})\ (J + 1) \tag{3}$$

In addition, the relative polarization of the pump laser and of the emitted far infrared radiation must be taken into consideration in the assignments. The rules

$$\Delta J_1 + \Delta J_2 \text{ even} \longrightarrow \| \text{ polarization}$$

$$\Delta J_1 + \Delta J_2 \text{ odd} \longrightarrow \perp \text{ polarization}$$

where ΔJ_1 is the change in the quantum number J of the IR transition and ΔJ_2 the change in J of the FIR transition have been applied in making the assignments. The polarization rules indicate that in situations where two FIR lines are found for the same pump frequency offset, then the infrared transition is a P or R transition when the high frequency line is parallel and Q transition when it is perpendicular. Once the Q-branch FIR emission line has been identified, the value of K can be calculated using equation (2) and rounding to the nearest integer value. Then the J quantum number can be determined from equation (3). The values of J and K refer to the lower level of the rotational transition.

Applying the symmetry requirements, Hezberg, 1951, allows the complete set of quantum numbers to be determined for the asymmetric levels. As an example, the complete assignment for the emission lines for the 9R(32) CO_2 pumpline are shown in Figure 5.

In order to check the assignment technique, the lines, which would be produced using the 9R(40) pumpline, were predicted based on the known coincidence of the pumpline with the Q-branch absorption line found by Kawaguchi and Tanaka, 1977. They confirmed the microwave transition $4_{1,\,4} \longrightarrow 3_{2,\,1}$ in the ν_9 excited state as shown in Figure 6. The predicted line at 1686.0 μm was observed using a rectangular metallic guide and in addition another line at 1089 μm was observed. The 1089 μm line was assigned to a refilling transition in the ground state and this nicely instilled

TABLE IV. Tentative Assignments of the Far Infrared Laser Lines from $^{12}CH_2F_2$ Pumped by the $^{12}C^{16}O_2$ Laser.

ABSORPTION			EMISSION			
CO_2 LINE	MODE	TRANSITION	λFIR(μm)	POL	1/λFIR (cm^{-1})	TRANSITION
$R_9(40)$	ν_9	$3_{2,2} \rightarrow 3_{2,1}$	1686.0	⊥	5.931 (a)	$3_{2,1} \rightarrow 2_{1,2}$
			1088.0	⊥	9.191 (a)	$4_{3,1} \rightarrow 3_{2,2}$ (b)
$R_9(34)$	ν_9	$18_{14,5} \rightarrow 18_{14,4}$	214.6	⊥	46.6029	$18_{14,4} \rightarrow 17_{13,5}$
			287.7	‖	34.7624	$18_{14,4} \rightarrow 18_{13,5}$
$R_9(32)$	ν_9	$18_{17,2} \rightarrow 18_{17,1}$	184.3	⊥	54.2576	$18_{17,1} \rightarrow 17_{16,2}$
			235.7	‖	42.4351	$18_{17,1} \rightarrow 18_{16,2}$
$R_9(22)'$	$\nu_4 + \nu_9$	$33_{23,11} \rightarrow 33_{23,10}$	122.5	⊥	81.6554	$33_{23,10} \rightarrow 32_{22,11}$
			166.7	‖	59.9964	$33_{23,10} \rightarrow 33_{22,11}$
$R_9(20)$	$\nu_4 + \nu_9$	$38_{23,16} \rightarrow 38_{23,15}$	117.7	⊥	84.9419	$38_{23,15} \rightarrow 37_{22,16}$
			166.6	‖	60.0128	$38_{23,15} \rightarrow 38_{22,16}$
$P_9(4)$	ν_9	$36_{6,31} \rightarrow 35_{6,30}$	289.5	‖	34.5423	$35_{6,30} \rightarrow 34_{5,29}$
			724.9	⊥	13.7946	$35_{6,30} \rightarrow 35_{5,31}$
$P_9(10)'$	ν_9	$41_{15,26} \rightarrow 40_{15,25}$	158.5	‖	63.0861	$40_{15,25} \rightarrow 39_{14,26}$
			272.3	⊥	36.7190	$40_{15,25} \rightarrow 40_{14,26}$

(a) Calculated Values (b) Ground State

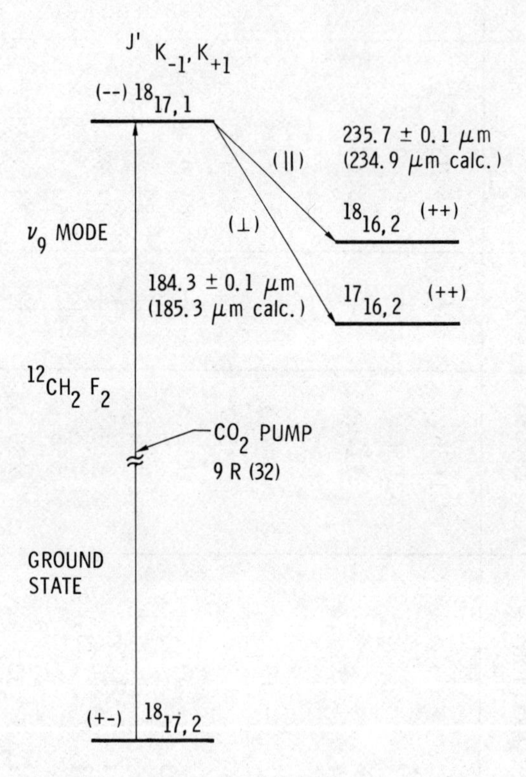

Figure 5. Assignments of the Far-Infrared emission lines pumped by the 9R(32) CO_2 Line.

confidence in the rest of the assignments. The four lines pumped by 9R(20) and 9R(22)' are assigned to transitions in the $\nu_4 + \nu_9$ combination band. Evidence that these lines are pumped on hot-band transitions was found by Galantowicz et al., 1979a, who measured increased absorption on these transitions after heating the absorption cell. This assignment explains the low value of absorption coefficient and higher operating pressures observed with these CO_2 pumplines. Similarly, Petersen and Duxbury, 1982, have assigned 9 of the lines pumped with the $^{12}C^{18}O_2$ laser as summarized in Table V.

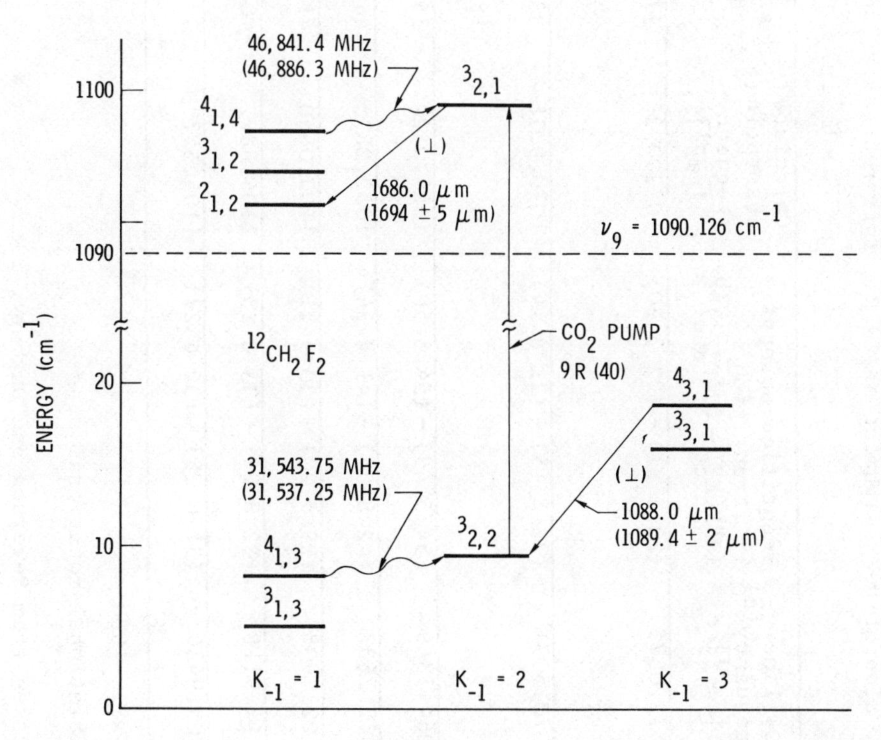

Figure 6. Assignments of the Far-Infrared emission lines pumped by the 9R(40) CO_2 line. The wavy transitions are microwave absorption lines observed by Kawaguchi and Tanaka, 1977.

TABLE V. Tentative assignment of CW Far Infrared Laser Lines from $^{12}CH_2F_2$ Pumped by the $^{12}C^{18}O_2$ Laser.

$^{12}C^{18}O_2$ pump line	CO_2 line frequency(a) [cm^{-1}]	Identified absorption transition $(J''K_a''K_c'') - (J'K_a'K_c')$ (b)	Calculated absorption frequency [cm^{-1}]	Measured FIR frequency [cm^{-1}]	Identified emission	Calculated FIR frequency [cm^{-1}]
9P(16)	1072.0525	(23 3 21) - (22 3 20)	1072.0926	17.04	(22 3 20) - (21 2 19)	17.02
				9.17	(22 3 20) - (22 2 21)	9.13
9P(18)	1070.5071	(23 8 15) - (22 8 14)	1070.9078	33.18	(22 8 14) - (21 7 15)	33.56
				18.92	(22 8 14) - (22 7 15)	19.21
9P(26)	1064.1349	(54 5 49) - (54 5 50)	1064.0427	35.35	(54 5 50) - (53 4 49)	35.56
9P(40)	1052.2632	(42 8 35) - (41 8 34)	1052.2875	44.72	(41 8 34) - (40 7 33)	44.92
9R(30)	1102.7908	(29 9 20) - (30 9 21)	1103.0650	41.14	(30 9 21) - (29 8 22)	41.10
9R(34)	1104.9118	(32 8 24) - (33 8 25)	1104.6827	40.44	(33 8 25) - (32 7 26)	40.35
9R(36)	1105.9436	(31 6 26) - (32 6 27)	1105.5579	33.51	(31 6 26) - (32 6 27)	33.69

(a) Freed et al., 1980

(b) All ν_9 vibrational-rotational states.

(c) Table V taken from Petersen and Duxbury, 1982

Many of the lines remain to be assigned because of the complications introduced by Coriolis-coupling between ν_3 and ν_9 and the overlap of the ν_9 and $\nu_4 + \nu_9$ modes. In particular, the 382.6 μm and 657.2 μm lines pumped by 9P(10)'' have not been easily assigned even though they fit the polarization requirements. The 1448.1 μm line pumped by 9R(44) would also be interesting, since this appears to be a rather strong transition. One other remaining question concerns the absorption on the 9R(18) $^{12}C^{16}O_2$ line which produces no FIR lines. This transition probably could best be investigated using IR-Microwave spectroscopic techniques.

V- Conclusion

Because of its many advantages including a large number of emission lines and broad spectral coverage, high output powers, high conversion efficiency, good stability, and ability for sealed-off operation, CH_2F_2 is one of the most useful CW optically pumped FIR laser molecules. For this reason it most certainly will find an increasing number of applications in Far Infrared systems. In the future, much work remains to complete the assignments and to refine the spectroscopic constants. Scaling to higher output power levels using the results of the theoretical models will be successful on all of the lines in properly optimized cavities. Finally, it seems apparent that the $^{13}CH_2F_2$ isotopic variety should be investigated, since the absorption spectrum overlaps well with the CO_2 pump laser bands. These challenges should keep the difluoromethane laser a subject of interesting investigation for many more years.

Acknowledgements

The Author would like to thank his colleagues, D. T. Hodges, N. C. Luhmann, J. R. Tucker, T. A. Galantowicz, C. O. Weiss, F. B. Foote, A. R. Calloway, R. D. Reel, H. Jones, E. L. Flectcher,

K. M. Evenson, and T. A. Detemple for their contributions on the CH_2F_2 Laser. Thanks are also extended to D. Antonini for assistance in preparing the manuscript.

References

Danielewicz, E. J., and Weiss, C. O., 1978: New Efficient CW Far-Infrared Optically Pumped CH_2F_2 Laser, IEEE J. Quantum Electron, QE-14, 705-707.

Danielewicz, E. J., Galantowicz, T. A., Foote, F. B., Reel, R. D., and Hodges, D. T., 1979: High Performance at new FIR wavelengths from Optically Pumped CH_2F_2, Opt. LETT., 4, 280 - 282.

Danielewicz, E. J., 1979: Molecular Parameters Determining the Performance of CW Optically Pumped FIR Lasers, in "Digest of the Fourth Int. Conf. on MM Waves and their Applications", Miami, Florida, 10-15 December, 1979, IEEE Cat. No. 79, CH-1384-7 MTT, pp.203-204.

Freed, C., Bradley, L.C. and O'Donnel, R. G., 1980: Absolute Frequencies of lasing transitions in seven CO_2 Isotopic Species, IEEE J. Quantum Electron, QE-16, 1195-1206

Galantowicz, T. A., Danielewicz, E.J., Foote, F. B. and Hodges, D.T. 1978: Characteristics of Non-Diffusion-Limited Optically Pumped CW Lasers -- experimental results for CH_2F_2 in "Proceedings of the International Conference of Laser 1978", V. J. Corcoran, ed., (STS Press, McLean, Virginia, 1979a).

Galantowicz, T. A., Tucker, J. R., and Danielewicz, E. J., 1979b: Pressure dependence of FIR Laser gain in collision relaxation dominated Molecular Systems, in "Digest of the Fourth Int. Conf. on MM Waves and Their Applications", Miami, Florida, 10-15-Dec.1979b, IEEE Cat. No. 79, CH-1384-7 MTT, Appendix pp.53-54.

Gamble, E. B. Jr., and Danielewicz, E. J., 1981: Rectangular Metallic Waveguide Resonator Performance Evaluation at Near Millimeter Wavelengths, IEEE J. Quantum Electron, QE-17, 2254-2256

Gamss, L. A. and Ronn, A. M., 1975: Vibrational Energy Transfer in CH_2F_2, Chem. Phys., 9, 319-326.

Heppner, J. and Weiss, C. O., 1980: Gain in the Quantum Mechanical Three-Level System of CW Laser Pumped FIR Laser Molecules, IEEE J. Quantum Electron, QE-16, 392-401.

Herzberg, G., 1951: Molecular Spectra and Molecular Structure II.

Infrared and Raman Spectra of Polyatomic Molecules (D. Van Nostrand Co., New York.

Hirota, E., Tanaka, T., Sakakibara, A., Ohashi, Y. and Morino, Y., 1970: Microwave Spectrum of Methylene Fluoride Centrifugal Distortion and Molecular Structure, J. Mol. Spectrosc., 34, 222-230.

Hirota, E., 1978a: Anharmonic Potential Function and Equilibrium Structure of Methylene Fluoride, J. Mol. Spectrosc., 71, 145-159.

Hirota, E., 1978b: Microwave Spectrum of Methylene Fluoride in Excited Vibrational States, J. Mole. Spectrosc., 69, 409-420.

Hodges, D.T., Foote, F. B. and Reel, R. D., 1976: Efficient High-Power Operation of the CW Far Infrared Waveguide Laser, Appl. Phys. Lett., 29, 662-664.

Julien, F. and Lourtioz, J. M., 1981: High Power Performances of a CW Optically Pumped 117 μm/ 184 μm CH_2F_2 Laser with Uniform Coupling, Opt. Commun., 38, 294-298.

Kawaguchi, K, and Tanaka, T., 1977: 9.4. μm CO_2 Laser Stark spectroscopic of the ν_9 fundamental Band of Methylene Fluoride, J. Mol. Spectrosc., 68, 125-133.

Lide, D. R., Jr., 1952: Microwave Spectrum of Methylene Fluoride, J. Amer. Chem. Soc., 74, 3548-3552.

Mead, D.G., 1979, High Resolution Spectroscopy of Some Gaseous Molecules with an Infrared Fourier Interferometer, in "Digest of the Fourth Int. Conf. on MM Waves and Their Applications," Miami,Florida, 10-15 Dec. 1979, IEEE Cat. No. 79, CH-1384-7 MTT, pp.217-218.

Peterson, F. R., Scalabrin, A. and Evenson, K. M., 1980: Frequencies of CW FIR Laser Lines from Optically Pumped CH_2F_2, INT. Journ. IR and MM Waves, 1, 111-115.

Peterson, J. C. and Duxbury, G., 1982: Optically Pumped Submillimeter Laser Lines from CH_2F_2 and CD_2Cl_2 using a $^{12}C^{18}O_2$ CW Laser. Int. Journ. IR and MM Waves, 3, pp.607-618.

Pugh, L. A., and Rao, K. N., 1976: Molecular Spectroscopy: Modern Research, Vol.II, K.N. Rao, ed., Academic Press, New York, ch.4.

Scalabrin, A., and Evenson, K.M., 1979: Additional CW FIR Laser Lines from Optically Pumped CH_2F_2, Opt. Lett., 4, 277-279.

Stewart, H. B. and Nielsen, H. H., 1979: Infrared Bands in the Spectrum of Difluoromethane, Phys. Rev., 75, 640-650.

Suzuki, I. and Shimanouchi, T., 1973: Vibration-Rotation Spectra and Molecular Force Field of Methylene Fluoride and Methylene Fluoride-d_2, J. Mol. Spectrosc., 46, 130-145.

Townes, C. H. and Schawlow, A., 1955: Microwave Spectroscopy, McGraw Hill, New York.

Tsunawaki, Y., Yamanaka, M. and Fujita, S., 1981: Doppler free OptoAcoustic Spectroscopy in CH_3OH and CH_2F_2, in "Digest of Sixth Int. Conf. on IR and MM Waves, 7-12 Dec. 1981, Miami, Florida, IEEE Cat. No. 81, CH 1645-1 MTT, pp F4-4.

FAR-INFRARED LASER LINES OBTAINED BY OPTICAL PUMPING OF FLUOROCARBON 12, CF_2Cl_2

J-M. Lourtioz

Institut d'Electronique Fondamentale, Bât. 220

Université Paris XI, 91405 ORSAY, France.

Introduction

Up to now only one work is known on optically pumped emissions in CF_2Cl_2 (Lourtioz et al.[1], 1981). We summarize in Section II these first laser results.

The infrared spectroscopy of CF_2Cl_2 is also at the beginning. Interest in CF_2Cl_2 stems from the important role that it is suspected to play in the chemistry of stratospheric ozone (Molina and Rowland, 1974). Unfortunately, this heavy asymmetric-top molecule presents an infrared spectrum of exceptionally high density which makes its interpretation a long and difficult task. As an example, nine CF_2Cl_2 absorption bands have been found to overlap in the spectral region of interest near 923 cm^{-1} (Morillon-Chapey et al., 1981).

Among these bands, only one, namely, the ν_6-fundamental of the main isotopic species ($CF_2{}^{35}Cl_2$), is known at the present time (Jones and Morillon Chapey, 1982). Moreover, from a first theoretical estimation, it can be shown that a maximum spectral density of approximately 10,000 lines/cm^{-1} is expected in the spectral region of the ν_6 Q-branch, which means an averaged number of 30 absorption lines falling within the frequency tuning range of a conventional CO_2 laser.

For all the reasons mentioned above, the definite assignment of a given FIR laser transition is generally excluded. The few tentative assignements previously reported (Lourtioz et al., 1981) must be considered with extreme care. In section III, we only give the CF_2Cl_2 spectroscopic data which may be presently taken with confidence.

CF_2Cl_2 laser results

The experimental set-up used for CF_2Cl_2 was the same as for CF_3Br (Lourtioz et al. 1981). The FIR emissions of these two heavy molecules were expected to be at very long wavelength ($\lambda_{FIR} > 1$ mm) which dictated the use of a low-loss metallic waveguide resonator (1.80 m long, 30 mm inner diameter). The pumping source was a CW regular $^{12}C\,^{16}O_2$ laser which was able to deliver ∿ 30 W output power on most of its emission lines.

The coincidences between $^{12}C^{16}O_2$ emission lines and CF_2Cl_2 absorptions were first investigated using the microphone cell technique (Walzer et al. 1978, Lourtioz et al., 1981). A strong microphone signal was detected on the P-branch at 10 μm from P28 to P44. An half-amplitude signal was obtained on the R-branch at 9 μm from R18 to R38. As compared to the CH_3OH molecule (one of the best FIR candidates), the microphone signals obtained with fluoro-carbon 12 were found to be almost three times larger under the same experimental conditions (pressure and IR power). These strong absorption signals are one of the characteristics of a multichannel pumping. Optical pumping of CF_2Cl_2 has been only investigated with the P-branch lines at 10 μm.

Table I summarizes the eleven laser lines emitted from CF_2Cl_2. The emitted wavelengths range from 614 μm to 1200 μm. Only the 684.7 μm line pumped by 10 P 42 may be considered as a strong line ($P_{FIR} > 0.1$ mW). Column 3 gives the relative intensity of the other lines.

TABLE 1
CF_2Cl_2 Laser Results

CO_2 pump	Measured FIR wavelengths (μm)	Relative FIR Intensity	CO_2 Laser power (W)	IR Frequency offset (MHz)	Optimum Pressure (m Torr)	Temperature Effect $P_{FIR} = P(T)$
10 P 30	751.4 ± 0.6	6	30	+ 10	100	↘
10 P 32	[a]614.3 ± 0.8	2,5	28	+ 33	75	↘↘
	684.74 ± 0.8	1.5		- 22		
10 P 34	858.73 ± 0.5	2.5	26	+ 12	55	} ↘↘
	1025 ± 1	1.5		+ 17		
	638.4 ± 0.6	3		+ 35		↘
	980 ± 10	1		—	45	
10 P 36	[a]1164 ± 2	1	23	- 31		↘
	1205 ± 2	2		- 31		↗↗
10 P 40			16			
	684.7 ± 0.5	15		- 13		→
10 P 42			12		80	
	[a]765.2 ± 0.5	5		+ 28		→

[a] = lines for which first tentative assignments have been proposed (Lourtioz et al., 1981)

Most of the FIR wavelengths have been measured by interferometry with an accuracy better than 10^{-3} (see column 2). The method of measurement is based on a comparison between the FIR wavelength and the pump wavelength used as a proper etalon (Lourtioz et al., 1980).

Due to the very dense CF_2Cl_2- absorption spectrum, several CF_2Cl_2 absorption transitions may be successfully pumped by the same CO_2 emission line. This situation occurs for the 10P34, 10P36 and 10P42 CO_2-pump lines. In each case, the FIR emissions respectively correspond to different absorption transitions of CF_2Cl_2.

Figure 1 illustrates this typical situation in the case of the P34 CO_2 pump line : the FIR cavity is respectively tuned on each FIR emission line and the FIR signal is monitored while PZT scanning the CO_2 frequency. The transferred Lamb-dips (Inguscio

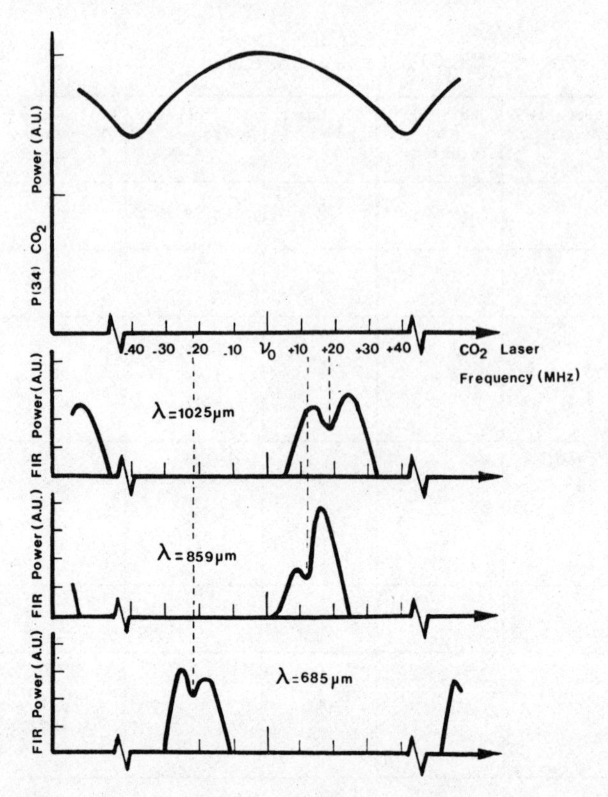

Fig. 1. Output powers of the 1025 μm, 859 μm and 685 μm CF_2Cl_2 laser lines as functions of the 10 P 34 CO_2 line frequency tuning. The IR frequency scale is corrected for the non-linearities of the CO_2 piézoceramic translator. The IR-FIR transferred Lamb-dips shown in the figure are recorded at the optimum laser pressure (55 mTorr). (Lourtioz et al., 1981).

et al., 1979) are clearly shown in the three curves $P_{FIR} = f(\nu_{CO_2})$ They allow a good estimation of the pump frequency offsets with respect to the CO_2 line-center. The pump frequency offsets corresponding to the different FIR lines are given in column five of Table I.

The influence of the temperature on the FIR signals is indicated in column 7. The most significant results are noted with double arrows (up or down). For instance a strong FIR power enhancement with temperature was observed for the 1205 μm line : the FIR signal was increased by a factor of three when the temperature was varied from 25°C up to 90°C. This positive effect suggests that we are probably concerned with an hot-band transition.

Spectroscopic information :

The CF_2Cl_2 molecule, commercially labelled as Freon 12, presents three main isotopic varieties, namely, $CF_2\,^{35}Cl_2$, $CF_2\,^{35}Cl\,^{37}Cl$ and $CF_2\,^{37}Cl_2$ in natural abundance ratio of approximately 9 : 6 : 1. This asymmetric-top molecule belongs to the C_{2V} point group when both chlorine atoms are of the same isotopic variety and to the C_S point group when the two chlorine atoms are different isotopes.

The large number of quasi coincident bands due both to the presence of these three isotopes and to numerous hot bands accounts for the exceptionally dense infrared spectrum of CF_2Cl_2.

Table II lists the CF_2Cl_2 absorption bands which are presently assigned in the spectral region around 923 cm^{-1}, except some $CF_2\,^{37}Cl_2$ bands which have not been detected yet. Contradictory assignments had been previously proposed for these rovibrational CF_2Cl_2 bands (Goldman et al., 1976, Giorgianni et al., 1979). In order to obtain the final assignments given in Table II, a pure sample of $CF_2\,^{35}Cl_2$ was synthesized and its IR spectra were compared to those of natural fluorocarbon-12 (Morillon-Chapey et al., 1981).

The fundamental ν_6 band is of A-type. Consequently all the absorption bands listed in Table II exhibit a typical (PQR) envelope. The 10 μm $^{12}C^{16}O_2$ emission lines from P_{28} to P_{40} may only coincide with R-branch transitions of these bands. The situation is different for the 10 P(J) $^{12}C^{16}O_2$ emission lines with $J > 40$. For instance, the 10 P 42 line may coincide either with Q-branch transitions of the fundamental ν_6 of the main isotopic species $CF_2\,^{35}Cl_2$ or with R-branch transitions of the other bands.

Among the rovibrational bands listed in Table II, only the ν_6-fundamental of the main isotopic species is presently known.

Isotopic and Vibrational Assignments of Fluorocarbon-12 in the 10.8 μm - Region. (Morillon-Cahpey et al. 1981).

FREQUENCIES cm^{-1}	Cl 35 - 35	Other Isotopic Species
	$923\ cm^{-1}$ range	
923.22	35 - 35 ν_6	
922.66	35 - 35 $\nu_5 + \nu_6 - \nu_5$	
921.79		35 - 37 ν_6
921.29	35 - 35 $\nu_4 + \nu_6 - \nu_4$	35 - 37 $\nu_5 + \nu_6 - \nu_5$
919.86		35 - 37 $\nu_4 + \nu_6 - \nu_4$
919.37	35 - 35 $2\nu_4 + \nu_6 - \nu_4$	
917.97		35 - 37 $2\nu_4 + \nu_6 - 2\nu_4$
917.52	35 - 35 hot band	

Parameters Determined [a] for the ν_6 Band of $CF_2{}^{35}Cl_2$ (in MHZ) (Jones and Morillon-Chapey, 1982).

	GROUND	$\nu_6 = 1$
A	4118.896(29)	4115.545(36)
B	2638.704(26)	2630.948(27)
C	2233.691(26)	2225.758(28)
$\Delta_J . 10^4$	4.80(56)	4.34(56)
$\Delta_{JK} . 10^4$	-4.09(28)	-5.86(31)
$\Delta_K . 10^3$	1.65(15)	1.61(16)
$\delta_j . 10^4$	1.192(18)	1.289(33)
δ_K	0^b	0^b

$$\nu_0 \begin{cases} = 27\ 678\ 026 \pm 4\ \text{MHz} \\ = 923.23956(12)\ \text{cm}^{-1} \end{cases}$$

[a] = from 116 data points, std. dev. = 0.9 MHz, numbers in parenthesis are 1 std. dev. in units of last decimal.

[b] = constrained at this value.

TABLE IV

Identified Infrared Transitions of the ν_6 Band of $CF_2{}^{35}Cl_2$. (Jones and Morillon-Chapey, 1982).

Laser Line [a]	Correction	Infrared Transition	Rel. Int.	Frequency
	(MHz)	J' - J"		(MHz)
N_2O P(24)	-30 ± 20	$30_{13,18}$ - $31_{13,19}$	591	27 517 854.8(2.6)
N_2O P(21)	-30 ± 20	$25_{10,16}$ - $26_{8,19}$	41	27 599 535.8(-4.5)
N_2O P(20)	-30 ± 20	$33_{6,28}$ - $33_{6,27}$	37	27 626 563.8(-14.2)
N_2O P(20)	-30 ± 20	$33_{5,28}$ - $33_{7,27}$	48	27 626 563.8(-3.3)
N_2O P(19)	0 ± 20	$41_{13,29}$ - $41_{13,28}$	52	27 653 522.3(-11.0)
13-18 10P(8)*	-7.5 ± 5	$34_{12,23}$ - $34_{12,22}$	123	27 667 367.1(2.7)
12-16 10P(42)*	+25 ± 10	$16_{5,12}$ - $16_{5,11}$	56	27 668 299.5(-4.7)
13-16 10R(12)*	-11 ± 5	$28_{10,18}$ - $28_{10,19}$	96	27 674 170.4(-0.8)
13-16 10R(12)*	+22 ± 5	$23_{10,13}$ - $23_{10,14}$	211	27 674 203.4(-0.5)
N_2O P(18)	-30 ± 20	$21_{7,14}$ - $21_{7,15}$	80	27 680 321.1(10.8)
N_2O P(17)	+30 ± 20	$41_{13,29}$ - $41_{11,30}$	30	27 707 110.0(1.0)
13-18 10P(6)*	0 ± 20	$7_{2,6}$ - $6_{2,5}$	174	27 711 180.1(1.8)
13-16 10R(14)*	+30 ± 20	$8_{3,6}$ - $7_{3,5}$	240	27 716 668.8(14.7)
12-16 10P(38)*	-30 ± 20	$14_{8,6}$ - $13_{6,7}$	2	27 790980.4(-0.8)

a = $\nu_{Transition} - \nu_{Laser}$

Rel.Int. = Relative intensities calculated at 300°K

The stars indicate the use of one among the different isotopic CO_2 lasers.

A first study on infrared Fourier transform and diode laser spectra (Nordstrom et al., 1979) and recently a much more powerful infrared microwave double resonance experiment (Jones and Morillon-Chapey, 1981 ; Jones and Morillon-Chapey, 1982) determined the molecular parameters of this band. The most accurate values obtained by Jones and Morillon-Chapey are reported in Table III. They allow to calculate accurately the rovibrational transitions within the ν_6 band of $CF_2{}^{35}Cl_2$ provided that the J and K_A values are not too high (i.e. J, K_A < 50), (J, K_A, K_c) being the rotational quantum numbers of the asymmetric-top molecule.

As an illustration, Table IV lists the assignments of several absorption transitions within the ν_6 fundamental of $CF_2{}^{35}Cl_2$ which have been found to coincide respectively with different emission lines of infrared lasers (N_2O laser, CO_2 isotopic lasers) (Jones and Morillon-Chapey, 1982).

In contrast, the J, K_A- values involved in the IR-FIR pumping cycles are notably high (J, K_A > 50) and the precision of the asymmetric-top model drastically degrades for such values. The limited number of experimental data obtained from these optical pumping experiments and the lack of accuracy in the FIR wavelength measurements do not allow a safe identification of the IR and FIR transitions within the ν_6 fundamental band of $CF_2{}^{35}Cl_2$.

Conclusions :

The eleven FIR emission lines obtained from CF_2Cl_2 by pumping with only five $^{12}C^{16}O_2$ laser lines have revealed this molecule as a good FIR candidate. As shown in Table IV, the N_2O laser and the CO_2 isotopic lasers can be also tried as pumping sources.

The spectroscopic data presently available on the different isotopes of CF_2Cl_2 do not allow to assign the FIR lines observed. Experiments with mono-isotopic species should be more fruitful for further investigations.

As compared to the FIR pumping experiments, IR-microwave double resonance experiments (Jones and Morillon-Chapey, 1982) appear to provide more safe informations about the rovibrational spectrum of this complex asymmetric-top molecule. Owing to the frequency tunability of the microwave sources, these latter experiments generally reveal the presence of several microwaves transitions (n > 2) branching from a given IR rovibrational transition. In many cases, this combined set of microwave frequencies allow to identify the IR transition without any ambiguity.

References

Giorgianni, S., Cambi, A., Franco, L., and Ghersetti, S. 1979, J. Mol. Spectrosc. 75, 389-405.

Goldman, A., Bonomo, F.S., and Murcray, D.G. 1976. Geophys. Res. Lett. 3, 309-312.

Inguscio, M., Moretti, A., and Strumia, F. 1979. Opt. Commun. 30, 355-360.

Jones, H., and Morillon-Chapey, M. 1981. 7th Colloquium on High Resolution Spectroscopy, Reading, Sept. 1981.

Jones, H., and Morillon-Chapey, M. 1982. J. Mol. Spectrosc. 91, 87-102

Lourtioz, J.M., Pontnau, J., and Julien, F. 1980, Infrared Phys. 20, 231-235.

Lourtioz, J.M., Pontnau, J., Morillon-Chapey, M., and Deroche, J.C. 1981. International Journal of Infrared and Millimeter Waves 2, 49-63.

Lourtioz, J.M., Pontnau, J., and Meyer, C. 1981 . International Journal of Infrared and Millimeter Waves 2, 525-532.

Morillon-Chapey, M., Diallo, A.O., and Deroche, J.C. 1981. J. Mol. Spectrosc. 88, 424-427.

Nordstrom, R.J., Morillon-Chapey, M., Deroche, J.C., and Jennings, D.E. 1979. J. Phys. Letter 40, L37-L40.

Walzer, K., Tacke, M., and Busse, G. 1978. Infrared Phys. 19, 175-177.

SUBMILLIMETER LASER LINES IN 1,1 DIFLUORETHYLENE, CF_2CH_2

G. Duxbury

Department of Natural Philosophy
University of Strathclyde
107 Rottenrow, Glasgow G4 ONG, Scotland

Introduction

1,1 difluorethylene, (1,1 difluoroethene), CF_2CH_2, was one of the first molecules to be used in optically pumped submillimetre lasers. Laser action was first observed by Dyubko et al. in 1972 [1], subsequently Hodges et al. [2] and Radford [3] discovered several new lines. Since many of the emission lines are strong, their frequencies have been measured by heterodyne methods by Radford et al. [4] and by Dyubko and Fesenko [5]. More recent experiments have extended the list of lines by the use of CO_2 laser hot bands [6], by the use of high pressure waveguide lasers [7], and by the use of isotopically labelled CO_2 in the pump laser [8].

Since a knowledge of the identity of the rotational transitions is a considerable aid to laser power optimisation, several attempts have been made to identify the pump and emission lines. The earliest attempt was made by Duxbury, Gamble and Herman [9], on the basis of laser Stark spectroscopy. Duxbury and Herman [10] then extended the Stark measurements, and used these to assign further transitions [11,12]. At approximately the same time Dyubko et al. [13] proposed assignments for some lines based upon double resonance experiments. A more complete understanding of the energy level pattern of the ν_4 and the ν_9 states of CF_2CH_2 has come from the very detailed diode laser spectroscopic studies of Sattler and his colleagues [6,14,15].

Submillimeter laser lines

The initial observations of laser action in CF_2CH_2 were made using open resonator cavities [1,2]. Subsequently it was shown that more power was obtainable if cylindrical metallic resonators were used [3,10], and also that the threshold pump powers needed were greatly reduced, and were less than two watts for many lines. Although dielectric waveguide resonators have largely superseded metallic resonators for wavelengths of 500 µm and shorter [16], no results of using CF_2CH_2 in dielectric resonators have appeared in the open literature. Very little work has been reported on pulsed operation of CF_2CH_2, and the paper by Bluyssen et al. [17] on the pulse shape of the 554 µm line remains the sole contribution so far.

Recently the number of lines observed has been considerably increased by the use of isotopic pump lasers containing $^{13}CO_2$ and $C^{18}O_2$ [8], by the use of hot bands [6], and by making use of the increased bandwidth of RF excited waveguide CO_2 lasers [7]. Since the waveguide CO_2 lasers typically emit from 5 to 10 watts on a single line compared to the 15-30 watts from a conventional CO_2 pump laser, it is essential to use cavities in which low threshold operation is possible. The waveguide pumping experiments have therefore used metallic waveguide FIR resonators. The strength of the absorption of some of the lines pumped in this way is shown by the observation of transferred Lamb dips in the output [18,11]. The main advantage of the open resonator and of the dielectric waveguide is that the polarisation of the FIR output is well defined, whereas the output from the metallic waveguides is depolarised unless steps are taken to control the polarisation by the use of a polarisation sensitive output coupler [19,20]. An alternative method is to use oversize rectangular waveguide [21,22].

The laser lines observed using CF_2CH_2 are summarised in tables 1 to 5. The majority of the wavelengths have been measured by interferometry, but most of the strong lines have had their frequencies measured using heterodyne methods [4,5].

Spectroscopic measurements and assignments

The observed submillimetre laser lines of CF_2CH_2 are due almost entirely to pure rotation transitions in the ν_4 and the ν_9 states. The centre of the A-type ν_4 band is at 925.7692 cm^{-1}, and that of the B-type ν_9 band is at 953.8057 cm^{-1} [10,15], and consequently the near coincidences occur with lines of the R and P branches of the 10 μm CO_2 laser. Although the ν_4 and ν_9 states lie very close together the Coriolis interaction between them appears to be small, and they can therefore be treated as isolated bands for the purpose of the energy level calculation. The laser Stark experiments of Duxbury and Herman [10] allowed the bandcentres to be accurately measured for the first time, and together with the microwave data of Ewart [23] allowed most of the FIR emission lines to be assigned [12]. Subsequently Sattler and his colleagues [6,14,15] used the detailed analysis of diode laser spectra of CF_2CH_2 to derive accurate centrifugal distortion constants for the excited states. The diode laser spectra have the advantage that, unlike the laser Stark spectra, they do not preferentially sample the spectra of levels with low J and K_a values. With the aid of the improved rotational and centrifugal distortion constants for the ν_4 and the ν_9 states, Sattler and Lafferty and their colleagues confirmed most of the assignments of Duxbury and Herman, and of Dyubko et al., and extended the analysis to include almost all known emission lines. Sattler has also predicted coincidences with $^{13}CO_2$ pumping [24]. The molecular constants for the ground,

Table I

$\lambda_{vac}/\mu m$	ν/GH_z	CO_2 pump line	Threshold Power/ watts [a]	Rel. Pol.	Rel. Strength [b]	Pressure/ m torr	Ref.
2070.192[d]	144.8141	10P[14]			VS		(5)
1018.258	294.4169	10P[14]	4		M	50	(2)(5)(12)
943.466	314.0943	10P[30]					(5)
918.148	326.4187	10P[22]					(5)
917.881	326.6138	10P[22]					(5)
889.086	337.1919	10P[22]	2	11	S	150	(1)(5)(7)(12)
888.862	337.2767	10P[22]	2	11	VS	150	(1)(5)(7)(12)
842.623	355.7847	10P[30]					(5)
764.1	393.	10P[10]		⊥	S		(25)
697.8	429.6	10P[24]					(11)
662.816	452.302	10P[24]	2	11	S	100	(1)(2)(4)(12)
605.	495.6	10P[48]			E		(6)
567.9	527.9	10P[24]	2.5		M	150 c	(2)(12)
554.365	540.7851(20)	10P[14]	1.5	⊥	VS	120 c	(1)(4)(5)(12)
524.	573.	10P[16]	2.5		W	80 c	(12)
486.7683	6.5.8833	10P[22]					(5)
484.7546	618.4175	10R[28]					(5)
463.6243	646.6281	10R[20]	1.5	11	S	100 c	(1)(5)(12)
458.		10P[30]	3	11	M	80 c	(1)(12)
407.2937	736.0596	10P[14]	2 CW	11	S	180 c	(2)(4)(12)
375.5449	798.2886	10P[22]	1.5 CW	11	S	250 c	(1)(4)(12)
288.5	1039.	10P[12]		11	W	300 c	(1)

a CW - copper waveguide, FP - Fabry-Perot VS > 1 mw

b S ≃ 01.-1 mw M 50-100μw W 10-50μw VW 10μw E - electrically chopped

c optimum pressure

d 591.44[1]μm line and 990.0μm line ommitted as unconfirmed

TABLE II - Lines pumped using an electrically chopped laser operating on the $01^11 - 11^10$ CO_2 hot band.

$\lambda_{vac}/\mu m$	ν/GH_2	CO_2 pump line ν/cm^{-1}	Reference
1062. [1]	282.3 [3]	P[14] 915.6277	(6)
890. [1]	336.8 [4]	P[16] 913.8803	(6)
618. [1]	485.4 [8]	P[22] 908.4873	(6)
535. [1]	560. [1]	P[25] 905.9502	(6)
497. [1]	603. [1]	P[27] 904.1021	(6)

TABLE III - Lines pumped using a $^{13}C^{16}O_2$ laser.

$\lambda_{vac}/\mu m$	ν/GH_2	CO_2 pump line	Rel. Pol.	Rel. Strength	CO_2 power used/watts	Pressure/ m torr	Ref.
950.6	315.4	10R [30]		S			(8)
615.9	486.8	10R [34]	11	S			(8)
584.8	512.6	10R [14]		S	12	80	(8)
546.0	549.1	10R [28]		S	11	80	(8)
468.4	640.0	10R [14]		S	12	80	(8)
427.7	700.9	10R [20]		S	12	80	(8)
347.6	862.5	10R [16]		S	12	80	(8)
326.	920.	10R [20]		S	12	80	(8)
324.4	924.1	10R [20]		S	12	80	(8)
259.5	1155.	10R [20]		S	12	80	(8)

TABLE IV - Lines pumped using a $^{12}C^{18}O_2$ laser.

$\lambda_{vac}/\mu m$	ν/GH_2	CO_2 pump line	Rel. Pol.	Rel. Strength	CO_2 power used watts	Pressure/ m torr	Ref.
591.7	506.7	10P [24]	11	S	10	80	(8)
469.5	638.5	10P [20]	11	S	12	80	(8)
437.6	685.1	10P [22]	11	S	12	80	(8)
403.6	742.8	10P [18]	11	S	12	80	(8)
399.3	750.8	10P [14]	11	S	8	80	(8)
373.4	802.9	10P [20]		S	12	80	(8)

TABLE V - Lines pumped using a CO_2 waveguide laser.

λ_{vac} /μm	ν/GH_2	CO_2 pump line	Ir offset /MHz	Rel. Pol.	Rel. Strength	Threshold CO_2 power/w	Pressure m torr	Ref.
557.7	537.5	10P 22	-116.9	11	VS	1	95	(7)
568.5	527.3	10P 22	86.9	11	M	1	65	(7)

TABLE VI - Molecular constants of the ν_4 and ν_9 bands of CF_2CH_2 in cm^{-1}.

	Ground State (a)	ν_4 Band (a)	ν_9 Band (a)
	0.367004867 [33]	0.63598861 [27]	0.36782001 [23]
	0.347873598 [16]	0.34740751 [24]	0.34812118 [21]
	0.178302671 [14]	0.17773941 [22]	0.17832848 [20]
$\Delta_K \times 10^7$	2.5859 [75]	2.596 [32]	2.8663 [87]
$\Delta_{JK} \times 10^8$	5.099 [91]	5.52 [20]	3.029 [86]
$\Delta_J \times 10^7$	1.6035 [26]	1.5970 [23]	1.6640 [24]
$\delta_K \times 10^7$	1.8898 [45]	1.9879 [66]	1.8370 [24]
$\delta_J \times 10^8$	6.687 [10]	6.8028 [69]	6.7707 [61]
$H_K \times 10^{13}$	4.81 [24]	164. [29]	4.81[b]
$H_{KJ} \times 10^{12}$	-	-6.4 [23]	-
$K_{JK} \times 10^{12}$	-	-2.65 [89]	-
$H_J \times 10^{13}$	-	-1.98 [85]	12.43 [75]
ν_0	-	925.771992 [44]	953.804922 [39]

a Uncertainties given for this work are one standard deviating, ref. [15].

b Fixed at ground state value.

TABLE VII - Assignments of optically pumped transitions in the ν_4 band of CF_2CH_2 [(a)].

CO_2 Line	Laser Freq.(cm^{-1})	Calculated IR Absorption Freq.(cm^{-})	Assignment	S(T)[(b)]	Calculated SMMW Emission Assignment	S[(c)]	Freq.(cm^{-})	Measured SMMW Emission Freq.(cm^{-})
10P[48]	916.5818	916.5825	$35_{11,24}-35_{13,23}$	0.5	$35_{11,24}-34_{11,23}$	25.1	16.5353	16.53±0.03
10P[48]	916.5818	916.5825	$35_{12,24}-35_{12,25}$	0.5	$35_{12,24}-34_{12,23}$	25.1	16.5353	16.53±0.03
10P 30	934.8945	934.8952	$15_{14,1}-14_{14,0}$	0.5	$15_{14,1}-14_{12,2}$	1.7	11.8677	11.86770
		934.8944	$45_{17,29}-45_{15,30}$	0.2	$45_{17,29}-44_{17,28}$	31.0	21.8531	21.83
		934.8944	$45_{16,29}-45_{16,20}$	0.2	$45_{16,29}-44_{16,28}$	31.0	21.8531	21.83
		934.8944	$45_{16,29}-45_{15,30}$	0.2	$45_{16,29}-45_{15,30}$	12.3	10.4767	10.47706
		934.8944	$45_{16,29}-45_{16,30}$	0.2	$45_{16,29}-45_{16,30}$	12.3	10.4767	10.47706
10P 24	940.4581	940.5468	$23_{18,5}-22_{18,4}$	1.5	$23_{18,5}-22_{18,4}$	13.1	15.0871	15.08716
			cascade	-	$22_{18,4}-21_{18,3}$	10.7	14.3321	14.33
		940.5469	$23_{18,5}-22_{18,4}$	1.5	$23_{18,5}-22_{16,6}$	4.0	17.7492	17.61
10P 14	949.4793	949.4778	$38_{29,9}-37_{29,8}$	0.5	$38_{29,9}-37_{29,8}$	19.8	24.5523	24.55230
10P 12	951.1923	951.1921	$43_{31,12}-42_{31,11}$	0.3	$43_{31,12}-42_{31,11}$	21.2	26.6279	26.6280
		951.1921	$43_{31,12}-42_{31,11}$	0.3	$43_{31,12}-42_{29,13}$	5.6	34.6328	34.66

(d)								
H P[14]	915.6277	915.6292	$23_{3,20}-24_{3,21}$	2.8	$23_{3,20}-22_{3,19}$	19.8	9.4247	9.416±0.009
H P[14]	915.6277	915.6292	$23_{4,20}-24_{4,21}$	2.8	$23_{4,20}-22_{4,19}$	19.8	9.4247	9.416±0.009
H P[16]	913.8803	913.8815	$20_{11,9}-21_{11,10}$	1.5	$20_{11,9}-19_{11,8}$	11.1	11.2228	11.233±0.013
H P[16]	913.8803	913.8815	$20_{12,9}-21_{12,10}$	1.5	$20_{12,9}-19_{12,8}$	11.1	11.2228	11.233±0.013
H P[22]	908.4873	908.4868	$32_{13,19}-33_{13,20}$	1.2	$32_{13,19}-31_{13,18}$	20.8	16.1847	16.19±0.03
H P[22]	908.4873	908.4868	$32_{14,19}-33_{14,20}$	1.2	$32_{14,19}-31_{14,18}$	20.8	16.1847	16.19±0.03
H P[24]	906.6393	906.6411	$33_{17,16}-34_{17,17}$	0.9	$33_{16,16}-32_{17,15}$	19.1	17.9676	Not observed
H P[24]	906.6393	906.6411	$33_{18,16}-34_{18,17}$	0.9	$33_{18,16}-32_{18,15}$	19.1	17.9676	Not observed
H P[25]	905.9502	905.9489	$32_{21,12}-33_{21,12}$	0.8	$32_{21,12}-31_{21,11}$	16.7	18.7033	18.69±0.03
H P[27]	904.1021	904.1025	$42_{15,28}-43_{15.29}$	0.6	$42_{15,28}-41_{15,27}$	29.6	20.0811	20.12±0.04
H P[27]	904.1021	904.1025	$42_{15,28}-43_{14,29}$	0.6	$42_{14,28}-41_{14,27}$	29.6	20.0811	20.12±0.04

(a) Reference (6).

(b) Relative integrated line intensity.

(c) Calculated submillimetre transition strength. Tabulated line intensities are calculated for a transition moment of 1 Debye, to facilitate future scaling, and 300K, but are not corrected for nuclear spin statistics.

(d) H-pumped by the 01´1-11´0 hot band lines.

TABLE VIII – Assignments of optically pumped transitions in the ν_9 band of $F_2C{=}CH_2$ [(a)(b)]

CO_2 Laser		Calculated IR Absorption			Calculated SMMW Emission			Measured SMMW Emission
Line	Freq. (cm^{-1})	Freq. (cm^{-})	Assignment	S(T)	Assignment	S	Freq. (cm^{-})	Freq. (cm^{-1})
10P [22]	942.3833	942.3826	$32_{8,15}-24_{9,16}$	1.9	$23_{8,15}-22_{8,14}$	15.8	11.2475	11.24751
		942.3826	$23_{9,15}-24_{8,16}$	1.9	$23_{9,15}-22_{9,14}$	15.8	11.2475	11.24751
			cascade	-	$22_{8,14}-21_{8,13}$	14.9	10.8915	10.89149
			cascade	-	$22_{9,14}-21_{9,13}$	14.9	10.8915	10.89149
		942.3841	$22_{10,13}-23_{9,14}$	1.7	$22_{10,13}-21_{10,12}$	14.2	11.2504	11.25034
		942.3841	$22_{9,13}-23_{10,14}$	1.7	$22_{9,13}-21_{9,12}$	14.2	11.2504	11.25034
			cascade	-	$21_{10,12}-20_{10,11}$	13.2	10.8947	10.89466
			cascade	-	$21_{9,12}-20_{9,11}$	13.2	10.8947	10.89466
		942.3841	$22_{10,13}-23_{9,14}$	1.7	$22_{10,13}-21_{8,14}$	0.8	20.5437	20.54365
		942.3841	$22_{9,13}-23_{10,14}$	1.7	$22_{9,13}-21_{7,14}$	0.8	20.5437	20.54356
10P [16]	947.7420	947.7410	$36_{18,19}-36_{19,18}$	0.6	$36_{18,19}-35_{18,18}$	21.9	19.0924	19.1
		947.7410	$36_{17,19}-36_{18,18}$	0.6	$36_{17,19}-35_{17,18}$	21.9	19.0924	19.1

10P [14]	947.4793	947.4795	$20_{8,13}-20_{9,12}$	1.0	$20_{8,13}-19_{8,12}$	13.7	9.8207	9.82069
		947.4795	$20_{7,13}-20_{8,12}$	1.0	$20_{7,13}-19_{7,12}$	13.7	9.8207	9.82069
		947.4795	$20_{7,13}-20_{8,12}$	1.0	$20_{7,13}-20_{7,14}$	5.4	4.8305	4.83047
		947.4798	$32_{18,14}-32_{19,13}$	0.8	$32_{18,14}-31_{18,13}$	17.7	18.0386	18.03865
		947.4798	$32_{19,14}-32_{20,13}$	0.8	$32_{19,14}-31_{19,13}$	17.7	18.0386	18.03865
10P [10]	952.8809	952.8799	$20_{16,4}-20_{17,3}$	2.0	$20_{16,4}-19_{16,3}$	10.7	13.1010	13.1
10R [20][a]	975.9304	975.9277	$48_{12,36}-47_{13,55}$	0.6	$48_{12,36}-47_{12,35}$	37.1	21.5686	21.56919
		975.9277	$48_{13,36}-47_{12,35}$	0.6	$48_{13,36}-47_{13,35}$	37.1	21.5686	21.56919
10R [28][b]	980.9132	980.9137	$33_{25,9}-32_{22,10}$	0.2	$33_{25,9}-32_{25,8}$	16.4	20.6282	20.62818

(a) Reference (6).

(b) See footnotes to table 7.

TABLE IX - Assignments of optically pumped laser transitions in the ν_4 and ν_9 bands of CF_2CH_2 using the $^{13}CO_2$ laser.(a)

CO_2 Laser		Calculated IR Absorption		Calculated SMMW Emission				Measured SMMW Emission
Line	Freq. (cm^{-1})	Freq. (cm^{-1})	Assignment	S(T)	Assignment	S	Freq. (cm^{-1})	Freq. (cm^{-1})
10R[14]	924.5276	924.5288	ν_4 $27_{22,6}-27_{22,5}$	0.95	$27_{22,6}-26_{22,5}$	12.0	17.0872	17.1
					$27_{22,6}-26_{20,7}$	4.1	21.3483	21.35
10R[20]	928.6567	928.6565	ν_4 $38_{27,11}-38_{27,12}$	0.32	$38_{27,11}-37_{27,10}$	-	22.4601	23.38
					$38_{27,11}-37_{25,12}$	-	30.83	30.83
10R[28]	933.8807	933.8795	ν_4 $38_{\substack{1\,3\\1\,4}}25-38_{\substack{1\,3\\1\,2}}26$	0.37	$38_{\substack{1\,3\\1\,4}}25-37_{\substack{1\,3\\1\,4}}24$	26.5	18.3098	18.31
10R[30]	935.1358	935.1368	ν_4 $28_{\substack{1\\2}}27-28_{\substack{0\\1}}28$	0.12	$28_{\substack{1\\2}}27-27_{\substack{1\\2}}26$	26.6	10.4870	10.52
10R[34]	935.1358	937.5843	ν_9 $32_{\substack{1\,4\\1\,3}}19-33_{\substack{1\,3\\1\,4}}20$	1.10	$32_{\substack{1\,4\\1\,3}}19-31_{\substack{1\,4\\1\,3}}18$	20.8	16.2400	16.27

(a) References (8), (24).

ν_4 and ν_9 states are given in table 6, and the assignments of the emission lines in tables 7,8 and 9.

It seems likely that the number of strong emission lines of CF_2CH_2 will be increased in the near future by a more thorough investigation using waveguide CO_2 lasers, and by the use of higher power isotopic CO_2 lasers, including the $^{13}C^{18}O_2$ and $C^{18}O^{16}$ lasers.

References

1. B.L. Bean and S. Perkowitz, "Complete Frequency Coverage for Submillimeter Laser Spectroscopy with Optically Pumped CH_3OH, CH_3OD, CD_3OD and CH_2CF_2", Optica Letts. 1, 202-204 (1977).

2. H.J.A. Bluyssen, R.E. McIntosh, A.F. van Etteyer and P. Wyder, "Pulsed Operation of an Optically Pumped Far-Infrared Molecular Laser", IEEE J. Quantum Electron. QE-11, 341-348 (1975).

3. G. Duxbury, T.J. Gamble and H. Herman, "Assignments of Optically Pumped Laser Lines of 1,1 Difluorethylene", IEEE Trans. on Microwave Theory and Tech., MTT-22, 1108-1109 (1974).

4. G. Duxbury and H. Herman, "The 10.4 μm CO_2 Laser Stark Spectra of the ν_4 and ν_9 Bands of 1,1 Difluorethylene", J. Mol. Spectrosc. 73, 444 (1978).

5. G. Duxbury and H. Herman, "Optically Pumped Millimeter Lasers", J. Phys. B: Atom. Molecl. Phys. 11, 935-949 (1978).

6. G. Duxbury and H. Herman, "Assignment of Optically Pumped Far-Infrared Laser Lines in 1,1 Difluorethylene", Infrared Phys. 18, 461-463 (1978).

7. G. Duxbury and H. Herman, "cw Optically Pumped Far-Infrared Waveguide Laser with Variable Output Michelson Coupler", J. Phys. E: Sci. Instrum. 11, 419-420 (1978).

8. S.F. Dyubko, B.S. Svich and L.D. Fesenko, "Submillimeter-band Gas Laser Pumped by a CO_2 Laser", J.E.T.P. Lett. 16, 418-419 (1972).

9. S.F. Dyubko and L.D. Fesenko, "Frequencies of Optically Pumped Submillimeter Laser", Conf. Digest of the Third Int. Conf. on Submillimeter Waves and Their Applications, Guildford, U.K., pp. 70-73 (1978).

10. S.F. Duybko, O.I. Baskakov, M.V. Moskienko and L.D. Fesenko, "Assignment of Some Optically Pumped FIR Laser Lines of $C_2H_2F_2$, HCOOD and CH_3I", Conf. Digest of the Third Int. Conf. on Submillimeter Waves and Their Applications, Guildford, U.K., pp. 68-69 (1978).

11. I.C. Ewart, Ph.D. Thesis, School of Chemistry, University of Bristol (1975).

12. E.B. Gamble and E.J. Danielewicz, "Rectangular Metallic Waveguide Resonator Performance Evaluation at Near Millimeter Wavelengths", IEEE J. Quantum Electron. QE-17, 2254-2256 (1981).

13. D.T. Hodges, R.D. Reel and D.H. Barker, "Low-Threshold cw Submillimeter- and Millimeter-Wave Action in CO_2-Laser-Pumped $C_2H_4F_2$, $C_2H_2F_2$, and CH_3OH, IEEE J. Quantum Electron QE-9, 1159-1160 (1973).

14. D.T. Hodges, F.B. Foote and R.D. Reel, "High-Power Operation and Scaling Behaviour of cw Optically Pumped FIR Waveguide Lasers", IEEE J. Quantum Electron. QE-13, 491-494 (1977).

15. W.J. Lafferty, J.P. Sattler, T.L. Worchesky and K.J. Ritter, "Diode Laser Heterodyne Spectroscopy on the ν_4 and ν_9 Bands of 1,1 Difluorethylene", J. Mol. Spectrosc. 87, 506-521 (1981).

16. J. McCombie, J.C. Petersen and G. Duxbury, "Submillimeter Laser Emission from CF_2CH_2, $^{12,13}CD_3I$ and $^{15}NH_3$ Using Isotopically Labelled CO_2 Pump Lasers", in Quantum Electronics and Electro-Optics, ed. P.L. Knight, Wiley, 251-254 (1983).

17. H.E. Radford, "New cw Lines from a submillimeter Waveguide Laser", IEEE J. Quantum Electron. QE-11, 213-214 (1975).

18. H.E. Radford, F.R. Peterson, D.A. Jennings and J.A. Mucha, "Heterodyne Measurements of Submillimeter Laser Spectrometer Frequencies", IEEE J. Quantum Electron. QE-13, 92-94 (1977).

19. J.P. Sattler, T.L. Worchesky, M.S. Tobin, K.J. Ritter, T.W. Daley and S.J. Lafferty, "Submillimeter-wave Emission Assignments for 1,1 difluorethylene", Int. J. Infrared and Millimeter Waves 1, 127-138 (1980).

20. J.P. Sattler, T.L. Worchesky, K.J. Ritter and W.J. Lafferty, "Technique for Wideband, Rapid, and Accurate Diode-Laser Heterodyne Spectroscopy: Measurements on 1,1 Difluorethylene", Optics Lett. 5, 21-23 (1981).

21. J.P. Sattler, private communication (1981).

22. M.S. Tobin, "Michelson Output Coupler with 1-dimensional Grid for an Optica-ly Pumped Near Millimeter Laser", in Proc. 4th Int. Conf. Infrared and Millimeter Waves and their Applications, F.L. IEEE Cat. 79 CH, 1384-1387 MTT (1979).

23. M.S. Tobin, "cw Submillimeter Wave Laser Pumped by an rf-excited CO_2 Waveguide Laser", in Proc. Soc. Photo-Opt. Instr. Eng. 259, 13-17.(1980).

24. M.S. Tobin, J.P. Sattler and T.W. Daley, "New SMMW Laser Transitions Optically Pumped by a Tunable CO_2 Waveguide Laser", IEEE J. Quantum Electron. QE-18, 79-86 (1982).

25. M. Yamanaka, H. Tsuda and S. Mitani, "Polarization Characteristics in Optically Pumped, Far-Infrared Rectangular Waveguide Laser", Opt. Commun. 15, 426-428 (1975).

OPTICALLY PUMPED FAR-INFRARED AMMONIA LASERS

C. O. Weiss

Physikalisch-Technische Bundesanstalt
Braunschweig, Federal Republic of Germany

M. Fourrier, C. Gastaud, and M. Redon

Laboratoire de Dispositifs Infrarouge
Tour 12, 2e étage
Université P. et M. Curie
4 place Jussieu 75230 Paris Cedex 05 - France

INTRODUCTION

NH_3 is one of the spectroscopically best studied molecules which also has properties uniquely favourable for optically pumped lasers :

1) high permanent dipole moment ,
2) high rotational constant,
3) inversion splitting of the ground state energy levels in the 1 cm wavelength range.

The high permanent dipole moment gives rise to high transition strengths for rotational as well as vibrational transitions. Thus the absorption of pump radiation is efficient, the FIR gain is much higher than in any other molecule (Marx et al., 1981) and even optically pumped, vibrational transitions exhibit gain high enough to be operated as CW lasers (Znotins et al., 1980). Besides the high dipole moment, the high rotational constant contributes to high gain vía a small partition function.

The inversion splitting of the ground state levels facilitates Stark tuning of the molecular transitions into resonance with pump laser lines (Fetterman et al., 1973) and permits Infrared-Microwave two-photon pumping (Willenberg et al., 1980-a).

The high rotational constant of NH_3 has as a consequence a rather rarefied rotational spectrum with few coincidences of absorption lines and lines of suitable pump lasers - prerequirement for optically excited laser emission. In view of the favourable properties of NH_3 as a laser molecule, work began early to excite NH_3 by more sophisticated techniques than straight resonant optical pumping by the $^{12}C^{16}O_2$ laser :

1. Optical pumping by N_2O-, isotopic CO_2- and Sequence band lasers (Chang et al., 1970 , Woods et al., 1980 ; Danielewicz and Weiss, 1978-a,b ; McCombie et al., 1981 ; Willenberg, 1981).
2. Stark tuning of transitions into resonance with $^{12}C^{16}O_2$ and N_2O lasers (Fetterman et al., 1973 ; Redon et al., 1979-a,b, 1980-a, b,c, ; Gastaud et al., 1981-a,b).
3. Off-resonant optical pumping i.e. pumping several Doppler widths away from line center in the "tail" of the vibrational pump line (Fetterman et al., 1972).
4. CW lasing through stimulated Raman scattering (Willenberg et al., 1980-b ; Willenberg, 1981).
5. CW two-photon pumping (Willenberg et al., 1980-a ; Willenberg, 1981).
6. Utilization of pulsed $^{12}C^{16}O_2$ lasers having larger oscillation bandwidths and thus providing better chances for absorption coincidences :

a) pulsed single photon pumping (Gullberg et al., 1973 ; Deka et al., 1980 ; Jones et al., 1978).

b) pulsed two-photon pumping (Jacobs et al., 1976 ; Lee et al., 1979 ; Pinson et al., 1981).

c) pulsed hot band pumping (Gullberg et al., 1973 ; Lee et al., 1979 ; Pinson et al., 1981).

The methods 2-5 are applicable due to the low partition functions, high dipole moment and ground state inversion splitting. They have required further, the precise spectroscopy of NH_3 allowing the calculation of transition frequencies e.g. within a fraction of a Doppler width. The best set of spectroscopic constants for the ν_2 band of NH_3 which is mostly involved in optically pumped NH_3 lasers is given in ref. (Shimoda et al., 1980). Due to the precision of the NH_3 spectroscopy, laser line assignments are no problem in NH_3 unless one is dealing with Stark effects, Microwave and recently coherent laser spectroscopy has provided data on relaxation times/line broadening parameters (Leite et al., 1977). Line intensities can be calculated according to classical textbooks (Townes and Schawlow,1955).

It is to be noted for NH_3 that the same laser transitions can frequently be pumped by different methods. Due to the amount of knowledge existing on NH_3, it also lends itself to quantitative comparisons with the theory of optically pumped lasers (Heppner et al., 1980 ; Marx et al., 1981) or more generally speaking the interaction of two coherent fields with 3-levels quantum mechanical systems. The adequacy of the semiclassical theory of 3-levels systems for optically pumped lasers was thus recently demonstrated for the "laser" and "stimulated Raman scattering" case (Marx et al., 1981).

Time varying Stark fields were applied by Fetterman et al. (1976) to a FIR NH_3 laser revealing a number of interesting effects. The FIR laser power was increased at high Stark modulation frequencies due to the then simultaneous pumping of all velocity groups of laser molecules, thus reducing pump saturation. When quasi statically sweeping the absorption transition through the pump frequency, a Lamb dip transferred from the pump to laser transition is observed. Coherent transient effects analogous to effects in 2-levels systems (Brewer and Shoemaker, 1971) were observed. Not much work in this interesting field of time dependent coherent effects in 3-levels systems has been reported in the FIR laser literature since then.

1. Stark tuning of IR transitions with $^{12}C^{16}O_2$ and N_2O lasers

In order to overcome the lack of exact coincidences between the frequencies of the IR absorption transitions of ammonia and those of the conventional CO_2 or N_2O laser lines, two main ways may be chosen : (i) tuning the frequencies of the CO_2 lines (high pressure lasers) or generating other frequencies (sequential or isotopic lasers), (ii) shifting the IR transitions of the gas which can be achieved by Stark tuning.

For resonant excitation of a gas, Stark effect has been used in IR-MW double resonance experiments (Shimizu and Oka, 1970 ; Fourrier and Redon, 1972). Its application to pumping for FIR lasing purposes has been investigated by Fetterman et al. (1973) in $^{14}NH_3$. Three emissions were so detected at 151.48 88.2 and 114 μm wavelength with the 10P(32). 10R(16) and 9P(12) CO_2 laser lines respectively. The feasibility of the method was thus demonstrated but the operating conditions were very critical : for pressures higher that 9.5 mTorr, electrical breakdown occurred and below 8.5 mTorr, the laser gain was under threshold. With an hybrid metallic dielectric waveguide, it has been possible to improve the operating conditions and thus to obtain 63 emissions on $^{14}NH_3$ (Redon et al., 1979-a,b ; 1980-a,b) and 41 emissions on $^{15}NH_3$ (Gastaud et al., 1981-a,b) in presence of Stark effect. Up to now, a great number of these emissions have not been lasing by other excitation techniques.

This type of systematic research in presence of Stark field for FIR emission can also be applied to other molecules (Gastaud et al.,1980).

The Stark resonator used for these systematic researches permits electric fields up to 70kV/cm with ammonia pressures up to 40 mTorr. When lasing occurs at low electric fields (some kV/cm), the pressure can be raised (for instance 80 mTorr for 8 kV/cm), and it is then possible to reach pressures higher than the optimum pressure of the NH_3 laser in zero field operation. Such high electric fields may be easily applied to the gas, because the gain of the FIR ammonia laser is so high that lasing occurs for a transverse section of the waveguide as small as 3 x 15 mm^2, 3 mm being the spacing between the 1.2 m long Stark electrodes.

The IR tuning of the ammonia transitions is obtained by the strong displacement of the levels belonging to the ground state of the ν_2 vibration, those of the first excited state being almost insensitive to the electric field. The Stark wave function mixing of the fundamental ν_2 state allows the pumping of the forbidden aa and ss transitions , increasing thus the number of possible pumping schemes in addition to those obtained by the allowed as and sa transitions.

The first step for seeking FIR lasing in presence of Stark effect is to know the electric field values corresponding to the IR resonant excitations by the pump lines. This may result from IR laser Stark spectroscopy. Shimizu had undertaken such spectroscopy for $^{14}NH_3$ (1970-a) and $^{15}NH_3$ (1970-b), but incompletely. Further investigations (Gastaud et al., 1981-b ; Shimoda et al., 1980 ; Redon, 1980) have been necessary. It should be emphasized that the FIR emission wavelength measurements make easier the identification of the IR laser Stark spectra (Redon et al., 1980-c).

Table Comments

Results are summarized in Tables 1 to 4.

The pump laser lines belong to the CO_2 laser if they are prefixed with 9 or 10 and to the N_2O laser otherwise ; moreover, there is one CO_2 sequential line, the 10R(3). The FIR transitions are named according to the absorption rules. The resonant field subscript $E_\perp$ (resp. $E_{//}$) means that the electric field of the linearly polarized pump radiation is perpendicular (resp. parallel) to the Stark field. The Stark values are given for the first M component (lower Stark field) giving rise to FIR laser action, generally the M = J component. The accuracy of the Stark field is estimated to be $\pm$ 0.1 % for values lower than 50 kV/cm and $\pm$ 0.8 % for higher values. Wavelengths are measured, with a precision of $\pm$ 0.3 % by means of a Perot-Fabry interferometer fitted with meshes. All the emissions belong to the $v_2 = 1$ state. IR transitions with an asterisk (*) are $\Delta k = -3$ absorption transitions.

Tables 1 and 2

The Stark frequency shifts appearing in the 5th column are calculated using the experimental value of the electric field, up to the second order, the zero field energy levels being obtained form data of ref. (Garing et al., 1959 ; Townes and Schawlow, 1955); Shimoda et al., 1980 ; Sasada, 1980).

<u>Tables 3 and 4</u>

The FIR polarization is reported in the 6th column. // (resp. ⊥) means that the electric field of the FIR radiation is parallel (resp. perpendicular) to the metallic planes of the Stark hybrid waveguide. Careful experimental checks show that for long wavelengths, the polarization is not always parallel as previously reported ; in some cases the polarization is parallel or perpendicular according to the cavity length for a given FIR transition. When no polarization appears, it becomes from the fact that either the wavelength is too short for our grids or the checking is too difficult due to other simultaneously lasing lines.

The FIR power is given for the maximum possible pressure (minimum 40 mTorr) compatible with the breakdown conditions. This power is estimated with a Scientech power meter 36.101. The indicated orders of magnitude are : W for $P<0.1$ mW ; M for $0.1 \text{ mW} \leqslant P<1$ mW ; S for $P\geqslant 1$ mW.

The powers are given for the pump selection rules leading to the maximum measured power. This corresponds to :
// (or $\Delta M=0$) for the Q IR transitions,
⊥ (or $\Delta M=\pm 1$) for the R IR transitions,
the exceptions being indicated by the label ⊥ for the Q IR transitions and // for the RIR transitions.

Powers for the 9R(30) pump line have not been checked because several FIR lines are lasing simultaneously.

Up to now, the Stark effect is the only mean that has permitted the obtention of a large number of FIR emissions in ammonia. The use of isotopic or sequence pump lasers could increase this number. It could perhaps play a role in the MIR coherent radiation generation.

2. CW resonant optical pumping

Since $^{14}NH_3$ does not possess absorption coincidences with $^{12}C^{16}O_2$ laser lines within a Doppler width (the closest coincidence is within 3.6 Doppler widths and leads to Raman emission) other IR lasers in the 10 μm wavelength region have been utilized for optical pumping, the first being the N_2O-laser (Chang et al., 1970) which from a 1.5 m discharge length (typical for flowing CW CO_2 lasers) delivers powers up to 10 W, quite adequate for optical pumping. Similar powers can be obtained from the Sequence band $^{12}C^{16}O_2$ laser which has consequently been used to pump $^{14}NH_3$ (Danielewicz and Weiss, 1978-a ; Frank et al., 1982) and $^{15}NH_3$ (Danielewicz and Weiss, 1978-b ; Frank et al., 1982). Quite successful has been optical pumping of $^{15}NH_3$ with $^{13}C^{16}O_2$ (Woods et al., 1980 ; Davis et al., 1981) which has yielded high power FIR emission.

A list of all coincidences between $^{14}NH_3$ and $^{15}NH_3$ absorption lines and the bands of all CO_2 lasers of the various isotopic compositions including mixed C isotopes, along with their Sequence bands as well as N_2O lasers and their Sequence bands has recently been computed, predicting a large number of possible efficient CW laser lines from NH_3 (Frank et al., 1982).

Due to the high absorption of NH_3 absorption lines, which is the prime reason for the high gain of NH_3 FIR lasers, the NH_3 FIR lasers exhibit remarkable amplitude and frequency stability because the pump radiation is typically absorbed over 1 m length so that feedback of pump radiation into the pump laser is largely eliminated.

Optical pumping in NH_3 leads to high population transfer, therefore, besides the directly pumped lines, also cascade lines have been observed (Chang et al., 1970 ; Redon et al., 1979-a) and these have been found to oscillate even in the absence of the directly pumped laser line indicating collisional pumping. Using mixtures of $^{14}NH_3$ and $^{14}ND_3$ FIR laser lines have been observed by pumping with a CW $^{12}C^{16}O_2$ laser which have been assigned to $^{14}ND_3$ and $^{14}NH_2D$ (Landsberg, 1980).

A number of FIR laser lines have been observed using $^{14}NH_3$ and $^{15}NH_3$ and pumping with a $^{13}C^{16}O_2$ laser (Davis et al., 1981). We have tried to assign these to the ν_2 bands of $^{14}NH_3$ and $^{15}NH_3$ without success. No absorption lines were found to match the quoted pump laser lines within 1 GHz for J values up to 16 and only one of the quoted FIR wavelengths occurs (within the quoted uncertainty) at all in the $\nu_2 = 1$ state (112.3 μm). However, this wavelength is obtained from $^{14}NH_3$ and $^{15}NH_3$ in comparable strength, similarly another wavelength is obtained from $^{14}NH_3$ and $^{15}NH_3$ under identical pumping conditions. We therefore conclude tentatively that these lines are not associated with NH_3 but rather with species like CH_2DOD produced as a result of an exchange reaction of NH_3 with CD_3OD which was present in some cases.

The resonators used for the NH_3 FIR lasers have been of the open resonator type (Chang et al., 1970 ; Frank et al., 1982) and of the narrow metallic waveguide type (Tanaka et al., 1977), the latter seemingly producing much higher output power. High pressure CW waveguide lasers are now available with tuning widths of 1 GHz. These lasers will allow to increase the number of CW laser lines from NH_3. They will also allow to experimentally study the transition from "laser" emission to stimulated Raman scattering (Table 5).

3. CW stimulated Raman scattering

CW laser emission was found from $^{14}NH_3$ when optically pumping by the 9R(30) $^{12}C^{16}O_2$ laser line even though the frequency mismatch of the 9R(30) line with the nearest NH_3 absorption line sR(5,0) is 180 MHz (3.6 Doppler widths) (Redon et al., 1979-a). It was later found that this emission is not due to the usual single-photon absorption-emission process but rather to stimulated Raman scattering, where absorption and emission form a two-photon transition between the lower pump level and the lower laser level (Willenberg et al., 1980-b) which is non-resonant with the intermediate (upper laser) level. After understanding this CW Raman emission process, additional Raman emissions could be excited in $^{14}NH_3$ (Willenberg et al., 1980-b ; Willenberg, 1981). Raman lasers of this type differ in many respects from the resonantly (single-photon) pumped lasers : i) The gain line widths is much larger than in the resonant systems. It is approximately equal to the Doppler width of the infrared pump transition. ii) The spectral position of the gain line can be shifted by a pump frequency shift. iii) Saturation of the Raman gain occurs only by two-photon saturation which leads to gain saturation intensities orders of magnitude higher than for resonantly pumped systems. iv) Since these Raman systems do not rely on population inversion in the usual sense, which in the case of FIR lasers is very much susceptible to collisions, the Raman systems can be operated at rather high pressures (Marx et al., 1981) (Table 6).

4. CW two-photon pumping

The lack of absorption coincidences of NH_3 with $^{12}C^{16}O_2$ laser lines can also be circumvented by infrared-microwave-two-photon pumping. This scheme is based on the IR-MW two-photon spectroscopy work of Oka (Freund and Oka, 1976) and the idea was first suggested by Fetterman (1978).

In a metallic waveguide resonator which serves both as a FIR laser resonator and MW resonator it was demonstrated on a number of FIR laser transitions that CW lasers can be pumped by two-photon absorption (Willenberg et al., 1980-a ; Willenberg, 1981). This type of laser has the important property that its amplitude and frequency can be electrically modulated by control of the microwave. Since IR-MW two-photon pumping provides a frequency tuneable pump source (the MW frequency can be tuned, the laser frequency can be step-tuned), there is prospect of combining two-photon pumping with Raman emission(Hyper-Raman-laser) to produce CW FIR radiation of at least limited tuneability (Table 7).

5. Pulsed off-resonant optical pumping

Fetterman et al. (1972) showed that optical pumping at a frequency ω is efficient even if ω differs from the pump transition angular frequency ω_o, if the Rabi-frequency on the pump transition is such that :

$$\mu E/\hbar \geq |\omega - \omega_o| .$$

where μ is the transition dipole moment and E the electric field strength of the pump field. The authors demonstrated FIR emission on NH_3 and other molecules using megawatt pulsed pump lasers with frequency offsets as large as 20 Doppler widths. It is however clear that under these conditions also stimulated Raman scattering can occur (Wiggins et al., 1978). It is therefore unclear which of the observed emission are actually "off resonance pumped" (resonant emission) and which are Raman (off-resonant) emission. In ref. (Fetterman et al., 1972), it is mentioned that more lines were observed than given in Table 8. However since their assignment was unclear they were not published.

6. Pulsed stimulated Raman scattering

In ref. (Wiggins et al., 1978), a more systematic investigation was made of the emission behaviour of pulsed off-resonantly excited NH_3. It was found that for the transitions investigated at low pressures the off-resonantly pumped resonant emission was predominant. At higher pressure Raman emission dominated for the same reasons as mentioned for the CW Raman case above.

7. Pulsed single photon pumping

Since the oscillation banwidth of high pressure TE CO_2 lasers are 20 times higher than for low pressure CW lasers, there is a better chance for absorption coincidences of NH_3 absorption lines with these pulsed pump lasers. The high pressure TE lasers are oscillating on many modes distributed over the gain bandwith (≈ 2GHz) unless special measures are taken to insure single mode operation (Fabry-Perot etalon or low pressure discharge cell inside the TE laser resonator). Gullberg et al. (1973) were therefore able to excite a number of FIR laser lines with a pulsed atmospheric pressure TE laser. Since in this work in some cases the pump frequency mismatches were > 1 GHz it it to be expected that the emission in these cases was of the "off resonantly pumped" or Raman type. It is noteworth that also FIR ground state transitions were observed to lase. There are of the "refilling" type.

It should be mentioned that CO_2 TE lasers are now available which can operate at a pressure sufficiently high to provide continuously tuneable pulsed radiation over much of the 9-11 μm wavelength region. This type of laser has been used so far for generation of mid-infrared radiation by optical pumping $^{14}NH_3$ (Deka et al. 1980). Similarly a HF laser has been used to generate mid-infrared laser lines by optically pumping $^{14}NH_3$ (Jones et al., 1978).

A rather elegant way of pulsed resonant optical pumping of $^{14}NH_3$ FIR laser lines was used by Yamabayashi et al. (1981). With a pulsed 9R(30) $^{12}C^{16}O_2$ laser, $^{14}NH_3$ was optically pumped. The population pumped into the excited vibrational ν_2 state is then distributed by collisions into various rotational levels leading to mid-infrared lasing on a number of $\nu_2 = 1 \rightarrow \nu_2 = 0$ transitions. This laser emission is obviously suited to resonantly pump FIR transitions in $^{14}NH_3$. Using this technique 33 FIR laser lines were excited (Table 9,9a).

8. Pulsed hot-band pumping

Even though the population of the $\nu_2 = 1$ vibrational state is only about 1% of the ground state population at room temperature, it has been possible to generale FIR radiation by pumping the $\nu_2=1 \rightarrow \nu_2=2$ band of $^{14}NH_3$. The absorber density was in this case increased by raising the NH_3 pressure to values up to 30 Torr. FIR lasing on rotational and inversion transitions of the $\nu_2=2$ state were reported by Gullberg et al. (1973). Lee et al. (1979) reported additional laser lines from the $\nu_2=2$ state and a line first reported in ref. (Tiee and Wittig, 1978) was assigned. In ref. (Pinson et al., 1981) levels in the $\nu_2=2$ state were pumped in a step-wise fashion. One CO_2 laser pumped a $\nu_2= 1$ level the population of which was then transferred by FIR lasing (or collisions ?) to other rotational levels in the $\nu_2= 1$ state. A second CO_2 laser was then used to pump from a $\nu_2= 1$ level to a $\nu_2= 2$ level from which lasing occurred. The experiment is interesting since a number of "refilling" lines are observed for which there cannot be a population inversion. Consequently they must be of a "refilling-Raman" type not reported before. Since in some cases a whole sequence of emissions and absorptions finally leads to laser emission a closer examination of the situation would probably reveal very interesting coherent multi-levels physics (Table 10).

9. Pulsed two-photon pumping

Two-photon optical pumping of $^{14}NH_3$ with photons of two different pulsed CO_2 laser lines was first demonstrated by Jacobs et al. (1976). It leads to optical pumping of the $\nu_2= 2$ state. Lasing can then occur on rotational and inversion transitions in the $\nu_2= 2$ state or on $\nu_2= 2 \rightarrow \nu_2= 1$ transitions.

Pump laser with n lines (n $\approx$ 100 for any isotopic species of CO_2) can provide n single-photon pump frequencies but $\simeq n^2/2$ two-photon pump frequencies. Thus the probability of coincidences is by a large factor higher for two-photon than for single photon pumping.

Additional laser lines were reported by Lee et al. (1979) using two-photon pulsed pumping also utilizing $^{13}C^{16}O_2$ pump lasers.

Also in ref. (Pinson et al., 1981) two-photon pumping was used to achieve lasing in the $\nu_2 = 2$ state as well as on $\nu_2 = 2 \rightarrow \nu_2 = 1$ transitions. In view of the overall pump mismatches in these experiments it is conceivable that emission also occurred due to Hyper-Raman scattering (Table 11).

The laser lines reported experimentally so far are apparently far from exploiting all possibilities of generation of coherent FIR radiation in NH_3. One particularly attractive feature of NH_3 in this respect is its well understood spectrosocpy which allows to predict the wavelengths generated and also allows to predict the feasibility of FIR generation by the various optical techniques for desired wavelengths.

Table 1 - FIR laser lines in $^{14}NH_3$ - observed by Stark tuning

FIR emission transition	Infrared laser line	$IR-NH_3$ pumped transition	Resonant field (kV/cm) $E_\perp$	M	$E_{//}$	M	Shift E(GHz) $\nu_{NH_3}-\nu_{CO_2}$	λ (μm) measured	λ (μm) calculated
saQ(1,1)	R(37)	saQ(1,1)	7.35	1	7.25	1	- 0.31	280.00	281.21
saR(1,1)	R(37)	saQ(2,1)	59.18	2	58.67	2	- 7.26	133.16	133.16
asR(2,0)	10R(8)	ssQ(3,3)*	40.92	3	41.02	3	-13.69	388.85	389.11
	10R(6)	asQ(3,3)*	36.01	2	24.24	3	6.13	389.67	
	R(36)	ssQ(3,3)*	17.51	2	11.65	3	- 1.65	389.30	
saR(2,1)	R(35)	aaQ(3,1)	-	-	49.54	3	3.16	107.18	106.26
saQ(2,2)	10R(6)	aaQ(2,2)	63.07	2	63.12	2	20.67	279.49	280.50
	R(37)	saQ(2,2)	32.39	2	32.19	2	- 8.16	279.79	
	R(35)	aaQ(2,2)	-	-	38.94	2	10.44	280.13	
saR(2,2)	9P(40)	saR(2,2)	38.10	2	38.33	2	-10.54	105.20	105.35
	10R(8)	saQ(3,2)	45.60	3	45.23	3	- 9.07	105.28	
	10R(6)	aaQ(3,2)	55.00	3	55.58	3	11.74	105.18	
	10P(32)	asQ(4,2)	42.23	3	41.90	3	3.50	105.35	
	R(35)	aaQ(3,2)	23.90	2	15.71	3	1.37	105.39	
asR(2,2)	10R(48)	asR(2,2)	20.50	2	20.63	2	3.77	404.08	404.69
	10P(32)	ssQ(3,2)	24.37	3	24.60	3	- 3.13	403.84	
	P(8)	asQ(3,2)	7.80	3	7.67	3	0.35	403.83	
saR(3,1)	10R(6)	aaQ(4,1)	49.47	4	49.73	4	2.23	88.70	88.69
asR(3,1)	P(5)	ssQ(4,1)	-	-	57.63	4	- 3.00	215.77	215.01

(continued)

Table 1 - $^{14}NH_3$ - (Continued)

FIR emission transition	Infrared laser line	IR-NH_3 pumped transition	Resonant field (kV/cm) $E_\perp$	M	$E_{//}$	M	Shift (GHz) $\nu_{NH_3} - \nu_{CO_2}$	λ(μm) measured	λ(μm) calculated
	P(6)	asQ(4,1)	37.43	4	37.97	4	1.40	215.50	
saQ(3,2)	9P(40)	saR(2,2)	38.10	2	38.33	2	-10.54	288.68	289.35
	10R(8)	saQ(3,2)	45.60	3	45.23	3	- 9.07	289.45	
	10R(6)	aaQ(3,2)	55.00	3	55.58	3	11.74	289.15	
saR(3,2)	10R(6)	aaQ(4,2)	7.02	4	7.07	4	0.20	88.10	88.06
	R(36)	saQ(4,2)	59.67	4	58.47	4	- 9.73	88.28	
asR(3,2)	10P(32)	asQ(4,2)	42.23	3	41.90	3	3.50	218.00	218.29
	P(7)	asQ(4,2)	34.37	4	34.17	4	4.04	218.55	
saQ(3,3)	10R(8)	saQ(3,3)	34.80	3	34.50	3	-10.86	279.00	279.33
	10R(6)	aaQ(3,3)	31.20	3	31.43	3	8.96	279.60	
	R(37)	saQ(3,3)	52.53	3	51.99	3	-19.89	278.41	
	R(34)	aaQ(3,3)	53.33	3	53.90	3	19.59	278.78	
saR(3,3)	9P(20)	saR(3,3)	42.23	3	42.10	3	-14.42	86.96	87.09
	9P(22)	aaR(3,3)	47.60	3	47.33	3	16.67	87.03	
	10P(32)	asQ(5,3)	-	-	12.33	5	0.95	87.03	
	R(36)	saQ(4,3)	-	-	45.33	4	-12.05	87.27	
	R(34)	aaQ(4,3)	46.33	3	46.67	3	7.70	87.27	
	P(7)	asQ(5,3)	-	-	15.90	5	1.56	87.21	
asR(3,3)	9P(12)	ssR(5,3)	26.30	4	-	-	- 2.49	223.68	223.91
	10P(32)	ssQ(4,3)	48.67	4	49.07	4	-12.90	223.64	
	P(7)	ssQ(4,3)	47.10	4	47.50	4	-12.30	224.10	

FIR emission transition	Infrared laser line	IR-NH_3 pumped transition	Resonant field (kV/cm) $E_\perp$	M	$E_{//}$	M	Shift (GHz) $\nu_{NH_3}-\nu_{CO_2}$	λ (μm) measured	calculated
	P(9)	asQ(4,3)	56.90	4	56.22	4	16.46	223.51	
asR(4,1)	9R(30)	saR(5,1)	21.08	5	21.33	5	- 0.35	147.73	147.62
saQ(4,2)	10R(6)	aaQ(4,2)	-	-	14.20	2	0.20	303.80	301.30
saR(4,2)	R(35)	saQ(5,2)	38.38	5	38.20	5	- 3.78	76.06	75.89
asR(4,2)	9R(30)	saR(5,2)	16.16	5	16.16	5	- 0.75	149.05	149.10
saQ(4,3)	9P(20)	saR(3,3)	42.23	3	42.10	3	-14.42	290.49	290.95
	9P(22)	aaR(3,3)	47.60	3	47.33	3	16.67	290.61	
	10P(32)	asQ(5,3)	12.47	5	12.33	5	0.95	290.42	
	R(36)	saQ(4,3)	-	-	45.33	4	-12.05	290.33	
	R(34)	aaQ(4,3)	46.33	3	46.67	3	7.70	289.69	
saR(4,3)	9R(2)	saR(4,3)	50.84	4	51.00	4	-14.16	75.10	75.17
	R(35)	saQ(5,3)	-	-	36.15	5	- 6.62	75.28	
asR(4,3)	9R(30)	saR(5,3)	14.27	5	14.27	5	- 1.25	151.45	151.49
	9R(28)	aaR(5,3)	55.15	5	54.68	5	12.08	151.74	
	9P(34)	ssR(4,3)	61.94	4	61.38	4	-17.95	151.23	
	10R(6)	saQ(6,3)	-	-	66.37	6	-13.79	151.78	
	10P(32)	asQ(5,3)	12.47	5	12.33	5	0.95	151.17	
	P(6)	ssQ(5,3)	22.20	5	22.33	5	- 2.78	151.26	
	P(7)	asQ(5,3)	16.00	5	15.90	5	1.54	151.82	

(continued)

Table 1 - $^{14}NH_3$ - (Continued)

FIR emission transition	Infrared laser line	$IR-NH_3$ pumped transition	Resonant field (kV/cm) $E_\perp$	M	$E_{//}$	M	Shift (GHz) $\nu_{NH_3}-\nu_{CO_2}$	λ (μm) measured	calculated
saQ(4,4)	10R(8)	saQ(4,4)	–	–	60.96	4	-26.78	276.80	276.78
	R(36)	saQ(4,4)	–	–	39.77	4	-14.71	276.67	
	R(34)	aaQ(4,4)	16.97	4	17.00	4	3.56	277.00	
saR(4,4)	9R(2)	saR(4,4)	39.03	4	39.20	4	-14.21	74.10	74.15
	10R(4)	aaQ(5,4)	64.74	5	65.40	5	22.07	74.07	
	R(35)	saQ(5,4)	44.77	4	35.95	5	- 9.96	74.24	
asR(4,4)	9R(30)	saR(5,4)	15.37	4	12.33	5	- 1.54	155.30	155.28
	9P(36)	asR(4,4)	26.33	4	26.20	4	- 7.48	155.03	
	9P(38)	asR(4,4)	65.85	4	66.38	4	29.40	154.94	
	10P(34)	asQ(5,4)	24.43	5	24.33	5	5.26	154.97	
	P(8)	ssQ(5,4)	28.43	4	22.57	5	- 4.46	155.47	
	P(10)	asQ(5,4)	69.63	5	67.50	5	24.55	155.60	
saR(5,1)	9R(30)	saR(5,1)	–	–	56.33	2	- 0.35	67.00	67.12
asR(5,1)	9P(10)	ssR(5,1)	46.42	5	46.03	5	- 1.53	112.26	112.22
saR(5,2)	9R(30)	saR(5,2)	16.16	5	16.16	5	- 0.75	66.72	66.80
asR(5,2)	P(4)	asQ(6,2)	37.12	6	37.19	6	2.91	113.11	112.98
saQ(5,3)	9R(2)	saR(4,3)	50.84	4	51.00	4	-14.16	305.55	306.28
	9P(12)	ssR(5,3)	20.97	5	–	–	- 2.49	307.00	
saR(5,3)	9R(30)	saR(5,3)	–	–	36.30	2	- 1.32	66.37	66.28

FIR emission transition	Infrared laser line	IR-NH_3 pumped transition	Resonant field (kV/cm) $E_\perp$	M	$E_{//}$	M	Shift (GHz) $\nu_{NH_3} - \nu_{CO_2}$	λ (μm) measured	calculated
asR(5,3)	9P(12)	ssR(5,3)	20.97	5	20.90	5	- 2.49	114.28	114.30
saQ(5,4)	9R(2)	saR(4,4)	39.03	4	39.20	4	-14.21	291.63	291.21
saR(5,4)	9R(30)	saR(5,4)	12.30	5	12.33	5	- 1.54	65.50	65.51
	9R(28)	aaR(5,4)	38.20	5	37.96	5	10.43	65.39	
	10R(6)	saQ(6,4)	61.17	6	61.17	6	-17.95	65.40	
	10R(4)	aaQ(6,4)	28.70	6	28.80	6	5.47	65.46	
	R(34)	saQ(6,4)	49.30	4	33.13	6	- 7.16	65.61	
asR(5,4)	P(6)	ssQ(6,4)	62.67	5	52.68	6	-13.77	116.59	116.27
	P(8)	asQ(6,4)	58.18	6	57.87	6	16.57	116.45	
saQ(5,5)	10R(6)	saQ(5,5)	14.60	5	14.60	5	- 2.99	273.37	273.37
	R(33)	aaQ(5,5)	47.00	2	-	-	4.47	273.71	
saR(5.5)	9R(30)	saR(5,5)	8.47	5	8.47	5	- 1.07	64.46	64.50
	9R(28)	aaR(5,5)	28.63	5	28.46	5	9.02	64.35	
	10R(6)	saQ(6,5)	59.13	6	59.13	6	-22.53	64.66	
	R(34)	saQ(6,5)	55.68	4	45.00	5	-11.73	64.22	
	R(32)	aaQ(6,5)	37.47	5	31.10	6	8.40	64.49	
asR(5,5)	9P(16)	ssR(5,5)	27.13	5	45.33	3	- 8.39	118.88	119.02
	9P(18)	asR(5,5)	48.73	5	49.01	5	20.53	119.21	
	P(9)	ssQ(6,5)	26.30	5	21.77	6	- 4.70	119.13	

(continued)

Table 1 - $^{14}NH_3$- (Continued)

FIR emission transition	Infrared laser line	IR-NH_3 pumped transition	Resonant field (kV/cm) $E_\perp$	M	$E_{//}$	M	Shift (GHz) $\nu_{NH_3}-\nu_{CO_2}$	λ(μm) measured	calculated
asR(6,2)	9R(16)	ssR(6,2)	49.63	6	49.43	6	- 4.63	90.98	90.93
saQ(6,3)	9R(30)	saR(5,3)	17.87	4	17.87	4	- 1.26	325.93	325.31
saR(6,3)	R(33)	saQ(7,3)	17.20	7	17.17	7	- 1.21	59.48	59.37
	R(32)	aaQ(7,3)	-	-	24.54	7	2.26	59.25	
asR(6,3)	10P(28)	ssQ(7,3)	56.27	7	56.60	7	- 8.95	91.67	91.71
saQ(6,4)	9R(30)	saR(5,4)	12.30	5	12.33	5	- 1.54	309.47	309.50
	9R(28)	aaR(5,4)	38.20	5	37.96	5	10.43	309.47	
asR(6,4)	10P(30)	ssQ(7,4)	8.83	7	8.83	7	- 0.53	92.40	92.88
saQ(6,5)	9R(30)	saR(5,5)	8.47	5	8.47	5	- 1.07	290.31	290.44
	9R(28)	aaR(5,5)	28.63	5	28.46	5	9.02	289.81	
saR(6,5)	R(33)	saQ(7,5)	57.25	5	41.27	7	-11.62	57.99	58.01
	R(31)	aaQ(7,5)	47.17	6	40.17	7	10.62	57.99	
asR(6,5)	9R(10)	ssR(6,5)	47.97	6	47.77	6	-16.02	94.53	94.45
	9R(8)	asR(6,5)	19.07	6	19.10	6	3.87	94.37	
	10P(32)	ssQ(7,5)	37.93	6	37.80	6	- 7.67	94.34	
saQ(6,6)	10R(6)	saQ(6,6)	-	-	57.70	6	-26.86	267.83	268.82
	R(34)	saQ(6,6)	60.39	4	40.07	6	-16.12	268.76	

FIR emission transition	Infrared laser line	$IR-NH_3$ pumped transition	Resonant field (kV/cm) $E_\perp$	M	$E_{//}$	M	Shift (GHz) $\nu_{NH_3}-\nu_{CO_2}$	λ (μm) measured	calculated
	R(32)	aaQ(6,6)	10.77	6	10.80	6	1.71	269.66	
	R(31)	aaQ(6,6)	64.72	5	54.32	6	23.23	268.91	
saR(6,6)	10R(4)	saQ(7,6)	21.67	7	21.57	7	- 5.19	57.08	57.04
	10R(2)	aaQ(7,6)	-	-	48.07	7	17.01	56.93	
	R(31)	aaQ(7,6)	17.77	6	15.17	7	2.70	57.11	
asR(6,6)	9R(6)	ssR(6,6)	22.07	6	22.00	6	- 6.10	96.58	96.67
	9R(4)	asR(6,6)	34.83	6	34.97	6	13.05	96.56	
	10P(36)	asQ(7,6)	19.47	7	19.43	7	4.33	96.47	
	P(10)	ssQ(7,6)	39.30	7	39.40	7	-12.85	96.87	
	P(12)	asQ(7,6)	45.20	7	44.93	7	16.34	96.76	
saR(7,3)	10R(4)	saQ(8,3)	60.40	8	60.23	8	- 3.89	53.84	53.86
saR(7,4)	10R(2)	aaQ(8,4)	-	-	60.53	8	12.73	53.34	53.40
asR(7,4)	10P(28)	ssQ(8,4)	52.83	7	45.93	8	- 8.59	77.29	77.26
saR(7,5)	10R(2)	aaQ(8,5)	33.73	6	25.17	8	4.50	52.68	52.80
asR(7,5)	P(6)	asQ(8,5)	10.33	8	10.13	8	0.90	78.37	78.27
saQ(7,6)	10R(4)	saQ(7,6)	21.67	7	21.57	7	- 5.19	287.71	288.52
	R(31)	aaQ(7,6)	17.77	6	15.17	7	2.70	289.10	

(continued)

Table 1 - $^{14}NH_3$ - (Continued)

FIR emission transition	Infrared laser line	$IR-NH_3$ pumped transition	Resonant field (kV/cm) $E_\perp$	M	$E_{//}$	M	Shift (GHz) $\nu_{NH_3}-\nu_{CO_2}$	λ (μm) measured	calculated
saR(7,6)	R(30)	aaQ(8,6)	43.10	3	16.00	8	2.61	51.99	52.03
asR(7,6)	9R(42)	ssR(7,6)	67.73	7	67.40	7	-26.93	79.70	79.62
	9R(38)	asR(7,6)	34.57	7	34.67	7	11.06	79.57	
	10P(34)	asQ(8,6)	18.87	8	18.83	8	3.62	79.49	
	P(8)	ssQ(8,6)	42.03	6	31.33	8	- 8.06	79.73	
	P(10)	asQ(8,6)	62.98	8	62.61	8	22.91	79.72	
saQ(7,7)	10R(2)	aaQ(7,7)	39.07	5	28.00	7	9.08	263.44	263.44
	R(32)	saQ(7,7)	8.40	7	8.47	7	- 1.13	263.60	
asR(7,7)	9R(36)	ssR(7,7)	54.00	7	-	-	-23.71	81.39	81.48
	9R(32)	asR(7,7)	39.80	7	-	-	16.08	81.46	
	P(12)	ssQ(8,7)	31.40	4	15.63	8	- 3.00	81.56	
saR(8,6)	R(30)	saQ(9,6)	12.83	8	11.43	9	- 1.33	47.94	47.91
asR(8,6)	10P(32)	asQ(9,6)	-	-	26.67	9	5.91	67.73	67.75
saQ(8,8)	R(29)	aaQ(8,8)	41.47	4	20.27	8	5.37	257.00	257.14
saQ(9,9)	R(29)	saQ(9,9)	15.87	9	15.80	9	- 3.57	250.26	250.06
	R(27)	aaQ(9,9)	52.36	6	-	-	12.76	250.68	
asR(10,9)	P(14)	asQ(11,9)	6.70	?	7.33	?	?	54.45	54.46
saQ(12,12)	R(23)	saQ(12,12)	21.33	?	-	-	?	224.00	225.07

Table 2 - FIR laser lines in $^{15}NH_3$ - Observed by Stark tuning

FIR emission transition	Infrared laser line	$IR-NH_3$ pumped transition	Resonant field (kV/cm) $E_\perp$	M	$E_{//}$	M	Shift (GHz) $\nu_{NH_3} - \nu_{CO_2}$	λ (μm) measured	calculated
saQ(1,1)	10R(8)	asR(1,1)	30.30	1	31.63	1	4.89	292.69	291.30
	10R(2)	saQ(1,1)	64.43	1	63.13	1	- 15.24	291.88	
	R(30)	saQ(1,1)	42.33	1	41.00	1	- 7.86	291.87	
saR(2,1)	R(29)	saQ(3,1)	13.93	3	13.83	3	- 0.31	107.59	107.77
saQ(2,2)	10R(3)	aaQ(2,2)	25.43	2	25.67	2	5.51	290.33	291.03
	R(30)	saQ(2,2)	48.13	2	47.37	2	-15.02	291.21	
	R(28)	aaQ(2,2)	27.03	2	27.33	2	6.11	290.83	
saR(2,2)	R(29)	saQ(3,2)	19.13	3	19.03	3	- 2.11	107.30	106.90
asR(2,2)	10R(40)	asR(2,2)	6.23	2	6.07	2	0.39	386.25	386.56
saQ(3,2)	R(29)	saQ(3,2)	19.13	3	19.03	3	- 2.11	301.03	301.19
saR(3,2)	9P(26)	saR(3,2)	37.73	3	37.90	3	- 6.93	89.37	89.17
	R(27)	aaQ(4,2)	57.03	4	58.90	4	9.64	89.43	
asR(3,2)	10P(36)	asQ(4,2)	50.17	3	38.10	4	5.03	213.59	213.37
saQ(3,3)	R(30)	saQ(3,3)	64.97	3	64.07	3	-26.76	289.23	289.60
	R(29)	saQ(3,3)	20.90	3	20.77	3	- 4.89	290.60	
	R(27)	aaQ(3,3)	45.93	3	46.50	3	16.27	290.30	
saR(3,3)	9P(26)	saR(3,3)	28.40	3	28.57	3	- 8.17	88.26	88.18
	9P(28)	aaR(3,3)	66.40	3	65.77	3	26.35	88.07	

(continued)

Table 2 - $^{15}NH_3$ - (Continued)

FIR emission transition	Infrared laser line	IR-NH_3 pumped transition	Resonant field (kV/cm) $E_\perp$	M	$E_{//}$	M	Shift (GHz) $\nu_{NH_3} - \nu_{CO_2}$	λ(μm) measured	calculated
	R(29)	saQ(4,3)	57.17	4	56.17	4	-16.81	88.18	
	R(27)	aaQ(4,3)	28.00	4	28.17	4	5.58	88.38	
asR(3,3)	10P(36)	ssQ(4,3)	47.13	4	47.57	4	-12.58	219.00	218.60
	P(12)	ssQ(4,3)	7.90	4	7.90	4	- 0.54	218.88	
	P(13)	asQ(4,3)	22.93	4	22.80	4	4.05	218.93	
saR(4,2)	9P(4)	saR(4,2)	57.00	4	57.23	4	- 9.64	76.72	76.70
	10R(3)	saQ(5,2)	49.17	5	48.93	5	- 5.93	76.67	
	R(28)	saQ(5,2)	45.97	5	45.83	5	- 5.33	77.03	
asR(4,2)	P(10)	asQ(5,2)	21.87	4	22.07	4	0.95	147.03	146.93
saQ(4,3)	9P(26)	saR(3,3)	28.40	3	28.57	3	- 8.17	301.93	301.80
	9P(28)	aaR(3,3)	66.40	3	65.77	3	26.35	302.77	
	10P(36)	asQ(5,3)	23.23	5	23.13	5	3.18	301.93	
	R(29)	saQ(4,3)	57.17	4	56.17	4	-16.81	298.90	
	R(27)	aaQ(4,3)	28.00	4	28.17	4	5.58	302.06	
saR(4,3)	9P(4)	saR(4,3)	42.27	4	42.47	4	-11.03	76.06	75.98
	9P(6)	aaR(4,3)	56.27	4	55.90	4	15.97	75.72	
	10R(3)	saQ(5,3)	45.23	5	45.00	5	- 9.58	75.85	
	R(28)	saQ(5,3)	43.33	5	43.13	5	- 8.97	76.22	
	R(26)	aaQ(5,3)	62.87	5	63.33	5	14.92	75.83	
asR(4,3)	9P(38)	ssR(4,3)	68.77	4	68.33	4	-21.07	149.43	149.24
	9P(40)	asR(4,3)	62.13	4	62.47	4	19.13	149.06	

FIR emission transition	Infrared laser line	IR-NH_3 pumped transition	Resonant field (kV/cm) $E_\perp$	M	$E_{//}$	M	Shift (GHz) $\nu_{NH_3}-\nu_{CO_2}$	λ (μm) measured	calculated
	10P(36)	asQ(5,3)	23.23	5	23.13	5	3.18	149.22	
	P(12)	asQ(5,3)	62.20	5	61.77	5	15.22	148.80	
saQ(4,4)	R(29)	saQ(4,4)	50.40	4	50.00	4	-20.75	285.33	287.02
	R(27)	aaQ(4,4)	3.87	4	3.87	4	0.22	288.00	
	R(26)	aaQ(4,4)	54.84	4	55.33	4	22.34	286.45	
saR(4,4)	9P(4)	saR(4,4)	34.87	4	35.03	4	-12.28	75.08	74.96
asR(4,4)	9P(40)	ssR(4,4)	29.73	4	29.63	4	- 9.32	152.90	152.74
	10P(38)	asQ(5,4)	32.87	5	32.63	5	8.74	153.01	
	P(14)	asQ(5,4)	42.93	5	42.67	5	13.16	152.69	
saR(5,2)	9R(20)	aaR(5,2)	23.00	5	22.93	5	1.51	60.60	67.45
	R(26)	aaQ(6,2)	54.67	6	54.53	6	5.52	67.33	
saQ(5,3)	9P(4)	saR(4,3)	42.27	4	42.47	4	-11.03	318.69	317.82
	9P(6)	aaR(4,3)	56.27	4	55.90	4	15.97	317.04	
saR(5,3)	R(27)	saQ(6,3)	23.93	6	23.83	6	- 2.72	66.97	66.91
	R(26)	aaQ(6,3)	22.17	3	11.00	6	0.62	67.00	
asR(5,3)	9P(18)	asR(5,3)	47.20	5	47.33	5	10.20	112.98	113.10
	P(10)	asQ(6,3)	36.33	6	36.20	6	5.55	113.27	
saQ(5,4)	9P(4)	saR(4,4)	34.87	4	35.03	4	-12.28	302.64	302.21

(continued)

Table 2 - $^{15}NH_3$ - (Continued)

FIR emission transition	Infrared laser line	IR-NH_3 pumped transition	Resonant field (kV/cm) $E_\perp$	M	$E_{//}$	M	Shift (GHz) $\nu_{NH_3}-\nu_{CO_2}$	λ(μm) measured	λ(μm) calculated
	9P(6)	aaR(4,4)	37.87	4	-	-	13.44	301.50	
saR(5,4)	9R(22)	saR(5,4)	57.10	5	-	-	-19.65	66.15	66.16
asR(5,4)	9P(18)	ssR(5,4)	52.77	5	52.53	5	-16.97	115.11	114.96
	9P(20)	asR(5,4)	48.37	5	48.57	5	15.65	115.08	
	P(11)	ssQ(6,4)	12.77	5	10.53	6	- 0.93	115.07	
	P(12)	asQ(6,4)	26.90	6	26.87	6	5.23	115.01	
saQ(5,5)	R(28)	saQ(5,5)	44.97	5	44.67	5	-18.62	282.50	283.35
	R(26)	aaQ(5,5)	12.07	5	12.10	5	2.12	283.21	
	10R(3)	saQ(5,5)	46.03	5	45.73	5	-19.25	283.33	
asR(5,5)	9P(22)	asR(5,5)	20.13	5	20.23	5	5.42	117.70	117.59
saR(6,3)	10P(2)	aaQ(7,3)	41.23	5	41.90	7	5.77	59.88	59.90
	R(25)	aaQ(7,3)	-	-	47.93	7	7.11	59.85	
asR(6,3)	P(8)	asQ(7,3)	10.50	7	10.57	7	0.50	91.18	91.00
saR(6,4)	R(26)	saQ(7,4)	25.00	7	24.93	7	-3.93	59.25	59.31

FIR emission transition	Infrared laser line	IR-NH_3 pumped transition	Resonant field (kV/cm) $E_\perp$	M	$E_{//}$	M	Shift (GHz) $\nu_{NH_3}-\nu_{CO_2}$	λ(μm) measured	calculated
asR(6,4)	P(10)	asQ(7,4)	17.40	7	17.33	7	2.06	92.17	92.09
saQ(6,5)	10P(2)	aaQ(6,5)	41.23	4	41.33	4	7.12	302.13	301.40
	R(25)	aaQ(6,5)	46.00	4	46.27	4	8.50	301.38	
saR(6,5)	R(26)	saQ(7,5)	39.70	7	39.53	7	-11.14	58.67	58.53
asR(6,5)	9R(2)	asR(6,5)	37.50	6	-	-	11.87	93.80	93.63
saR(6,6)	R(26)	saQ(7,6)	49.33	7	49.07	7	-18.86	57.69	57.55
	R(24)	aaQ(7,6)	-	-	18.37	7	3.92	57.54	
asR(6,6)	10P(40)	asQ(7,6)	33.47	7	33.30	7	10.73	95.71	95.73
	P(16)	asQ(7,6)	25.53	7	25.47	7	7.04	95.80	
saR(7,4)	10P(2)	saQ(8,4)	18.03	8	18.03	8	- 1.96	53.96	53.84
	R(25)	saQ(8,4)	9.57	8	9.50	8	- 0.59	53.88	
asR(7,4)	9R(38)	asR(7,4)	63.33	7	63.60	7	16.43	76.91	76.80
saQ(7,6)	R(26)	saQ(7,6)	56.80	6	-	-	-18.66	301.03	299.40
	R(24)	aaQ(7,6)	-	-	18.37	7	3.92	299.25	
saR(7,6)	10P(2)	saQ(8,6)	54.77	8	54.70	8	-19.42	52.51	52.46
	10P(4)	aaQ(8,6)	32.13	8	32.27	8	8.65	52.52	
	R(23)	aaQ(8,6)	-	-	28.37	8	7.09	52.47	

(continued)

Table 2 - $^{15}NH_3$ - (Continued)

FIR emission transition	Infrared laser line	$IR-NH_3$ pumped transition	Resonant field (kV/cm) $E_\perp$	M	$E_{//}$	M	Shift (GHz) $\nu_{NH_3}-\nu_{CO_2}$	λ(μm) measured	calculated
asR(7,6)	9R(30)	asR(7,6)	45.30	7	45.37	7	16.73	79.09	79.08
saQ(7,7)	10P(2)	saQ(7,7)	19.47	7	19.40	7	- 5.33	273.03	272.93
	R(25)	saQ(7,7)	-	-	28.90	4	- 4.03	272.73	
saR(8,6)	R(22)	aaQ(9,6)	-	-	37.77	9	9.92	48.25	48.27
saQ(9,9)	R(22)	saQ(9,9)	19.50	8	17.33	9	- 4.36	259.00	258.96

Table 3 - Emitting properties of FIR lines in $^{14}NH_3$ (increasing fields classification for each pump per line)

IR laser line	Resonant field (kV/cm)	IR Stark Component	Measured FIR wavelength (μm)	FIR transition	FIR polarization	FIR power
9R(42)	67.73	ssR(7,6,7)	79.70	asR(7,6)	//	W
9R(38)	34.57	asR(7,6,7)	79.57	asR(7,6)	//	S
9R(36)	54.00	ssR(7,7,7)	81.39	asR(7,7)	//	W
9R(32)	39.80	asR(7,7,7)	81.46	asR(7,7)	//	W
9R(30)	8.47	saR(5,5,5)	290.34	saQ(6,5)		
			64.46	saR(5,5)		
	12.30	saR(5,4,5)	309.47	saQ(6,4)		
			65.50	saR(5,4)		
	12.33//		155.30	asR(4,4)		
	14.27	saR(5,3,5)	151.45	asR(4,3)		
	16.16	saR(5,2,5)	66.72	saR(5,2)		
			149.05	asR(4,2)		
	17.87	saR(5,3,4)	325.93	saQ(6,3)		
	21.08	saR(5,1,5)	147.73	asR(4,1)		
	36.30//	saR(5,3,2)	66.37	saR(5,3)		
	56.33//	saR(5,1,2)	67.00	saR(5,1)		
9R(28)	28.63	aaR(5,5,5)	289.81	saQ(6,5)	⊥	M
			64.35	saR(5,5)	//	M
	38.20	aaR(5,4,5)	309.47	saQ(6,4)	⊥	W
			65.39	saR(5,4)	//	S
	55.15	asR(5,3,5)	151.74	asR(4,3)	//	S

(continued)

Table 3 - $^{14}NH_3$ - (Continued)

IR laser line	Resonant field (kV/cm)	IR Stark component	Measured FIR Wavelength (μm)	FIR transition	FIR polarization	FIR power
9R(16)	49.63	ssR(6,2,6)	90.98	asR(6,2)	//	S
9R(10)	47.97	ssR(6,5,6)	94.53	asR(6,5)	//	M
9R(8)	19.07	asR(6,5,6)	94.37	asR(6,5)	//	S
9R(6)	22.07	ssR(6,6,6)	96.58	asR(6,6)	//	S
9R(4)	34.83	asR(6,6,6)	96.56	asR(6,6)	//	S
9R(2)	39.03	saR(4,4,4)	291.63	saQ(5,4)	⊥	W
			74.10	saR(4,4)	//	M
	50.84	saR(4,3,4)	305.55	saQ(5,3)	⊥	W
			75.10	saR(4,3)	//	M
9P(10)	46.42	ssR(5,1,5)	112.26	asR(5,1)	//	M
9P(12)	20.97	ssR(5,3,5)	114.28	asR(5,3)	//	S
			307.00	saQ(5,3)		W
	26.30	ssR(5,3,4)	223.68	asR(3,3)		W
9P(16)	27.13	ssR(5,5,5)	118.88	asR(5,5)	//	S
9P(18)	48.73	asR(5,5,5)	119.21	asR(5,5)	//	S
9P(20)	42.23	saR(3,3,3)	290.49	saQ(4,3)	//	M
			86.96	saR(3,3)	//	S

IR laser line	Resonant field (kV/cm)	IR Stark component	Measured FIR wavelength (μm)	FIR transition	FIR polarization	FIR power
9P(22)	47.60	aaR(3,3,3).	290.61	saQ(4,3)		M
			87.03	saR(3,3)	//	S
9P(34)	61.94	ssR(4,3,4)	151.23	asR(4,3)	//	S
9P(36)	26.33	ssR(4,4,4)	155.03	asR(4,4)	//	S
9P(38)	65.85	asR(4,4,4)	154.94	asR(4,4)	//	W
9P(40)	38.10	saR(2,2,2)	288.68	saQ(3,2)	⊥ or //	M
			105.20	saR(2,2)	//	S
10R(48)	20.50	asR(2,2,2)	404.08	asR(2,2)	//	S
10R(8)	34.50	saQ(3,3,3)	279.00	saQ(3,3)	⊥	S
	41.02	ssQ(3,3,3)*	388.85	asR(2,0)	//	M
	45.23	saQ(3,2,3)	289.45	saQ(3,2)	⊥	W
			105.28	saR(2,2)	//	S
	60.96	saQ(4,4,4)	276.80	saQ(4,4)	⊥	W
10R(6)	14.20	aaQ(4,2,2)	303.80	saQ(4,2)		W
	7.07	aaQ(4,2,4)	88.10	saR(3,2)	//	W
	14.60	saQ(5,5,5)	273.37	saQ(5,5)	⊥	W
	24.24	asQ(3,3,3)*	389.67	asR(2,0)	//	W
	31.43	aaQ(3,3,3)	279.60	saQ(3,3)	//	S
	49.73	aaQ(4,1,4)	88.70	saR(3,1)	//	M
	55.58	aaQ(3,2,3)	289.15	saQ(3,2)	⊥	W
			105.18	saR(2,2)	//	S

(continued)

Table 3 - $^{14}NH_3$ - (Continued)

IR laser line	Resonant field (kV/cm)	IR Stark component	Measured FIR wavelength (μm)	FIR transition	FIR polarization	FIR power
	57.70	saQ(6,6,6)	267.83	saQ(6,6)	⊥	W
	59.13	saQ(6,5,6)	64.66	saR(5,5)	//	M
	61.17	saQ(6,4,6)	65.40	saR(5,4)	//	W
	63.12	aaQ(2,2,2)	279.49	saQ(2,2)	⊥ or //	M
	66.37	saQ(6,3,6)	151.78	asR(4,3)	//	M
10R(4)	21.57	saQ(7,6,7)	287.71	saQ(7,6)	⊥	M
			57.08	saR(6,6)		S
	28.80	aaQ(6,4,6)	65.46	saR(5,4)		M
	60.23	saQ(8,3,8)	53.84	saR(7,3)		W
	65.40	aaQ(5,4,5)	74.07	saR(4,4)	//	M
10R(2)	25.17	aaQ(8,5,8)	52.68	saR(7,5)		W
	28.00	aaQ(7,7,7)	263.44	saQ(7,7)	⊥	W
	48.07	aaQ(7,6,7)	56.93	saR(6,6)		M
	60.53	aaQ(8,4,8)	53.34	saR(7,4)		W
10P(28)	45.93	ssQ(8,4,8)	77.29	asR(7,4)	//	W
	56.60	ssQ(7,3,7)	91.67	asR(6,3)	//	M
10P(30)	8.83	ssQ(7,4,7)	92.40	asR(6,4)	//	W
10P(32)	12.33	asQ(5,3,5)	151.17	asR(4,3)		S
			290.42	saQ(4,3)		M
			87.03	saR(3,3)		M
	24.60	ssQ(3,2,3)	403.84	asR(2,2)	//	M
	26.67	asQ(9,6,9)	67.73	asR(8,6)	//	W
	37.80	ssQ(7,5,6)	94.34	asR(6,5)	//	W

IR laser line	Resonant field (kV/cm)	IR Stark component	Measured FIR wavelength (μm)	FIR transition	FIR polarization	FIR power
	41.90	asQ(4,2,3)	218.00	asR(3,2)	//	S
			105.35	saR(2,2)	//	W
	49.07	ssQ(4,3,4)	223.64	asR(3,3)	//	S
10P(34)	18.83	asQ(8,6,8)	79.49	asR(7,6)	//	M
	24.33	asQ(5,4,5)	154.97	asR(4,4)	//	M
10P(36)	19.43	asQ(7,6,7)	96.47	asR(6,6)	//	M
R(37)	7.25	saQ(1,1,1)	280.00	saQ(1,1)	⊥ or //	M
	32.19	saQ(2,2,2)	279.79	saQ(2,2)	⊥	M
	51.99	saQ(3,3,3)	278.41	saQ(3,3)	⊥	W
	58.67	saQ(2,1,2)	133.16	saR(1,1)	//	W
R(36)	11.65	ssQ(3,3,3)*	389.30	asR(2,0)	//	W
	39.77	saQ(4,4,4)	276.67	saQ(4,4)	⊥	W
	45.33	saQ(4,3,4)	87.27	saR(3,3)	//	W
			290.33	saQ(4,3)	⊥	W
	58.47	saQ(4,2,4)	88.28	saR(3,2)	//	M
R(35)	15.71	aaQ(3,2,3)	105.39	saR(2,2)	//	W
	36.15	saQ(5,3,5)	75.28	saR(4,3)	//	W
	35.95	saQ(5,4,5)	74.24	saR(4,4)	//	W
	38.20	saQ(5,2,5)	76.06	saR(4,2)	//	W
	38.94	aaQ(2,2,2)	280.13	saQ(2,2)	⊥	W
	49.54	aaQ(3,1,3)	107.18	saR(2,1)	//	W

(continued)

Table 3 - $^{14}NH_3$- (Continued)

IR laser line	Resonant field (kV/cm)	IR Stark component	Measured FIR wavelength (μm)	FIR transition	FIR polarization	FIR power
R(34)	17.00	aaQ(4,4,4)	277.00	saQ(4,4)	⊥	W
	33.13	saQ(6,4,6)	65.61	saR(5,4)	//	W
	40.07	saQ(6,6,6)	268.76	saQ(6,6)	⊥	W
	45.00	saQ(6,5,5)	64.22	saR(5,5)	//	W
	46.67	saQ(4,3,3)	289.69	saQ(4,3)	//	W
			87.27	saR(3,3)	//	W
	53.90	aaQ(3,3,3)	278.78	saQ(3,3)	⊥ or //	W
R(33)	17.17	saQ(7,3,7)	59.48	saR(6,3)		M
	41.27	saQ(7,5,7)	57.99	saR(6,5)		W
	47.00 ⊥	aaQ(5,5,2)	273.71	saQ(5,5)	⊥	W
R(32)	8.47	saQ(7,7,7)	263.60	saQ(7,7)	⊥	W
	10.80	aaQ(6,6,6)	269.66	saQ(6,6)	⊥	W
	24.54	aaQ(7,3,7)	59.25	saR(6,3)		W
	31.10	aaQ(6,5,6)	64.49	saR(5,5)		W
R(31)	15.17	aaQ(7,6,7)	289.10	saQ(7,6)	⊥	W
			57.11	saR(6,6)		W
	40.17	aaQ(7,5,7)	57.99	saR(6,5)		W
	54.32	aaQ(6,6,6)	268.91	saQ(6,6)	⊥	W
R(30)	11.43	saQ(9,6,9)	47.94	saR(8,6)		W
	16.00	aaQ(8,6,8)	51.99	saR(7,6)		W
R(29)	15.80	saQ(9,9,9)	250.26	saQ(9,9)	⊥	W
	20.27	aaQ(8,8,8)	257.00	saQ(8,8)	⊥	W

IR laser line	Resonant field (kV/cm)	IR Stark component	Measured FIR wavelength (μm)	FIR transition	FIR polarization	FIR power
R(27)	52.36 ⊥	aaQ(9,9,6)	250.68	saQ(9,9)		W
R(23)	21.33 ⊥	saQ(12,12,?)	224.00	saQ(12,12)		W
P(4)	37.19	asQ(6,2,6)	113.11	asR(5,2)	//	W
P(5)	57.63	ssQ(4,1,4)	215.77	asR(3,1)	//	W
P(6)	10.13	asQ(8,5,8)	78.37	asR(7,5)	//	W
	22.33	ssQ(5,3,5)	151.26	asR(4,3)	//	M
	37.97	asQ(4,1,4)	215.50	asR(3,1)	//	M
	52.68	ssQ(6,4,6)	116.59	asR(5,4)	//	W
P(7)	15.90	asQ(5,3,5)	151.82	asR(4,3)	//	S
			87.21	saR(3,3)		
	34.17	asQ(4,2,4)	218.55	asR(3,2)	//	M
	47.50	ssQ(4,3,4)	224.10	asR(3,3)	//	M
P(8)	7.67	asQ(3,2,3)	403.83	asR(2,2)	//	M
	22.57	ssQ(5,4,5)	155.47	asR(4,4)	//	M
	31.33	ssQ(8,6,8)	79.73	asR(7,6)	//	M
	57.87	asQ(6,4,6)	116.45	asR(5,4)	//	W
P(9)	21.77	ssQ(6,5,6)	119.13	asR(5,5)	//	M
	56.22	asQ(4,3,4)	223.51	asR(3,3)	//	M

(continued)

Table 3 - $^{14}NH_3$ - (Continued)

IR laser line	Resonant field (kV/cm)	IR Stark component	Measured FIR wavelength (μm)	FIR transition	FIR polarization	FIR power
P(10)	39.40	ssQ(7,6,7)	96.87	asR(6,6)	//	M
	62.61	asQ(8,6,8)	79.72	asR(7,6)	//	W
	67.50	asQ(5,4,5)	155.60	asR(4,4)	//	W
P(12)	15.63	ssQ(8,7,8)	81.56	asR(7,7)	//	W
	44.93	asQ(7,6,7)	96.76	asR(6,6)	//	W
P(14)	7.33	asQ(11,9,?)	54.45	asR(10,9)		W

Table 4 - Emitting properties of FIR lines in $^{15}NH_3$ (increasing fields classification for each pump laser line)

IR laser line	Resonant field (kV/cm)	IR Stark component	Measured FIR wavelength (μm)	FIR transition	FIR polarization	FIR power
9R(38)	63.33	asR(7,4,7)	76.91	asR(7,4)	//	W
9R(30)	45.30	asR(7,6,7)	79.09	asR(7,6)	//	W
9R(22)	57.10	saR(5,4,5)	66.15	saR(5,4)	//	W
9R(20)	23.00	aaR(5,2,5)	67.60	saR(5,2)	//	W
9R(2)	37.50	asR(6,5,6)	93.80	asR(6,5)	//	W
9P(4)	34.87	saR(4,4,4)	75.08	saR(4,4)	//	W
			302.64	saQ(5,4)	⊥ or //	W
	42.27	saR(4,3,4)	76.06	saR(4,3)	//	S
			318.69	saQ(5,3)	⊥ or //	W
	57.00	saR(4,2,4)	76.72	saR(4,2)	//	M
9P(6)	37.87	aaR(4,4,4)	301.50	saQ(5,4)	⊥	W
	56.27	aaR(4,3,4)	317.04	saQ(5,3)	⊥	W
			75.72	saR(4,3)	//	M
9P(18)	47.20	asR(5,3,5)	112.98	asR(5,3)	//	S
	52.77	ssR(5,4,5)	115.11	asR(5,4)	//	S
9P(20)	48.37	asR(5,4,5)	115.08	asR(5,4)	//	S
9P(22)	20.13	asR(5,5,5)	117.70	asR(5,5)	//	S

(continued)

Table 4 - $^{15}NH_3$ - continued

IR laser line	Resonant field (kV/cm)	IR Stark component	Measured FIR wavelength (μm)	FIR transition	FIR polarization	FIR power
9P(26)	28.40	saR(3,3,3)	301.93	saQ(4,3)	⊥ or //	M
			88.26	saR(3,3)	//	S
	37.73	saR(3,2,3)	89.37	saR(3,2)	//	S
9P(28)	66.40	aaR(3,3,3)	88.07	saR(3,3)	//	S
			302.77	saQ(4,3)	⊥	W
9P(38)	68.77	ssR(4,3,4)	149.43	asR(4,3)	//	S
9P(40)	29.73	ssR(4,4,4)	152.90	asR(4,4)	⊥	S
	62.13	asR(4,3,4)	149.06	asR(4,3)	//	S
10R(40)	6.23	asR(2,2,2)	386.25	asR(2,2)	//	S
10R(8)	30.30	asR(1,1,1)	292.69	saQ(1,1)	//	M
10R(2)	63.13	saQ(1,1,1)	291.88	saQ(1,1)	//	M
10R(3)	25.67	aaQ(2,2,2)	290.33	saQ(2,2)	⊥ or //	M
	45.00	saQ(5,3,5)	75.85	saR(4,3)	//	W
	45.73	saQ(5,5,5)	283.33	saQ(5,5)	⊥	W
	48.93	saQ(5,2,5)	76.67	saR(4,2)	//	W
10P(2)	18.03	saQ(8,4,8)	53.96	saR(7,4)		W
	19.40	saQ(7,7,7)	273.03	saQ(7,7)	⊥	W
	41.33	aaQ(6,5,4)	302.13	saQ(6,5)	//	W

IR laser line	Resonant field (kV/cm)	IR Stark component	Measured FIR wavelength (μm)	FIR transition	FIR polarization	FIR power
	41.90	aaQ(7,3,7)	59.88	saR(6,3)		W
	54.70	saQ(8,6,8)	52.51	saR(7,6)		W
10P(4)	32.27	aaQ(8,6,8)	52.52	saR(7,6)		W
10P(36)	23.13	asQ(5,3,5)	149.22	asR(4,3)	//	S
			301.93	saQ(4,3)	⊥ or //	M
	38.10	asQ(4,2,4)	213.59	asR(3,2)	//	W
	47.57	ssQ(4,3,4)	219.00	asR(3,3)	//	S
10P(38)	32.63	asQ(5,4,5)	153.01	asR(4,4)	//	M
10P(40)	33.30	asQ(7,6,7)	95.71	asR(6,6)	//	W
R(30)	42.33 ⊥	saQ(1,1,1)	291.87	saQ(1,1)	//	M
	48.13 ⊥	saQ(2,2,2)	291.21	saQ(2,2)	//	S
	64.07	saQ(3,3,3)	289.23	saQ(3,3)	⊥ or //	M
R(29)	13.93 ⊥	saQ(3,1,3)	107.59	saR(2,1)	//	M
	19.13 ⊥	saQ(3,2,3)	107.30	saR(2,2)	//	S
	19.03	saQ(3,2,3)	301.03	saQ(3,2)	⊥	W
	20.90 ⊥	saQ(3,3,3)	290.60	saQ(3,3)	//	S
	50.00	saQ(4,4,4)	285.33	saQ(4,4)	⊥ or //	W
	56.17	saQ(4,3,4)	88.18	saR(3,3)	//	S
			298.90	saQ(4,3)	//	M
R(28)	27.33	aaQ(2,2,2)	290.83	saQ(2,2)	//	M

(continued)

Table 4 - $^{15}NH_3$ - (Continued)

IR laser line	Resonant field (kV/cm)	IR Stark component	Measured FIR wavelength (μm)	FIR transition	FIR polarization	FIR power
	43.13	saQ(5,3,5)	76.22	saR(4,3)	//	M
	44.67	saQ(5,5,5)	282.50	saQ(5,5)	⊥	W
	45.83	saQ(5,2,5)	77.03	saR(4,2)	//	W
R(27)	3.87	aaQ(4,4,4)	288.00	saQ(4,4)	⊥	M
	23.83	saQ(6,3,6)	66.97	saR(5,3)	//	W
	28.17	aaQ(4,3,4)	302.06	saQ(4,3)	⊥	M
			88.38	saR(3,3)	//	S
	45.93 ⊥	aaQ(3,3,3)	290.30	saQ(3,3)	⊥ or //	S
	58.90	aaQ(4,2,4)	89.43	saR(3,2)	//	M
R(26)	11.00	aaQ(6,3,6)	67.00	saR(5,3)	//	W
	12.10	aaQ(5,5,5)	283.21	saQ(5,5)	⊥	W
	24.93	saQ(7,4,7)	59.25	saR(6,4)	//	M
	39.53	saQ(7,5,7)	58.67	saR(6,5)	//	W
	49.07	saQ(7,6,7)	57.69	saR(6,6)		W
	54.53	aaQ(6,2,6)	67.33	saR(5,2)	//	W
	55.33	aaQ(4,4,4)	286.45	saQ(4,4)	//	W
	56.80 ⊥	saQ(7,6,6)	301.03	saQ(7,6)	//	W
	63.33	aaQ(5,3,5)	75.83	saR(4,3)	//	W
R(25)	9.50	saQ(8,4,8)	53.81	saR(7,4)		W
	28.90	saQ(7,7,4)	272.73	saQ(7,7)	//	W
	46.27	aaQ(6,5,4)	301.38	saQ(6,5)	⊥ or //	W
	47.93	aaQ(7,3,7)	59.85	saR(6,3)	//	W

IR laser line	Resonant field (kV/cm)	IR Stark component	Measured FIR wavelength (μm)	FIR transition	FIR polarization	FIR power
R(24)	18.37	aaQ(7,6,7)	57.54	saR(6,6)		W
			299.25	saQ(7,6)	⊥	W
R(23)	28.37	aaQ(8,6,8)	52.47	saR(7,6)		W
R(22)	17.33	saQ(9,9,9)	259.00	saQ(9,9)	⊥	W
	37.77	aaQ(9,6,9)	48.25	saR(8,6)		W
P(8)	10.57	asQ(7,3,7)	91.18	asR(6,3)	//	M
P(10)	17.33	asQ(7,4,7)	92.17	asR(6,4)	//	W
	22.07	asQ(5,2,4)	147.03	asR(4,2)	//	M
	36.20	asQ(6,3,6)	113.27	asR(5,3)	//	M
P(11)	10.53	ssQ(6,4,6)	115.07	asR(5,4)	//	W
P(12)	7.90	ssQ(4,3,4)	218.88	asR(3,3)	//	M
	26.87	asQ(6,4,6)	115.01	asR(5,4)	//	M
	61.77	asQ(5,3,5)	148.80	asR(4,3)	//	S
P(13)	22.80	asQ(4,3,4)	218.93	asR(3,3)	⊥ or //	S
P(14)	42.67	asQ(5,4,5)	152.69	asR(4,4)	//	W
P(16)	25.47	asQ(7,6,7)	95.80	asR(6,6)	//	W

Table 5 - List of CW resonantly pumped FIR laser lines from NH_3

Molecule	λ Laser (μm)	Transition in absorption	Pump laser line	Pump transition	Pol.	Optimum pressure	Power pump/ laser	Offset (MHz)	Ref.
$^{14}NH_3$	81.5	aR(7,7)	N_2O P(13)	aQ(8,7)	⊥	70mTorr	3W/40mW	4	1,2,3
$^{14}NH_3$	263.3	sQ(7,7)	N_2O P(13)	aQ(8,7)	//	50mTorr	3W/0.5mW	4	1,2c
$^{14}NH_3$	42.6	aR(13,13)	N_2O P(29)	aQ(14,13)	⊥	100mTorr	3W/1mW	10	4
$^{15}NH_3$	218.6	aR(3,3)	$^{12}C^{16}O_2$ Seq.10P(35)	aQ(4,3)	⊥	50mTorr	7W/23mW	17	3,5
$^{15}NH_3$	111.9	aR(5,2)	$^{12}C^{16}O_2$ Seq.10P(31)	aQ(6,2)	⊥	50mTorr	8W/11mW	27	3,5
$^{14}NH_3$	290.9	sQ(4,3)	$^{12}C^{16}O_2$ Seq. 9P(17)	sR(3,3)	⊥	100mTorr	3.5W/0.05mW	150	6
$^{14}NH_3$	87.1	sR(3,3)	$^{12}C^{16}O_2$ Seq. 9P(17)	sR(3,3)	//	100mTorr	3.5W/1mW	150	6
$^{15}NH_3$	373.4	aR(2,0)	$^{12}C^{16}O_2$ 10R(42)	aR(2,0)	//	50mTorr	16W/23mW	26	3,5
$^{14}NH_3$	112.3	aR(5,1)	$^{12}C^{16}O_2$ Seq. 9P(7)	aR(5,1)	//	50mTorr		90	3
$^{15}NH_3$	93.6	aR(6,5)	$^{12}C^{16}O_2$ Seq. 9R(7)	aR(6,5)	//	50mTorr		122	3
$^{15}NH_3$	524.9	sQ(13,7)	$^{12}C^{16}O_2$ Seq.10P(5)	sQ(13,7)	//	50mTorr		6	3

Molecule	λ Laser (μm)	Transition in absorption	Pump laser line	Pump transition	Pol.	Optimum pressure	Power pump/ laser	Offset (MHz)	Ref.
$^{15}NH_3$	273.4	sQ(12,11)	$^{12}C^{16}O_2$ Seq.10P(7)	sQ(12,11)	//	50mTorr		3	3
$^{15}NH_3$	152.7	aR(4,4)	$^{13}C^{16}O_2$ 10R(18)	aQ(5,4)	⊥	80mTorr	18W/180mW	31	7,8
$^{15}NH_3$	287.7	sQ(4,4)	$^{13}C^{16}O_2$ 10R(18)	aQ(5,4)	//	60mTorr		31	8c
$^{14}NH_3$	147.3	aR(4,0)	$^{12}C^{16}O_2$ 9R(30)	sR(5,0)	//	100mTorr		175	9d
$^{14}ND_3$	87	pR(10,1)	$^{12}C^{16}O_2$ 9R(40)	pR(12,1)	//				10
$^{14}NH_2D$	77		$^{12}C^{16}O_2$ 10R(30)		//				10e
$^{14}NH_2D$	108		$^{12}C^{16}O_2$ 10R(26)		⊥				10e
$^{14}NH_2D$	77		$^{12}C^{16}O_2$ 10R(14)		//				10e
$^{14}NH_2D$	124		$^{12}C^{16}O_2$ 10R(14)		//				10e
$^{14}NH_2D$	86		$^{12}C^{16}O_2$ 10R(40)		//				10e
$^{14}NH_2D$	113		$^{12}C^{16}O_2$ 10R(40)		//				10e

(continued)

Ref. 1 Chang et al., 1970.

Ref. 2 Tanaka et al., 1977.

Ref. 3 Frank et al., 1982.

Ref. 4 Willenberg, 1981.

Ref. 5 Danielewicz and Weiss, 1978-a.

Ref. 6 Danielewicz and Weiss, 1978-b.

Ref. 7 Woods et al., 1980.

Ref. 8 McCombie et al., 1981.

Ref. 9 Redon et al., 1979-a.

Ref. 10 Landsberg, 1980.

c : cascade from line directly above

d : cascade from 67μm Raman line, table 6

e : tentatively ascribed to $^{14}NH_2D$

Table 6 - List of CW Raman emission lines from NH_3

Molecule	$\lambda(\mu m)$	Nearest laser trans. in absorption	Pump laser line	Nearest pump trans. in absorption	Pol.	Opt. pressure	Pump/ Laser power	Offset (MHz)	Ref.
$^{14}NH_3$	67.2	sR(5,0)	$^{12}C^{16}O_2$ 9R(30)	sR(5,0)	//	500mTorr	10W/10mW	175	1,2,3
$^{14}NH_3$	370.4	sQ(9,5)	$^{12}C^{16}O_2$ 10R(2)	sQ(9,5)	//	500mTorr		178	2
$^{14}NH_3$	234.4	sQ(11,11)	$^{12}C^{16}O_2$ 10P(2)	sQ(11,11)	//	500mTorr		451	4
$^{14}NH_3$	291.0	sQ(5,4)	$^{12}C^{16}O_2$ 10R(6)	sQ(5,4)	//	500mTorr		560	4

Ref. 1 Redon et al., 1979-a.

Ref. 2 Willenberg et al., 1980-b.

Ref. 3 Landsberg, 1980.

Ref. 4 Willenberg, 1981.

Table 7 - List of CW two-photon pumped FIR laser lines from NH_3

Molecule	λ laser (μm)	Transition in absorption	Pump laser line	μ-wave pump frequency (GHz)	2-photon pump-transition	Offset (MHz)	Ref.
$^{14}NH_3$	87	sR(3,3)	$^{12}C^{16}O_2$ Seq. 9P(17)	24.020	aaR(3,3)	130	1
$^{14}NH_3$	291	sQ(4,3)	$^{12}C^{16}O_2$ Seq. 9P(17)	24.020	aaR(3,3)	130	1
$^{14}NH_3$	88	sR(3,2)	$^{12}C^{16}O_2$ Seq. 9P(17)	22.585	aaR(3,2)	260	1
$^{14}NH_3$	301	sQ(4,2)	$^{12}C^{16}O_2$ Seq. 9P(17)	22.585	aaR(3,2)	260	1
$^{14}NH_3$	279	sQ(3,3)	$^{12}C^{16}O_2$ Seq. 10R(11)	24.485	aaQ(3,3)	610	1
$^{14}NH_3$	291	sQ(5,4)	$^{12}C^{16}O_2$ Seq. 10R(6)	22.090	aaQ(5,4)	560	1
$^{14}NH_3$	281	sQ(2,2)	$^{12}C^{16}O_2$ 10R(8)	22.795	aaQ(2,2)	940	1
$^{14}NH_3$	234	sQ(11,11)	$^{12}C^{16}O_2$ 10R(2)	29.463	aaQ(11,11)	450	2
$^{14}NH_3$	225	sQ(12,12)	$^{12}C^{16}O_2$ 10P(4)	30.282	aaQ(12,12)	1150	2
$^{14}NH_3$	312	sQ(7,5)	$^{12}C^{16}O_2$ 10R(4)	20.173	aaQ(7,5)	630	2
$^{14}NH_3$	273	sQ(5,5)	$^{12}C^{16}O_2$ 10R(6)	27.515	aaQ(5,5)	2980	2

Molecule	λ laser (μm)	Transition in absorption	Pump laser line	μ-wave pump frequency (GHz)	2-photon pump-transition	Offset (MHz)	Ref.
$^{14}NH_3$	288	sQ(7,6)	$^{12}C^{16}O_2$ Seq.10R(7)	20.672	aaQ(7,6)	2250	2
$^{14}NH_3$	370	sQ(9,5)	$^{12}C^{16}O_2$ 10R(2)	16.620	aaQ(9,5)	180	2

Ref. 1 Willenberg et al., 1980-a.

Ref.2 Willenberg, 1981.

Table 8 - List of pulsed NH_3FIR laser lines optically pumped off resonance

Molecule	λ Laser (μm)	Transition in absorption	Pump laser line	Pump trans. in absorption	Offset (MHz)	Ref.
$^{14}NH_3$	280.5	sQ(2,2)	$^{12}C^{16}O_2$ 10R(8)	sQ(2,2)	940	1
$^{14}NH_3$	291.2	sQ(5,4)	$^{12}C^{16}O_2$ 10R(6)	sQ(5,4)	560	1
$^{14}NH_3$	72.6	sR(4,4)	$^{12}C^{16}O_2$ 10R(6)	sQ(5,4)	560	1
$^{14}NH_3$	58.0	sR(6,5)	$^{12}C^{16}O_2$ 10R(4)	sQ(7,5)	630	1
$^{14}NH_3$	151.9	aR(4,3)	$^{12}C^{16}O_2$ 10P(32)	aQ(5,3)	951	1

Ref. 1 Fetterman et al., 1972.

Table 9 - List of pulsed single photon pumped FIR laser lines from NH_3

Molecule	λ Laser (μm)	Transition in absorption	Pump laser line	Pump transition in absorption	Offset (MHz)	Pol.	Ref.
$^{14}NH_3$	151.8	aR(4,3)	$^{12}C^{16}O_2$ 10P(32)	aQ(5,3)	951	⊥	1,2
$^{14}NH_3$	291.3	sQ(4,3)	$^{12}C^{16}O_2$ 10P(32)	aQ(5,3)	951	//	1,2c
$^{14}NH_3$	84.7	aR(5,3)	$^{12}C^{16}O_2$ 10P(32)	aQ(5,3)	951		2 G
$^{14}NH_3$	216.4	sQ(13,13)	$^{12}C^{16}O_2$ 10P(6)	sQ(13,13)	1028	//	1
$^{14}NH_3$	225.3	sQ(12,12)	$^{2}C^{16}O_2$ 10P(4)	sQ(12,12)	1150	//	1
$^{14}NH_3$	311.7	sQ(7,5)	$^{2}C^{16}O_2$ 10R(4)	sQ(7,5)	630		1
$^{14}NH_3$	58.0	sR(6,5)	$^{2}C^{16}O_2$ 10R(4)	sQ(7,5)	630		1
$^{14}NH_3$	119.0	aR(5,5)	$^{2}C^{16}O_2$ 10R(4)	sQ(7,5)	630		1c
$^{14}NH_3$	74.2	sR(4,4)	$^{2}C^{16}O_2$ 10R(6)	sQ(5,4)	560	⊥	1

(continued)

Table 9 - (Contunued)

Molecule	λ Laser (μm)	Transition in absorption	Pump laser line	Pump transition in absorption	Offset (MHz)	Pol.	Ref.
$^{14}NH_3$	291.9	sQ(5,4)	$^{12}C^{16}O_2$ 10R(6)	sQ(5,4)	560	//	1
$^{14}NH_3$	155.1	aR(4,4)	$^{12}C^{16}O_2$ 10R(6)	sQ(5,4)	560	⊥	1c
$^{14}NH_3$	281.4	sQ(2,2)	$^{12}C^{16}O_2$ 10R(8)	sQ(2,2)	940	//	1
$^{14}NH_3$	281.3	sQ(1,1)	$^{12}C^{16}O_2$ 10R(14)	aR(1,1)	1446	⊥	1cc
$^{14}NH_3$	256.7	aR(1,1)	$^{12}C^{16}O_2$ 10R(14)	aR(1,1)	1446		1G
$^{14}NH_3$	90.6	aR(6,1)	$^{12}C^{16}O_2$ 9R(16)	aR(6,1)	4916	//	1+
$^{14}NH_3$	67.3	sR(5,0)	$^{12}C^{16}O_2$ 9R(30)	sR(5,0)	175		1
$^{14}NH_3$	147.1	aR(4,0)	$^{12}C^{16}O_2$ 9R(30)	sR(5,0)	175	//	1c
$^{14}NH_3$	88.9	sR(3,0)	$^{12}C^{16}O_2$ 9R(30)	sR(5,0)	175		1c

Molecule	λ Laser(μm)	Transition in absorption	Pump laser line	Pump transition in absorption	Offset (MHz)	Pol.	Ref.
$^{14}NH_3$	388.0	aR(2,0)	$^{12}C^{16}O_2$ 9R(30)	sR(5,0)	175		1c
$^{14}NH_3$	83.8	sR(5,0)	$^{12}C^{16}O_2$ 9R(30)	sR(5,0)	175		1G

Ref. 1 Gullberg et al., 1973.

Ref. 2 Pinson et al., 1981.

c : cascade from line directly above

cc : collisional cascade

G : ground state transition

\+ : it appears more likely that with 9R(16) pumping aR(6,0) would be excited (offset 1.220GHz) leading to aR(6,0) emission at 90.4 μm parallel polarization.

Table 9a - Pulsed FIR laser lines obtained by pumping $^{14}NH_3$ with vibrational laser lines from $^{14}NH_3$ from a $^{12}C^{16}O_2$ 9R(30) pumped $^{14}NH_3$ laser (Ref. : Yamabayashi et al., 1981)

FIR wavelength (μm)	Transition in absorption	Laser pumped, $G \to \nu_2$ vibrational trans.
59.42	sR(6,3)	sP(8,3)
67.24	sR(5,0)	sP(7,0)
76.32	sR(4,1)	sP(6,1)
76.52	aR(7,3)	aP(9,3)
79.63	aR(7,6)	aP(9,6)
87.07	sR(3,3)	sP(5,3)
88.83	sR(3,0)	sP(5,0)
90.38	aR(6,0)	aP(8,0)
90.90	aR(6,2)	aP(8,2)
91.68	aR(6,3)	aP(8,3)
92.87	aR(6,4)	aP(8,4)
94.44	aR(6,5)	aP(8,5)
96.63	aR(6,6)	aP(8,6)
106.2	sR(2,1)	sP(4,1)
112.3	aR(5,1)	aP(7,1)
113.0	aR(5,2)	aP(7,2)
114.3	aR(5,3)	aP(7,3)
116.3	aR(5,4)	aP(7,4)
119.0	aR(5,5)	aP(7,5)
133.1	sR(1,0)	sP(3,0)
147.3	aR(4,0)	aP(6,0)
147.8	aR(4,1)	aP(6,1)
149.1	aR(4,2)	aP(6,2)
151.6	aR(4,3)	aP(6,3)
155.3	aR(4,4)	aP(6,4)
215.2	aR(3,1)	aP(5,1)
218.5	aR(3,2)	aP(5,2)
223.8	aR(3,3)	aP(5,3)
296.0	sQ(3,1)	sP(4,1)
343.3	sQ(7,3)	sP(8,3)
389.1	aR(2,0)	aP(4,0)
393.5	aR(2,1)	aP(4,1)
404.6	aR(2,2)	aP(4,2)

Table 10 - List of pulsed FIR laser lines generated by hot band ($\nu_2 \rightarrow 2\nu_2$) pumping of NH_3

Molecule	λlaser(μm)	Transition in absorption	Pump laser line	Pump transition in absorption	Offset (MHz)	Pol.	Ref.
$^{14}NH_3$	56.9	sP(5,2)	$^{12}C^{16}O_2$ 10P(20)	sQ(4,2)	280		1,2
$^{14}NH_3$	35.1	sQ(1,1)	$^{12}C^{16}O_2$ 10P(14)	sQ(1,1)	1008	//	1,2
$^{14}NH_3$	34.3	sQ(7,7)	$^{12}C^{16}O_2$ 10P(12)	sQ(7,7)	336	//	1,2
$^{14}NH_3$	64.7	sP(6,3)	$^{12}C^{16}O_2$ 9P(24)	sR(4,3)	294	//	1,2+
$^{14}NH_3$	36.2	sQ(5,3)	$^{12}C^{16}O_2$ 9P(24)	sR(4,3)	294	⊥	1,2+
$^{14}NH_3$	25.5	sR(5,3)	$^{12}C^{16}O_2$ 10P(26)	sQ(6,3)	1197		1
$^{14}NH_3$	25.9	sR(5,1)	$^{12}C^{16}O_2$ 9P(10)	sR(5,1)	2208		1
$^{14}NH_3$	26.4	sR(4,3)	$^{12}C^{16}O_2$ 9P(24)	sR(4,3)	294		1+
$^{14}NH_3$	27.8	sR(3,2)	$^{12}C^{16}O_2$ 10P(20)	sQ(4,2)	280		1,3
$^{14}NH_3$	49.0	aR(3,3)	$^{12}C^{16}O_2$ 9P(24)	sR(4,3)	294		1
$^{14}NH_3$	33		$^{12}C^{16}O_2$ 10R(4)				3

(continued)

Ref. 1 Gullberg et al., 1973.
Ref. 2 Lee et al., 1979.
Ref. 3 Tiee and Wittig, 1978.

+ the same lines were observed in ref. (Pinson et al., 1981) using two-step pumping with 10P(32) and 9P(24) into the $\nu_2 = 2$ state.

Table 11 - List of pulsed two-photon pumped ($G \rightarrow 2\nu_2$) FIR laser lines of NH_3

Molecule	λlaser (μm)	Transition in absorption	Pump laser lines	Pump transition in absorption	Inter offset (MHz)	Total offset (MHz)	Ref.
$^{14}NH_3$	26.1	sR(4,4)	$^{12}C^{16}O_2$ 10P(34)+10P(18)	aaQ(5,4)	5181	51	1,2
$^{14}NH_3$	26.7	sR(4,2)	$^{12}C^{16}O_2$ 9P(34)+10P(24)	aaR(4,2)	5181	138	2
$^{14}NH_3$	35.1	sQ(1,1)	$^{12}C^{16}O_2$ 10P(24)+10P(24)	aaQ(1,1)	267513	585	2
$^{14}NH_3$	35.5	sQ(5,4)	$^{12}C^{16}O_2$ 10P(34)+10P(18)	aaQ(5,4)	5181	51	1,2
$^{14}NH_3$	23.9	sR(6,5)	$^{12}C^{16}O_2$ 9R(8) +10P(24)	aaR(6,5)	3346	1538	3
$^{14}NH_3$	36.2	sQ(7,5)	$^{12}C^{16}O_2$ 9R(8) +10P(24)	aaR(6,5)	3346	1538	3
$^{14}NH_3$	87.61	sP(8,5)	$^{12}C^{16}O_2$ 9R(8) +10P(24)	aaR(6,5)	3346	1538	3

Ref. 1 Jacobs et al., 1976.
Ref. 2 Lee et al., 1979.
Ref. 3 Pinson et al., 1981.

References

Brewer, R.G. and Shoemaker, R.L., 1971 : "Photo Echo and Optical Nutation in Molecules", Phys. Rev. Lett. 27, 631-634.

Chang, T.Y., Bridges, T.J. and Burkhardt, E.G., 1970 : "CW Laser Action at 81.5 and 263.4 μm in Optically Pumped Ammonia Gas", Appl. Phys. Lett. 17, 357-358.

Danielewicz, E.J. and Weiss, C.O., 1978-a : "Far Infrared Laser Emission from $^{15}NH_3$ Optically Pumped by a CW Sequence Band CO_2 Laser", IEEE J. Quantum Electron. QE-14, 222-223.

Danielewicz, E.J. and Weiss, C.O., 1978-b : "New CW Far Infrared D_2O, $^{12}CH_3F$ and $^{14}NH_3$ Laser Lines", Opt. Commun. 27, 98-100.

Davis, B.W., Vass, A., Pidgeon, C.R. and Allan, G.R., 1981 : "New FIR Laser Lines from an Optically Pumped Far-Infrared Laser with Isotopic $^{13}C^{16}O_2$ Pumping", Opt. Comm. 37, 303-305.

Deka, B.K., Dyer, P.E. and Winfield, R.J., 1980 : "Optically Pumped NH_3 Laser Using a Continuously Tunable CO_2 Laser", Opt. Commun. 33, 206-208.

Fetterman, H.R., Schlossberg, H.R. and Waldman, J., 1972 : "Submillimeter Lasers Optically Pumped Off-Resonance", Opt. Commun. 6, 156-159.

Fetterman, H.R., Schlossberg, H.R. and Parker, C.D., 1973 : "CW Submillimeter Laser Generation in Optically Pumped Stark-Tuned NH_3", Appl. Phys. Lett. 23, 684-686.

Fetterman, H.R., Parker, C.D., Tannenwald, P.E., 1976 : "Enhancement of Optically Pumped Far Infrared Lasing by Stark Modulation", Opt. Commun. 18, 10-12.

Fetterman, H.R., 1978 : private communication.

Fourrier, M. and Redon, M., 1972 : "Infrared-Microwave Double Resonance of NH_3 in the Presence of a High Field Stark Effect", Appl. Phys. Lett. 21, 463-464.

Frank, E.M., Weiss, C.O., Siemsen, K., Grinda, M. and Willenberg, G.D., 1982 : "Predictions of Far-Infrared Laser Lines from $^{14}NH_3$ and $^{15}NH_3$", Opt. Lett. 7, 96-98.

Freund, S.M. and Oka, T., 1976 : "Infrared-Microwave Two-Photon Spectroscopy : The ν_2 Band of NH_3", Phys. Rev. A-13, 2178-2190.

Garing, J.S., Nielsen, H.H. and Rao, K.N., 1959 : "The Low-Frequency Vibration Rotation Bands of the Ammonia Molecule", J. Mol. Spectr. 3, 496-527.

Gastaud, C., Sentz, A., Redon, M. and Fourrier, M., 1980 : "New CW FIR Laser Action by Stark Tuning from Optically Pumped CH_3OH and CH_3OD", IEEE J. Quantum. Electron. QE-16, 1285-1287.

Gastaud, C., Sentz, A., Redon, M. and Fourrier, M., 1981-a : "Continuous-wave Stark Far Infrared Lasing Lines in $^{15}NH_3$ Optically Pumped by a CO_2 or N_2O Laser", Opt. Lett. 6, 449-451.

Gastaud, C., Sentz, A., Redon, M. and Fourrier, M., 1981-b : "Nouvelles Raies d'Emission en Infrarouge Lointain dans $^{15}NH_3$ Pompé par les Lasers à CO_2 et N_2O en Présence d'Effet Stark", Colloque O.H.D. Toulouse, 335-337.

Gullberg, K., Hartmann, B. and Kleman B., 1973 : "Submillimeter Emission from Optically Pumped $^{14}NH_3$", Physica Scripta 8, 177-182.

Heppner, J., Weiss, C.O., Hübner, U. and Schinu, G., 1980 : "Gain in CW Laser Pumped FIR Laser Gases", IEEE J. Quantum Electron. QE-16, 392-402.

Jacobs, R.R., Prosnitz, D., Bischel, W.K. and Rhodes, C.K., 1976 : "Laser Generation from 6 to 35 μm Following Two-Photon Excitation of Ammonia", Appl. Phys. Lett. 29, 710-712.

Jones, C.R., Buchwald, M.I., Gundersen, M. and Bushnell, A.H., 1978 : "Ammonia Laser Optically Pumped with an HF Laser", Opt. Commun. 24, 27-30.

Landsberg, B.M., 1980 : "New CW FIR Laser Lines from Optically Pumped Ammonia Analogues", Appl. Phys. 23, 127-130.

Lee, W., Kim, D., Malk, E. and Leap, J., 1979 : "Hot-Band Lasing in NH_3", IEEE J. Quantum Electron. QE-15, 838-839.

Leite, J.R.R., Ducloy, M., Sanchez, A., Seligson, D. and Feld, M.S., 1977 : "Laser Saturation Resonances in NH_3 observed in the Time-Delayed Mode", Phys. Rev. Lett. 39, 1469-1472.

Marx, R., Hübner, U., Abdul-Halim, I., Heppner, J., Ni, Y.C., Willenberg, G.D. and Weiss, C.O., 1981 : "Far-Infrared CW Raman and Laser Gain of $^{14}NH_3$", IEEE J. Quantum Electron. QE-17, 1123-1127.

McCombie, J., Peterson, J.C. and Duxbury, G., 1981, 5th National Quantum Electronics Conference of G.B., Hull University, paper 62.

Pinson, P., Delage, A., Girard, G. and Michon, M., 1981 : "Characteristics of Two-Step and Two-Photon Excited Emissions in $^{14}NH_3$", J. Appl. Phys. 52, 2634-2637.

Redon, M., Gastaud, C. and Fourrier, M., 1979-a : "New CW FIR Far-Infrared Lasing in $^{14}NH_3$ Using Stark Tuning", IEEE J. Quantum Electron. QE-15, 412-414.

Redon, M., Gastaud, C. and Fourrier, M., 1979-b : "Far-Infrared Emissions in NH_3 Using Forbidden Transitions Pumped by a CO_2 Laser", Opt. Commun. 30, 95-98.

Redon, M., 1980 : Thesis, Université Pierre et Marie Curie, Paris.

Redon, M., Gastaud, C. and Fourrier, M., 1980-a : "Far-Infrared Emissions in Ammonia by Infrared Pumping Using a N_2O Laser", Inf. Phys. 20, 93-98.

Redon, M., Gastaud, C. and Fourrier, M., 1980-b : "New CW FIR Laser Lines Obtained in Ammonia Pumped by a CO_2 Laser, Using the Stark Tuning Method", Int. J. Infrared and Millimeter Waves 1, 95-109.

Redon, M., Gastaud, C. and Fourrier, M., 1980-c : "New CW FIR Laser Action in Stark Tuned Ammonia and Stark FIR Lasing Spectroscopy in the ν_2 Band", Opt. Commun. 34, 455-459.

Sasada, H., 1980:"Microwave Inversion Spectrum of $^{15}NH_3$", J. Mol. Spectr. 83, 15-20.

Shimizu, T. and Oka, T., 1970 : "Infrared-Microwave Double Resonance of NH_3 Using an N_2O Laser", Phys. Rev. A2, 1177-1181.

Shimizu, F., 1970-a : "Stark Spectroscopy of NH_3 ν_2 Band by 10 μm CO_2 and N_2O Lasers", J. Chem. Phys. 52, 3572-3576.

Shimizu, F., 1970-b : "Stark-Spectroscopy of $^{15}NH_3$ ν_2 Band by 10 μm Lasers", J. Chem. Phys. 53, 1149-1151.

Shimoda, K., Ueda, Y. and Iwakori, J., 1980 : "Infrared Laser Stark Spectroscopy of Ammonia", Appl. Phys. 21, 181-189 ; 22, 439.

Tanaka, A., Tanimoto, A., Murata, N., Yamanaka, M. and Yoshinaga, H., 1977 : "CW Efficient Optically-Pumped Far-Infrared Waveguide NH_3 Lasers", Opt. Commun. 22, 17-21.

Tiee, J.J. and Wittig, C., 1978 : "Optically Pumped Molecular Lasers in the 11-17 μm Region", J. Appl. Phys. 49, 61-64.

Townes, C.H. and Schawlow, A.L., 1955 : "Microwave Spectroscopy", McGraw-Hill, New-York.

Wiggins, J.D., Drozdowicz, Z. and Temkin, R.J., 1978 : "Two-Photon Transitions in Optically Pumped Submillimeter Lasers", IEEE J. Quantum Electron. QE-14, 23-30.

Willenberg, G.D., Weiss, C.O. and Jones, H., 1980-a : "Two-Photon Pumped CW Laser", Appl. Phys. Lett. 37, 133-135.

Willenberg, G.D., Hübner, U. and Heppner, J., 1980-b : "Far-Infrared CW Raman Lasing in NH_3", Opt. Commun., 33, 193-196.

Willenberg, G.D., 1981 : "Continuous-wave Far-Infrared Two-Photon-Pumped, Single-Photon-Pumped, and Raman Emission from $^{14}NH_3$", Opt. Lett. 6, 372-373.

Woods, R.A., Davis, B.W , Vass, A. and Pidgeon, C.R , 1980 : "Application of an Isotopically Enriched $^{13}C^{16}O_2$ Laser to an Optically Pumped Far-Infrared Laser", Opt. Lett. 5, 153-154.

Yamabayashi, N., Fukai, K., Miyazaki, K. and Fujisawa, K., 1981 : "Resonant Pumping Far-Infrared NH_3 Laser", Appl. Phys. B26, 33-36.

Znotins, T.A., Reid, J., Garside, B.K. and Ballik, E.A., 1980 : "12 μm NH_3 Laser Pumped by a Sequence CO_2 Laser", Opt. Lett. 5, 528-530.

LIST OF THE OPTICALLY PUMPED LASER LINES OF D_2O

T.A. Detemple

Department of Electrical Engineering
University of Illinois
Urbana, Illinois 61801

Introduction

Deuterated water, D_2O, is a light, asymmetric top molecule which has the singular distinction of being the most powerful emitter in the far-infrared spectral region when pumped by high pressure, pulsed CO_2 (TEA) lasers (Evans et al., 1976 and 1977). Since it is the lightest molecule to be optically pumped, the origin of its emission strength can be traced directly to the large individual ground state populations caused by the small rotational partition function (DeTemple and Lawton, 1978). The strength in emission is further documented by the fact that the first identifiable pure rotational stimulated Raman emission was seen in this molecule on a number of different transitions (Petuchowski et al., 1977; Temkin, 1977; Wiggins et al., 1978). Because of the strength of emission of this molecule in the FIR and because of the possibility of laser-isotope separation for use as a nuclear fuel, the infrared spectroscopy is of more than passing interest and will be reviewed here.

Emission Lines

The use of infrared lasers as the excitation source for FIR lasers was first demonstrated in 1970 in CH_3F and later extended to other methyl halide compounds (Chang and Bridges, 1970 ; Chang and McGee, 1971). The original experiments were first performed using conventional CO_2 lasers emitting relatively low power (Chang, 1970) and later extended to the case of TEA laser pumping at much higher powers (DeTemple, et al., 1973). The initial experiments using the latter source were performed in CH_3F and were very spectacular in that the conversion appeared to be near the Manley-Rowe limit and the gain was so high that the system operated in a mirrorless mode (DeTemple, et al., 1973; Brown, et al., 1973).

Paralleling this period of development were investigations on tunable infrared spectroscopy by either frequency shifting the source by some modulation means or frequency shifting the molecule by the Stark effect. Initial studies on acousto-optic modulated CO_2 lasers revealed the presence of nearby coincidences in D_2O vapor which was suggestive of its possible use as a laser molecule (Keilmann, 1974). Subsequent experiments demonstrated strong emission associated with a number of pump lines some of which presented a spectroscopic puzzle in terms of assignments (Plant et al., 1974; Keilmann, et al., 1975). As it emerged, the origin of the puzzle lay in a few assignment errors in, at-the-time, the most recent spectroscopic data (Williamson, 1969). During this initial discovery period of the FIR lasing properties in D_2O, separate studies on the microwave ground state (Steenbeckeliers and Bellet, 1973) and ν_2-band infrared absorption were both used to provide revised assignments (Lin and Shaw, 1977) which ultimately permitted the assignment of all FIR lines (Petuchowski, et al., 1977). At a later date, tunable diode laser studies on selected transitions (Worchesky et al., 1978) were used to reconfirm the identification of near resonant stimulated Raman emission on pure rotational transitions. This effect, which has been seen on a number of other transitions in D_2O and other molecules, has been verified in tunable pump (Nicholson, 1979) and tunable gain studies (Drozdowicz et al., 1978).

The known pump and emission lines are tabulated in Table I along with the known pump detuning (offset = $\nu_{laser} - \nu_{molecule}$), the transition assignments, the emitted wavelength, rounded to the nearest micron, the emitted wavenumber based on either measurements (M) or calculated from the spectra (C), the emission polarization and the dominant emission mode (L=laser, R=Raman, C= cascade). The asymmetric rotor designation for the states is J_{K^-,K^+} where J is the rotational quantum number and $K^{-(+)}$ is the K-quantum number in the prolate (oblate) symmetric top limit (Schalow and Townes, 1955). An alternate representation is J_τ where $\tau = K^- - K^+$. All infrared transitions in Table I are associated with the ν_2 or rocking mode which has its transition dipole moment (μ_{ν_2} = .12D) along the symmetry axis as is the permanent dipole moment. Both infrared and FIR transitions thus belong to 'type-b' subbranches with selection rules of $\Delta J=0,\pm 1$ and $\Delta\tau=0,\pm 2$ for the strongest transition. Other transitions are observable and the line strengths are tabulated elsewhere (Schawlow and Townes, 1955).

The strongest emission lines are at 66 μm and 385 μm, the latter of which is being developed for Thompson scattering diagnostics (Woskoboinikow et al., 1981). Other strong lines at 111 μm and 113 μm are also of interest (Dodel and Douglas, 1982). In Fig. 1 are shown linkage diagrams for two pumping transitions (Petuchowski et al., 1977; Dodel and Douglas, 1982; Brown et al., 1979). As seen in this figure, the emission lines are not part of a classical optically pumped three-level system but are much more complicated in terms of the photon kinetics and are probably influenced by some collisional kinetics. With present spectroscopic data, the origin of the emission lines can be identified but the detailed reasons for the various line characteristics has not been explored in great detail. Further laser system characteristics are outlined elsewhere (DeTemple, 1979).

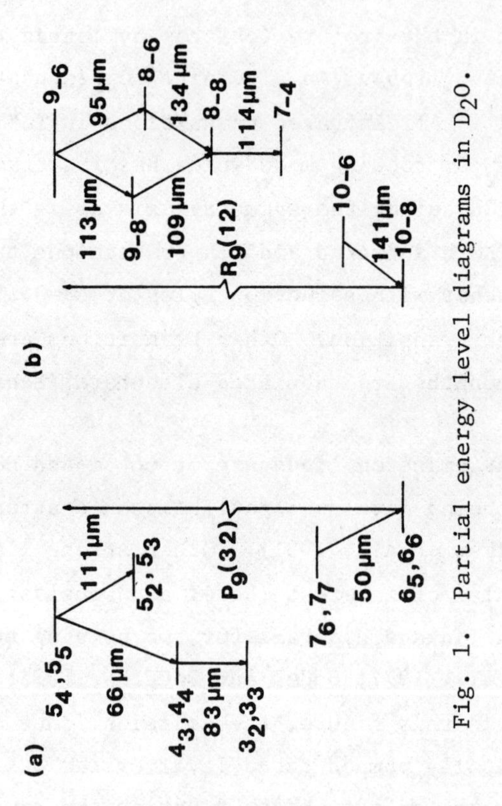

Fig. 1. Partial energy level diagrams in D_2O.

000

The ground state was first investigated in the FIR in the late 1930's (Fuson et al., 1939). The low-J microwave transitions were subsequently investigated much later (Benedict et al., 1970) and the pioneering work of Steenbeckelier led to a very precise determination of ground state transitions (36 in number) up to 600 GHz (Steenbeckeliers and Bellet, 1973). The accuracy of this work was further verified by high resolution (100 MHz) FIR Fourier transform spectrometer (FTS) measurements up to 40 cm^{-1} (1200 GHz) (Fleming and Gibson, 1976). Because of the lightness of this molecule and the implied centrifugal distortions, a 22 constant fit to Watson's Hamiltonian (Watson, 1968) was needed to characterize the microwave measurements (Steenbeckeliers and Bellet, 1973).

100, 020, 001

Interest in the shorter wavelength infrared transitions stems from atmospheric transmission considerations, from simple contamination problems associated with isotopic scaling reactions and from electrically excited laser transitions. Earlier studies of these bands were those of Benedict et al. (1956) and were used to identify many electrically excited laser transitions (Benedict et al., 1969). An unpublished analysis was also available which was based on measurements taken at the National Bureau of Standards (Benedict, 1976). Most recently, the Orsay group has been mapping the spectra of H_2O and its isotopic derivatives (Flaud et al., 1981). Their most recent work was concerned with D_2O, in particular, the ν_1, $2\nu_2$ and ν_3 bands (Papineau et al., 1981; 1980). Based on high resolution IR-FTS measurements (5×10^{-3} cm^{-1} resolution), they were able to derive from the measured spectra, a complete set of ground and excited rotational levels. These data were subsequently fit to Watson's Hamiltonian using 24 ground state constants, 15 constants per vibrational mode plus five coupling constants needed to account for Coriolis-type and Fermi-type interactions between excited states. Since the J,K range in this work

(Papineau, 1981) was much higher than in the ground state microwave studies (Fuson et al., 1939) the resulting constants represent an improvement over the previous set. Although the constants and levels are listed elsewhere (Papineau et al., 1981), for convenience the experimentally determined rotational energy levels up through J=11 are listed in Table II. For the majority of these levels, the statistical error at 68% confidence level is less than 10^{-3} cm^{-1} (30 MHz) and is about a factor of five better than the energy levels calculated from the determined constants (Papineau et. al., 1981).

010

The ν_2 band, the band of main interest here, has been investigated much less than the others mainly because of source and detector problems. Earlier studies were extended by Williamson (1969) using conventional grating spectroscopy. Williamson's work was extended and improved, particularly in the assignments, by Lin and Shaw (1977) who supplemented his data with IR-FTS measurements (.1 cm^{-1} resolution). Earlier use of their data and assignments yielded agreement with known matches and emission lines but illustrated the limits on the accuracy of their transitions (.04 cm^{-1}) (Petuchowski et al., 1977). The inaccuracy was also apparent in acousto-optic CO_2 spectroscopy performed by Keilmann et al. (1975) and in the tunable diode measurements of Worchesky et al. (1978). Because of the paucity of higher resolution data, the data of Williamson (1969) and Lin (1977) still represent the most complete and collectively accurate band data for ν_2. In an effort to improve their level positions, we have redetermined the ν_2 rotational state energies by using their data for the infrared transitions (Lin, 1976), Papineau's et al. (1981) data for the ground state levels and performing a simple weighted average (Lin and Shaw, 1977) to arrive at the revised positions listed in Table II.

The accuracy of the predictions of line positions from the data in Table II can be illustrated with the ν_2-band-measurements shown in Table IV. The measurements reflect the tunable diode laser studies of Worchesky et al. (1978) and one wing absorption measurement of Petuchowski et al. (1977) ($R_9(34)$). Except for the latter data the revised set of constants in Table II yields a slight improvement over the previous set of data. Two other measurements of note are the 385 μm frequency and the splitting of the ν_2 $5_2 \rightarrow 5_4$ and $5_3 \rightarrow 5_5$ FIR transitions. The former transition is predicted at 25.9931 cm^{-1} whereas the heterodyne measurements of Fetterman et al. (1979) yielded a value of 25.980 cm^{-1} indicating an error of .013 cm^{-1}. For the latter case, the splitting is found from Table I to be .0467 cm^{-1} and is to be compared with the measured value by Dodel and Douglas (1981) of .0522 cm^{-1} indicating an error of .0065 cm^{-1}. Hence between the infrared and ν_2 FIR transitions, the 010 energy levels in Table II appear accurate to about .01 cm^{-1}.

The accuracy in the position of the remaining levels (000, 100, 020 and 001) is discussed more fully in Papineau et al. (1981) and Papineau (1980). As an example of the accuracy, the recently measured electrically excited D_2O laser line (Dommin et al., 1980) $100{:}12_{1,12} \rightarrow 020{:}11_{4,7}$ was found to be 118.65366 cm^{-1} and is to be compared with the value of 118.6529 cm^{-1} from the data of Papineau (1981). The difference of 22.8 MHz is consistent with the 10^{-3} cm^{-1} accuracy of the levels and illustrates the utility of the present set of energy levels.

Other Features

From a systems viewpoint, the spectroscopy and energy transfer features are about of equal importance in understanding and assessing the laser/Raman transitions in any optically pumped system (DeTemple, 1979). The main energy transfer features can be separated into J-changing and ν-changing kinetics and will be discussed later.

V-T-R Relaxation

Since D_2O is an asymmetric top, there appears to be no equivalent ΔK-changing relaxation bottleneck which exists in symmetric top molecules and effects the long pulse and cw laser performance (DeTemple and Danielewicz, 1976; Dangoisse et al., 1980). Hence the next stage bottleneck for long pulse operation is the vibrational or V-T-R relaxation. Although early relaxation data indicated an extraordinarily fast decay (Burnett and North, 1969), only recently has the magnitude of the decay been quantified. Two separate experimental approaches, IR-IR double resonance (Sheffield et al., 1980) and laser induced fluorescence (Miljanic and Moore, 1980) have yielded ν_2 self-relaxation rates of 1.03×10^6/torr-sec. This value is about 10^2-10^3 times larger than in the methyl halides and is only a factor of 10 less than the average rotational relaxation rate implying a minimal bottleneck effect in D_2O.

J_τ Relaxation

In highly polar molecules such as D_2O, the dipole-dipole interaction is of such a strength that almost every D_2O - D_2O collision causes a change in J_τ and hence the lifetime and linewidth. Because of the obvious difficulties of measuring the line broadening in the FIR and IR, reliance on the predictive abilities of line broadening models has been common. One such model is based on the work of Anderson (1949), extended by Tsao and Curnette (1962) and referred to as the ATC model. This model was first employed by Benedict and Kaplan (1964) for line broadening estimates in H_2O. The predictions of this model along with others was recently compared with high-resolution measurements in H_2O and shown to be of reasonable accuracy; 85% of the broadening rates could be predicted with $<25\%$ error (Mandin et al., 1980). The measured broadening rates ranged between 13 and 40 MHz/torr (FWHM) (Mandin et al., 1980). The ATC method has also been applied to the D_2O pump transitions listed in Table I and compared with empirically determined broadening rates extracted from the pressure dependence of the absorption (Nicholson, 1979). The comparison essentially verified the $<25\%$ error of the method and suggests is utility in the absence of measurements.

Contaminents

Although usually not a problem, H-bearing contaminents are thought to be detrimental to D_2O laser operation caused by isotopic mixing reactions producing HDO. The most likely source of the impurity is adsorbed H_2O in the cell walls and can be minimized by heat/purge cycles. The contaminent is manifested in a long-time decay of the output (Dodel and Douglas, 1982) and in gas absorption measurements at the FIR emission frequencies (Petuchowski et al., 1977). For example, recent high resolution studies of HDO (Papineau, 1980) show that the ground state transition $4_0 \to 5_0$ is within 5 GHz of the strong 66 μm transition which is much closer than the nearest ground state transition in H_2O $(6_{-3} \to 7_{-1})$ or D_2O $(6_{-1} \to 7_1)$. Hence system-to-system differences in laser behavior may exist depending on the degree of original D_2O purity and cell contamination.

Summary

The basic conclusions which can be deduced at this state are that because of the sparseness of the IR-spectrum the transition assignments and frequencies of the ν_2-band and ν_2-FIR transitions can be obtained with accuracies of about .01 cm^{-1}. This accuracy is sufficient for identifying candidate (DeTemple and Lawton, 1978) and actual (Dodel and Douglas, 1983) TEA-laser pumped transitions (DeTemple and Lawton, 1978) but insufficient for continuous wave operation for which one such transition is already known (Danielewicz and Weiss, 1978).

The line broadening and rotational relaxation rates appear to be a more serious problem. Anamolies in the FIR-polarization (Brown et al., 1979) and time-dependence (Dodel and Douglas, 1982) have been noted along with the possibility of very long-lived rotational states (Sheffield et al., 1980). These are suggestive of the importance of collisional processes in actual laser systems and point to future areas of study.

Table I. D_2O Laser Assignments.

LINE	LOWER	UPPER	OFFSET	λ(μm)	1/λ(cm^{-1})	MODE	LOWER	UPPER	POL	OPER
$P_9(44)$	7_5	6_3	0 C	47	211.203C	0	7_4	8_6	‖	
	7_4	6_4	566MHzC		211.203C		7_5	8_5		
				112	89.797C	ν_2	6_1	6_3	‖	
					89.599C		6_2	6_4		
				61	164.050C	ν_2	5_3	6_3	‖	
					164.098C		5_2	6_4		
				140	71.294C	ν_2	5_0	5_2	⊥	
$P_9(32)$	6_6	5_5	1.2GHz	66	151.702C	ν_2	4_4	5_4	‖	R
	6_5	5_5			151.665C		4_3	5_5		
				83	120.570C	ν_2	3_2	4_4	‖	C
					120.543C		3_3	4_3		
				111	89.932C	ν_2	5_2	5_4	⊥	R?
					89.885C		5_3	5_5		
				50	198.764C	0	6_6	7_6	‖	R
							6_5	7_7		
$P_9(32)$	7_{-1}	6_{-3}	-554GHz	116	85.846C	ν_2	5_{-3}	6_{-3}	‖	
$P_9(18)$	11_{-6}	10_{-6}	-1.2GHz	93	107.956C	ν_2	9_{-4}	10_{-6}	‖	
				134	74.631C	ν_2	10_{-8}	10_{-6}	⊥	
				91	110.321C	ν_2	9_{-6}	10_{-8}	‖	
$P_9(12)$	12_{-8}	11_{-6}	-786MHz	91	109.979C	ν_2	11_{-10}	11_{-8}	⊥	
				106	94.386C	ν_2	11_{-8}	11_{-6}	⊥	

LINE	LOWER	UPPER	OFFSET	λ(μm)	1/λ(cm^{-1})	MODE	LOWER	UPPER	POL	OPER
$R_9(12)$	10_{-8}	9_{-6}	565MHz	113	88.828C	ν_2	9_{-8}	9_{-6}		
				95	105.683C	ν_2	8_{-6}	9_{-6}	‖	
				134	74.826C	ν_2	8_{-8}	8_{-6}	‖	
				114	87.628C	ν_2	7_{-4}	8_{-6}	‖	
				109	91.680C	ν_2	8_{-8}	9_{-8}	‖	
				141	70.883C	0	10_{-8}	10_{-6}	⊥	
$R_9(22)$	5_0	4_0	-326MHz	385	25.980M	ν_2	4_{-2}	4_0	⊥	R,L
				358	27.897C	ν_2	4_{-4}	4_{-2}	⊥	
				240	41.735C	0	5_0	6_{-2}	⊥	
				276	36.181C	0	6_{-2}	6_0	⊥	
$R_9(30)$	10_{-7}	10_{-9}	-651MHz	99	101.107C	ν_2	9_{-9}	10_{-9}	‖	R,L
$R_9(32)$	9_{-9}	8_{-7}	589MHz	122	82.304C	ν_2	7_{-7}	8_{-7}	‖	
				98	101.723C	0	9_{-9}	10_{-9}	‖	
$R_9(34)$	4_2	3_0	-2.3GHz	253	39.464C	ν_2	3_{-2}	3_0	⊥	
$R_{10}(22)$	11_{-1}	10_{-3}	-138MHzC	57	175.125C	ν_2	9_{-3}	10_{-3}	‖	
				124	80.788C	ν_2	10_{-5}	10_{-3}	⊥	
$R_9(17)$ SEQUENCE BAND	10_{-8}	9_{-6}	-32MHz	113	88.828C	ν_2	9_{-8}	9_{-6}	⊥	L

Table II. D_2O Energy Levels (cm^{-1}).

J	K^-	K^+	000	010	020	100	001
0	0	0	0.0000	1178.3780	2336.8394	2671.6455	2787.7186
1	0	1	12.1170	1190.5119	2348.9716	2683.6056	2799.7587
1	1	1	20.2590	1199.7940	2359.6904	2691.6068	2807.3938
1	1	0	22.6843	1202.3396	2362.3481	2694.0113	2809.8428
2	0	2	35.8780	1214.2975	2372.7913	2707.0527	2823.3308
2	1	2	42.0693	1221.5017	2381.3005	2713.1212	2829.0296
2	1	1	49.3394	1229.1291	2389.2677	2720.3295	2836.3704
2	2	1	73.6762	1256.8474	2421.1961	2744.2458	2859.1949
2	2	0	74.1420	1257.3107	2421.6314	2744.7103	2859.6954
3	0	3	70.4474	1248.9202	2407.4905	2741.1545	2857.5524
3	1	3	74.5062	1253.7876	2413.4518	2745.1166	2861.1850
3	1	2	88.9711	1268.9812	2429.3183	2759.4569	2875.7815
3	2	2	110.0343	1293.2520	2457.6137	2780.1221	2895.3391
3	2	1	112.2515	1295.4390	2459.7104	2782.3328	2897.7087
3	3	1	156.6057	1345.5833	2516.9530	2825.9476	2939.4863
3	3	0	156.6629	1345.6278	2516.9990	2826.0057	2939.5524
4	0	4	114.9868	1293.5107	2452.1786	2785.0781	2901.5803
4	1	4	117.3120	1296.4209	2455.8886	2787.3358	2903.5903
4	1	3	141.0871	1321.4073	2482.0494	2810.9025	2927.5333
4	2	3	158.1110	1341.3739	2505.7920	2827.5607	2943.1009
4	2	2	164.1780	1347.4004	2511.6388	2833.6043	2949.5210
4	3	2	205.8860	1394.9372	2566.3038	2874.5570	2988.5739
4	3	1	206.2769	1395.2642	2566.6165	2874.9522	2989.0188
4	4	1	269.3754	1466.1705	2646.8436	2937.0369	3048.6239
4	4	0	269.3813	1466.1535	2646.8481	2937.0383	3048.6308
5	0	5	169.0384	1347.5432	2506.2585	2838.3753	2955.0101
5	1	5	170.2433	1349.1211	2508.3600	2839.5376	2955.9963
5	1	4	204.9378	1385.6988	2546.7815	2873.9146	2990.7820
5	2	4	217.5859	1400.9417	2565.4245	2886.2431	3002.1360

J	K^-	K^+	000	010	020	100	001
5	2	3	229.9923	1413.3537	2577.6463	2898.5870	3015.1307
5	3	3	267.5306	1456.6298	2628.0573	2935.3555	3049.9545
5	3	2	269.0103	1457.9833	2629.2551	2936.8569	3051.6273
5	4	2	331.0721	1527.9236	2708.6406	2997.8938	3110.2172
5	4	1	331.1236	1527.9500	2708.6758	2997.8939	3110.2797
5	5	1	411.5418	1617.8554	2809.7249	3077.0113	3186.2225
5	5	0	411.5423	1617.8351	2809.7256	3077.0077	3186.2235
6	0	6	232.5221	1410.8766	2569.5359	2900.9705	3017.8555
6	1	6	233.1061	1411.6637	2570.6510	2901.5297	3018.2187
6	1	5	279.5651	1460.9044	2622.6106	2947.5395	3064.4600
6	2	5	288.0943	1471.5448	2636.1575	2955.8073	3072.0629
6	2	4	309.2656	1492.9561	2657.4561	2976.8428	3094.0261
6	3	4	341.3888	1530.6042	2702.0992	3008.1322	3123.4521
6	3	3	345.4470	1534.3727	2705.4559	3012.3010	3127.9750
6	4	3	405.2835	1602.2022	2782.9558	3071.2053	3184.2828
6	4	2	405.5319	1602.4227	2783.1269	3071.2525	3184.5797
6	5	2	485.5938	1691.9998	2883.9455	3149.8708	3260.6871
6	5	1	485.5998	1692.0219	2883.9481	3149.8685	3260.6943
6	6	1	582.4087	1799.6933	3004.157*	3244.8750	3351.6511
6	6	0	582.4087	1799.6948	3004.1605	3244.8746	3351.6511
7	0	7	305.4954	1483.5683	2642.0348	2972.9227	3089.2658
7	1	7	305.7673	1483.9525	2642.6028	2973.1813	3090.1592
7	1	6	364.0467	1546.0702	2708.5526	3030.8605	3147.7316
7	2	6	369.2665	1552.8624	2717.6204	3035.8848	3152.5063
7	2	5	401.2623	1585.4810	2750.4364	3067.6064	3185.3928
7	3	5	427.1989	1616.5757	2788.2098	3093.5458	3208.7719
7	3	4	436.0603	1624.9056	2795.7693	3101.7476	3218.4808
7	4	4	492.0216	1689.0036	2869.8141	3156.9230	3270.8085
7	4	3	492.8803	1689.7592	2870.4120	3157.3886	3271.8170
7	5	3	572.1305	1778.6611	2970.6447	3235.0837	3347.5119
7	5	2	572.1647	1778.6643	2970.6673	3235.0770	3347.5515

Table II. D_2O Energy Levels (cm^{-1}). (Continued)

J	K^-	K^+	000	010	020	100	001
7	6	2	668.8514	1886.2358	3090.8636	3329.4578	3434.9245
7	6	1	668.8521	1886.2428	3090.8726	3329.4537	3434.9256
7	7	1	781.1724	2010.5214	3228.6279	3444.3860	3544.1705
7	7	0	781.1724	2010.5409	3228.6279	3444.3857	3544.1705
8	0	8	388.0189	1565.6660	2723.8276	3054.2923	3171.2861
8	1	8	388.1420	1565.8727	2724.1094	3054.4092	3172.6797
8	1	7	457.8234	1640.4918	2803.8316	3123.3326	3240.2202
8	2	7	460.7654	1644.4803	2809.4572	3126.1413	3243.1528
8	2	6	505.0490	1690.0557	2855.7410	3169.0034	3288.2046
8	3	6	524.6087	1714.2429	2886.0753	3189.2563	3305.5406
8	3	5	540.8819	1729.7847	2900.4755	3205.1908	3323.0498
8	4	5	591.2184	1788.3725	2969.1802	3254.8749	3369.6963
8	4	4	593.5869	1790.4075	2970.8567	3256.7176	3372.4261
8	5	4	671.1951	1877.8096	3069.8576	3332.7511	3446.7247
8	5	3	671.3355	1877.9296	3069.9429	3332.7483	3446.8820
8	6	3	767.7187	1985.2953	3189.9884	3426.4120	3531.2277
8	6	2	767.7202	1985.2929	3189.9907	3426.3906	3531.2305
8	7	2	880.0546	2109.6614	3327.8754	3544.0189	3641.3909
8	7	1	880.0548	2109.6698	3327.8755	3544.0174	3641.3910
8	8	1	1006.9616	2249.5480	3481.497*	3665.5449	3762.9765
8	8	0	1006.9616	2249.5629	3481.497*	3665.5449	3762.9765
9	0	9	480.1254	1657.2527	2814.9640	3145.1117	3262.4312
9	1	0	480.1806	1657.3462	2815.1030	3145.1631	3262.4054
9	1	8	560.7538	1743.9552	2908.0736	3224.8271	3341.8337
9	2	8	562.3170	1746.1743	2911.3617	3226.3058	3343.8526
9	2	7	619.5618	1805.6137	2972.3690	3283.2974	3401.3291
9	3	7	633.2168	1823.1708	2995.3210	3296.3742	3413.3508
9	3	6	659.4169	1848.6233	3019.3659	3322.0089	3441.0639
9	4	6	702.7022	1900.0268	3080.9327	3364.9135	3480.7429
9	4	5	708.1703	1904.7919	3084.9081	3369.7400	3486.9181
9	5	5	782.8087	1989.4996	3181.6074	3442.9548	3558.3304

J	K^-	K^+	000	010	020	100	001
9	5	4	783.2682	1989.9093	3181.8852	3443.0443	3558.8310
9	6	4	879.0281	2096.7384	3301.5378	3535.8067	3640.0980
9	6	3	879.0479	2096.7662	3301.5526	3535.7278	3640.1149
9	7	3	991.2966	2221.1510	3439.4651	3655.6857	3750.8668
9	7	2	991.2971	2221.1420	3439.477*	3655.6758	3750.8675
9	8	2	1118.3355	2361.3050	3593.377*	3776.6297	3872.8697
9	8	1	1118.3355	2361.3058	3593.377*	3776.6273	3872.8697
9	9	1	1258.8695	2515.0235	–	3913.9273	4007.2254
9	9	0	1258.8695	2515.0235	–	3913.9273	4007.2254
10	0	10	581.8242	1758.3629	2915.4725	3245.3889	3363.0377
10	1	10	581.8479	1758.3592	2915.5395	3245.4111	3363.0402
10	1	9	672.9202	1856.4955	3021.2442	3335.4303	3452.6350
10	2	9	673.7174	1857.6863	3023.0859	3336.1758	3456.3797
10	2	8	743.8033	1931.1268	3099.2637	3405.7672	3523.8030
10	3	8	752.6102	1942.9602	3115.5413	3414.1586	3531.8414
10	3	7	790.8205	1980.6498	3151.7940	3449.3887	3571.5693
10	4	7	826.1962	2023.7480	3204.865*	3486.7936	3603.6510
10	4	6	837.0509	2023.4321	3213.0565	3496.8144	3615.6418
10	5	6	906.9443	2113.7922	3305.886*	3565.7020	3682.2930
10	5	5	908.2113	2114.7943	3306.6513	3566.2056	3683.6427
10	6	5	1002.8070	2220.6600	3425.5236	3657.7203	3761.4370
10	6	4	1002.8803	2220.7901	3425.5622	3657.5088	3761.4970
10	7	4	1114.8946	2345.0071	3563.3453	3779.5300	3872.6058
10	7	3	1114.8973	2345.0435	3563.3738	3779.4878	3872.6093
10	8	3	1242.0058	2485.4338	3717.530*	3899.8938	3994.9475
10	8	2	1242.0059	2485.4343	3717.530*	3899.8935	3994.9476
10	9	2	1382.7869	2639.5551	–	4037.1506	4129.6681
10	9	1	1382.7869	2639.5544	–	4037.1506	4129.6681
10	10	1	1535.9674	–	–	4187.5933	4276.0503
10	10	0	1535.9674	–	–	4187.5933	4276.0503
11	0	11	693.1099	1868.8919	3025.3583	3355.0915	3473.1228

Table II. D_2O Energy Levels (cm^{-1}). (Continued)

J	K^-	K^+	000	010	020	100	001
11	1	11	693.1207	1868.8853	3025.3923	3355.1297	3473.1249
11	1	10	794.4325	1978.2511	3143.4500	3455.2559	3572.7193
11	2	10	794.8294	1978.8646	3144.4520	3455.6183	3572.6319
11	2	9	877.1159	2065.7696	3235.524*	3537.2038	3655.1280
11	3	9	882.4112	2073.2502	3246.3313	3542.2076	3660.8679
11	3	8	934.0513	2124.8878	3296.864*	3594.2668	3713.4354
11	4	8	961.3383	2159.2440	3340.666*	3620.1667	3738.0672
11	4	7	980.2564	2176.4832	3355.6440	3637.7197	3758.4896
11	5	7	1043.5037	2250.4645	3442.629*	3700.9389	3818.5099
11	5	6	1046.5295	2252.9472	3444.497*	3702.5446	3821.6927
11	6	6	1139.0599	2357.1007	3561.9353	3792.2297	3915.5054
11	6	5	1139.2943	2357.3119	3562.067*	3791.7788	3915.6420
11	7	5	1250.8473	2481.1773	3699.3966	3895.3744	4006.6038
11	7	4	1250.8576	2481.1500	3699.572*	3895.1997	4006.6173
11	8	4	1377.9454	2621.952*	3853.906*	4035.3136	4129.1934
11	8	3	1377.9461	2621.952*	3853.906*	4035.3116	4129.1939
11	9	3	1518.9337	-	-	4172.4907	4264.2338
11	9	2	1518.9337	-	-	4172.4907	4264.2338
11	10	2	1672.4803	2945.204*	-	4323.1929	4411.0962
11	10	1	1672.4803	2945.204*	-	4323.190*	4411.0962
11	11	1	-	-	-	4485.535*	4568.531*
11	11	0	-	-	-	4485.535*	4568.531*

All energy levels are taken from measured lines unless donated by an * in which case calculated values are shown. The sources are: 000, 020, 100, 001; Refs. N. Papineau, et al., J. Molec. Spec. 87, 219-232 (1981) and N. Papineau, These de Doctorat 3eme Cycle, Universite Paris-Sud, Centre D'Orsay, 1980; 010, Refs. Lin, C. L. and Shaw, J. H., J. Molec. Spec. 66, 441-447, (1977) and Lin, C. L., Ph.D. Thesis, The Ohio State University, 1976.

Table III. Rotational Constants.

STATE	$E_v(cm^{-1})$	$A(cm^{-1})$	$B(cm^{-1})$	$C(cm^{-1})$	$Z_R(293K)$
000[a]	0	15.51997	7.27297	4.84529	220.24
010[b]	1178.3780	16.63492	7.33869	4.78978	213.00
020[a]	2336.8394	18.14266	7.40173	4.73405	204.28
100[a]	2671.6455	15.18014	7.18024	4.77997	225.65
001[a]	2787.7186	14.88703	7.24521	4.79256	226.54

[a] N. Papineau, et al., J. Molec. 87, 219-232 (1981)

[b] C. L. Lin, et al., J. Molec. Spec. 66, 441-447 (1977).

Table IV. Selected ν_2 Line Positions Relative to CO_2 Laser Lines.

LINE $C^{12}O_2^{16}$	TRANSITION ν_2	DETUNINGS[a] LIN & SHAW[b]	TABLE II	MEAS.[c]
$P_9(32)$	$6_{6,1} \rightarrow 5_{5,0}$	1.7GHz	1.4GHz	1.16GHz
	$6_{6,0} \rightarrow 5_{5,1}$	1.7GHz	807MHz	1.16GHz
	$7_{3,4} \rightarrow 6_{2,5}$	-52MHz	-327MHz	-555MHz
$R_9(12)$	$10_{k,9} \rightarrow 9_{2,8}$	833MHz	731MHz	564MHz
$R_9(22)$	$5_{3,3} \rightarrow 4_{2,2}$	-82MHz	-509MHz	-325MHz
$R_9(30)$	$10_{2,9} \rightarrow 10_{1,10}$	-295MHz	-200MHz	-651MHz
$R_9(32)$	$9_{0,9} \rightarrow 8_{1,8}$	772MHz	544MHz	588MHz
$R_9(34)$	$4_{3,1} \rightarrow 3_{2,2}$	-2.8GHz	-3.1GHz	±2.3GHz[d]
$R_9(17)$[e]	$10_{1,9} \rightarrow 9_{2,8}$	237MHz	136MHz	-32MHz

[a] $(\nu_{laser} - \nu_{molecule})$

[b] C. L. Lin, Ph.D. Thesis, The Ohio State University, 1976.

[c] T. L. Worchesky, et al., Opt. Lett. 2, 70-71 (1978).

[d] S. J. Petuchowski, et al., IEEE J. Quan. Electron. QE-13, 476-481 (1977).

[e] CO_2 laser sequence band 002→021.

Acknowledgements

It is a pleasure to acknowledge the contributions of Professor G. Benedict, Professors C. Camy-Perot and J.-M. Flaud, Dr. G. Dodel, Dr. S. J. Petuchowski and Professor J. Shaw to this work. The support of the National Science Foundation and the University of Illinois Industrial Affiliates Program is gratefully acknowledged.

References

Anderson, P.W., 1949: "Pressure Broadening in the Microwave and Infra-Red Region", Phys. Rev. 76, 647-661.

Benedict, W.S., Gailar, N., and Plyler, E.K., 1956: "Rotation-vibration Spectra of Deuterated Water Vapor", J. Chem. Phys. 24, 1139-1165.

Benedict, W.S., and Kaplan, L.D., 1964: "Calculation of Line Widths in H_2O-H_2O and H_2O-O_2 Collisions", J. Quant. Spectros. Radiat. Transfer, 4, 453-469.

Benedict, W.S., Pollack, M.A., and Tomlinson, W.J., 1969: The Water-Vapor Laser", IEEE J. Quan. Electron. QE-5, 108-123.

Benedict, W.S., Clough, S.A., Frenkel, L., and Sullivan, T.E., 1970: "Microwave Spectrum and Rotational Constants for the Ground State of D_2O", J. Chem. Phys. 53, 2565-2570.

Benedict, W.S., 1976: private communication.

Brown, F., Horman, S.R., Palevsky, A., and Button, K.J., 1973: "Characteristics of a 30 kW-peak, 496 μm, Methyl Fluoride Laser", Opt. Comm. 9, 28-30.

Brown, F., Davidson, S., and Muehlner, D., 1979: Post Deadline Paper, Fourth Int. Conf. on Infrared and Millimeter Waves, Miami.

Burnett, G.M., and North, A.M., eds., 1969: Transfer and Storage of Energy by Molecules Vol. 2 Vibrational Energy, (Wiley-Interscience, London).

Chang, T.Y. and Bridges, T.J., 1970: "Laser Action at 452, 496 and 541 μm in Optically Pumped CH_3F", Opt. Comm. 1, 423-425.

Chang, T.Y. and McGee, J.D., 1971: "Millimeter and Submillimeter Wave Laser Action in Symmetric Top Molecules Optically Pumped via Parallel Absorption Bands", Appl. Phys. Lett. 19, 103-105.

Dangoisse, D., Glorieux, P., and Wascat, J., 1980: Paper D4.4, Fifth Int. Conf. on Infrared and Submillimeter Waves, Wurtzberg, F.R.G.

Danielewicz, E.J., and Weiss, C.D., 1978: "New CW Far Infrared D_2O, $^{12}CH_3F$ and $^{14}NH_3$ Lasers", Opt. Comm. 27, 98-100.

DeTemple, T.A., Plant, T.K., and Coleman, P.D., 1973: "Intense Superradiant Emission at 496 μm From Optically Pumped Methyl Fluoride", Appl. Phys. Lett. 22, 644-646.

DeTemple, T.A. and Danielewicz, E.J., 1976: "Continuous-Wave CH_3F Waveguide Laser at 496 μm: Theory and Experiment", IEEE J. Quan. Electron. QE-12, 40-47.

DeTemple, T.A. and Lawton, S.A., 1978: "The Identification of Candidate Transitions for Optically Pumped Far Infrared Lasers: Methyl Halides and D_2O", IEEE J. Quan. Electron. QE-14, 762-768.

DeTemple, T.A., 1979: Infrared and Millimeter Waves, Vol. I, K.J. Button, ed. (Academic Press, NY) pp 129-184.

Dodel, G. and Douglas, N.G., 1981: "Separation of the 5_5-5_3/5_4-5_2 Doublet in the ν_2 Band of $D_2{}^{16}O$", J. Molec. Spec. 87, 297.

Dodel, G. and Douglas, N.G., 1982: "Investigation of D_2O Laser Emission at 50, 66, 83, 111 and 116 μm", IEEE J. Quan. Electron. QE-18, 1294-1301.

Dodel, G. and Douglas, N.G., 1983: "New Pulsed Far Infrared Laser Lines in D_2O", in press.

Domnin, Yu.S., Tatarinkov, V.M., and Shumyatskii, P.S., 1980: "Phase Locking of a D_2O Laser to a Frequency Standard", Sov. J. Quan. Elec. 10, 116-117.

Drozdowicz, Z., Lax, B., and Temkin, R., 1978: "Gain Spectrum of a Pulsed Laser-Pumped Submillimeter Laser", Appl. Phys. Lett. 33, 154-156.

Evans, D.E., Peebles, W.A., Sharp, L.E., and Taylor, G., 1976: "Far-Infrared Superradiant Laser Action in Heavy Water:, Opt. Comm. 18, 479-484.

Evans, D.E., Guinee, R.A., Huckridge, D.A., and Taylor, G., 1977: "Time Resolved Pulses and Wavelength Measurements for the 114 μm and 66 μm Emissions in the FIR Superradiant D_2O Laser", Opt. Comm. 22, 337-342.

Fetterman, H.R., Tannewald, P.E., Parker, C.D., Melngailis, J., Williamson, R.C., Woskoboinikow, P., Praddaude, H.C., Mulligan, W.J., 1979: "Real-Time Spectral Analysis of Far Infrared Laser Pulses Using a Saw Dispersive Delay Line", Appl. Phys. Lett. 34, 123-125.

Flaud, J.-M., Camy-Peyret, C., and Toth, R.A., 1981: Water Vapor Line Parameters from Microwave to Medium Infrared, Vol. 19, Int. Tables of Selected Constants, (Pergamon, Oxford).

Fleming, J.W. and Gibson, M.J., 1976: "Far Infrared Absorption Spectra of Water Vapor $H_2{}^{16}O$ and Isotopic Modifications", J. Molec. Spec. 62, 326-337.

Fuson, N., Randall, H.M., and Dennison, D.M., 1939: "Infrared Absorption Spectrum of Heavy Water", Phys. Rev. 56, 982-1000.

Keilmann, F., 1974: private communication.

Keilmann, F., Sheffield, R.L., Leite, J.R.R., Feld, M.S., and Javan, A., 1975: "Optical Pumping and Tunable Laser Spectroscopy of the ν_2 Band of D_2O", Appl. Phys. Lett. 26, 19-22.

Lin, C.L., 1976: Ph.D. Thesis, "Measurement and Analysis of the ν_2 Band of Heavy Water", The Ohio State University.

Lin, C.L. and Shaw, J.H., 1977: "Measurement and Analysis of the ν_2 Band of $D_2{}^{16}O$", J. Molec. Spec. 66, 441-447.

Mandin, J.-Y., Flaud, J.-M., Camy-Peyret, C., and Guelachuili, G., 1980: Measurements and Calculations of Self-Broadening Coefficients of Lines Belonging to the ν_2 Band of $H_2{}^{16}O$", J. Quant. Spectros. Radiat. Transfer, 23, 351-370.

Miljanic, S.S. and Moore, C.B., 1980: "Vibrational Relaxation of $D_2O(\nu_2)$", J. Chem. Phys. 73, 226-229.

Nicholson, J.P., 1979: "Direct Observation of Raman Shifts in a FIR Superradiant Laser", Opt. Comm. 29, 49-50.

Papineau, N., 1980: These de Doctorat 3eme Cycle, "Etude des Bandes $2\nu_2$, ν_1, ν_3 de D_2O et $2\nu_2$, ν_1 de HDO. Determination des Constants Rotationnelles de l'etat Fondamental de H_2O et HDO", Universite Paris-Sud Centre, D'Orsay.

Papineau, N., Flaud, J.-M., and Camy-Peyret, C., 1981: "The $2\nu_2$ ν_1 and ν_3 Bands of $D_2{}^{16}O$. The Ground State (000) and the Triad of Interacting States {(020), (100), (001)}", J. Molec. Spec. 87, 219-232.

Petuchowski, S.J., Rosenberger, A.T., and DeTemple, T.A., 1977: "Stimulated Raman Emission in Infrared Excited Gases", IEEE J. Quan. Electron. QE-13, 476-481.

Petuchowski, S.J. and DeTemple, T.A., 1980: "A Quantum Mechanically Based Propagation Model for Serial Three-Wave Interactions in Optically Pumped Lasers: 50 μm and 66 μm in D_2O", Int. J. of Infrared and Millimeter Waves, Vol. 1, No. 3, (Plenum Publishing Corporation) 387-414.

Plant, T.K., Newman, L.A., Danielewicz, E.J., DeTemple, T.A., and Coleman, P.D., 1974: "High Power Optically Pumped Far Infrared Lasers", IEEE Trans. Microwave Theory and Techniques, MTT-22, 988-900.

Schalow, A.L., and Townes, C.A., 1955: Microwave Spectroscopy, (MacGraw Hill, NY) Chapters 3 and 4.

Sheffield, R.L., Boyer, K., and Javan, A., 1980: "Study of Vibrational and Rotational Relaxation in D_2O", Opt. Lett. 5, 10-11.

Steenbeckeliers, G. and Bellet, J., 1973: "Application of Watson's Centrifugal Distortion Theory to Water and Light Asymmetric Tops. General Methods. Analysis of the Ground State and the ν_2 State of D_2O^{16}", J. Molec. Spec. 45, 10-34.

Temkin, R.J., 1977: "Theory of Optically Pumped Submillimeter Lasers", IEEE J. Quan. Electron. QE-13, 450-454.

Tsao, C.J. and Curnutte, B., 1962: "Line-Widths of Pressure Broadened Spectral Lines", J. Quant. Spectros. Radiat. Transfer, 2, 41-91.

Watson, J.K.G., 1968: "Determination of Centrifugal Distortion Coefficients of Asymmetric-Top molecules. III. Sextic Coefficients", J. Chem. Phys. 48, 4517-4524.

Wiggins, J.D., Drozdowicz, Z., and Temkin, R.J., 1978: "Two Photon Transitions in Optically Pumped Submillimeter Lasers", IEEE J. Quan. Electron. QE-14, 23-30.

Williamson, J.G., 1969: Ph.D. Thesis, "ν_2 Bands of $H_2{}^{18}O$, $H_2{}^{16}O$, and $D_2{}^{16}O$", The Ohio State University.

Worchesky, T.L., Ritter, K.J., Sattler, J.P., and Riessler, W.A., 1978: "Heterodyne Measurements of Infrared Absorption Frequencies of D_2O', Opt. Lett. 2, 70-71.

Woskoboinikow, P., Praddaude, H.D., Mulligan, W.J., and Cohn, D.R., 1981: "Submillimeter-Wave Thomson-Scattering ion Temperature Diagnostic Using a Pulsed D_2O Laser at 385 μm", Paper ThB18, Conf. on Laser Eng. and Applications, Washington, D.C.

THE SULFUR DIOXIDE SUBMILLIMETER WAVE LASERS

Joseph P. Sattler

U.S. Army Electronics Research and Development Command
Harry Diamond Laboratories, DELHD-RT-CA,
Adelphi, MD 20783

Walter J. Lafferty

Molecular Spectroscopy Division
National Bureau of Standards
Washington, D.C. 20234

A. Introduction

Coherent far-infrared or submillimeter-wave (SMMW) radiation has been obtained by using sulfur dioxide (SO_2) as the active medium of optically pumped lasers (Bugaev and Shliteris; Calloway and Danielewicz, 1981A; Simonis, et al.; Tobin; Calloway and Danielewicz, 1981B) and of electric discharge lasers (Dyubko et al.; Hard; Hassler and Coleman; Hubner et al.; Hassler et al.). This article will primarily review the results for optical pumping with a CO_2 laser, but will also discuss briefly the mechanisms proposed for the discharge laser.

The molecule SO_2 was recognized in 1979 by Danielewicz as a good candidate for optical pumping because it satisfied certain molecular selection criteria. The SO_2 SMMW laser operates in a manner similar to many other optically pumped cw lasers because CO_2-laser radiation of a frequency closely coincident with an SO_2 absorption selectively populates a rotational level of the ν_1 excited vibrational state of SO_2.

The cw laser action in the SMMW spectral region occurs between the rotational levels of the ν_1 state. The detailed interaction of the molecules with high-power laser pulses involves more complicated concepts (Temkin; Chang), but the underlying molecular spectroscopy is still necessary for an understanding of the results.

B. Spectroscopic details

The structure of the sulfur dioxide molecule has long been established from Raman, IR, and electron diffraction experiments (listed by Herzberg) and early microwave studies (Dailey et al.; Crable and Smith; Sirvetz). The equilibrium O-S-O angle is $119.33(1)^{\circ}$ and the equilibrium O-S distance is 1.43076(13) Å (Saito). Of the 24 natural species, $^{32}S^{16}O_2$ is 94.5% abundant. This nonlinear triatomic molecule (C_{2v} symmetry) is a near-prolate asymmetric rotor (Ray's $\kappa \simeq -0.94$) with a ground state permanent dipole moment of 1.63305(4) D lying in the direction of the symmetry axis which is also the axis of the intermediate inertial moment (Patel, et al.; Brown et al.).

A compilation of microwave data from many sources has been prepared by Lovas. The infrared spectrum between 500 and 5000 cm^{-1} was investigated at about 0.3 cm^{-1} resolution with a grating-prism spectrometer (Shelton et al.). The fundamental vibration bands are the type-B $\nu_1(a_1)$ symmetric stretch centered near 1152 cm^{-1}, the type-B $\nu_2(a_1)$ bend centered near 518 cm^{-1}, and the type-A $\nu_3(b_1)$ asymmetric stretch centered near 1362 cm^{-1}. These bands are also Raman active. Citations to earlier IR and Raman work may be found in Herzberg, and Shelton et al. Vibrational term values for overtone and combination bands may be obtained from Corice et al. Energy transfer between states lying below 3000 cm^{-1} has been discussed by Siebert and Flynn. At present, the best values for the fundamental band centers are: for ν_1, 1151.71352(32) (Sattler et al., 1981A); for ν_2, 517.75(10) cm^{-1} (Dana and Fontenella); for ν_3, 1362.0295(11) cm^{-1} (Brand et al.).

The rotational energy levels for an asymmetric top molecule may be denoted by the three numbers J, K_p, and K_o, where J is the total rotational angular momentum quantum number, and K_p and K_o are the quantum numbers associated with the projection of J on the unique axis for the limiting prolate and oblate symmetric tops, respectively (Herzberg; Townes and Schawlow). The selection rules for pure rotational transitions and for the type-B IR transitions of the ν_1 band are: $\Delta J = 0, \pm 1$; $J = 0 \nleftrightarrow J = 0$; ΔK_p, $\Delta K_o = \pm 1, \pm 3 \ldots$ For the ground and ν_1 states of $^{32}S^{16}O_2$, owing to nuclear spin statistics, K_p and K_o must both be even or odd.

The ν_1 IR band has been studied with tunable lasers. A waveguide CO_2 laser was used to find five near-coincidences with SO_2 absorptions; some identifications were proposed and frequency offsets were measured (Herlemont et al.). Three studies of SO_2 have been made by use of the lead-salt diode laser (F. Allario et al.; Hinkley et al.; Sattler et al., 1981A). From measurements in the region between 1110 to 1170 cm^{-1} which used grating-measured ammonia lines for absolute frequency calibration, the ν_1 band center was determined to be 1151.71(1) cm^{-1} (Hinkley et al.). Recently the frequencies of 55 lines in the extreme P-branch of the ν_1 band were measured to within 2×10^{-4} cm^{-1} (6 MHz) by use of diode laser heterodyne techniques (Sattler et al., 1981A). These data were combined with all available ground state and ν_1-state microwave data (Lovas) to determine the set of Watson spectroscopic constants presented in Table I (Sattler et al., 1981A). From the constants one can predict, to an accuracy that exceeds that of the interferometric measurements by over three orders of magnitude, all the observed optically pumped SMMW laser lines of $^{32}S^{16}O_2$.

The 9R(J) emission lines of $^{12}C^{16}O_2$ overlap the absorption lines of $^{32}S^{16}O_2$ in the extreme P-branch of the ν_1 band. The FWHM doppler width of the absorptions near 1075 cm^{-1} is about 50 MHz at 300 K. Near-coincidences, where the offset, $|\nu(SO_2) - \nu(CO_2)|$, is less than 60 MHz (a condition favorable for pumping by a cw CO_2 laser) occur for the 9R(J) $^{12}C^{16}O_2$ lines where J = 14, 18, 28, and 40. The offsets are about 95 MHz for the 9R(26) and 9R(34) lines, requiring, for example, a TEA or waveguide CO_2 laser for optical pumping.

It is now useful to consider what SMMW transitions one might expect. The CO_2 laser induces transitions from the ground state and selectively populates a rotational level, J', of the ν_1 vibrational manifold. From this level (using absorption notation where, for example, the emission transition, J' → J", from the selectively populated level J', is designated as R branch when J' = J" + 1), strong lasing can occur on R- or Q-branch rotational lines in the ν_1 state of SO_2. It is notable that Bugaev and Shliteris have observed weak lasing on a P-branch line. From the lower level, J", after the primary transition, cascade transitions are again possible. Finally, one might expect to observe refilling transitions in the ground state. In Fig. 1 is presented the energy level diagram appropriate for pumping with the 9R(28) CO_2 laser line.

C. The optically pumped $^{32}S^{16}O_2$ SMMW laser

The behavior of SO_2 under cw pumping by a $^{12}C^{16}O_2$ laser was investigated independently, but approximately at the same time, by Calloway and Danielewicz, 1981A and by Bugaev and Shliteris, and subsequently by Tobin (private communication). A TEA $^{12}C^{16}O_2$ laser pump was used by Simonis et al. A brief analysis with the work of Tobin and Simonis has been presented (Sattler, et al., 1981B). The measured output wavelengths are listed in Table II, along with other experimental data.

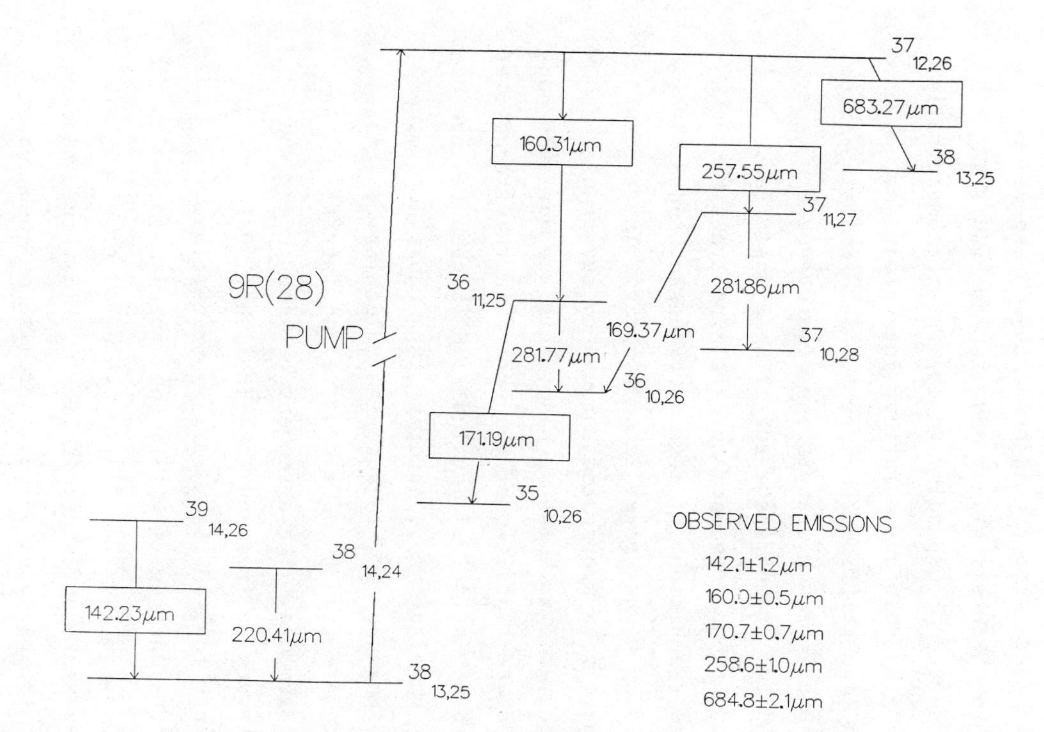

Fig. 1. Energy levels of $^{32}S^{16}O_2$ affected by optical pumping with the 9R(28) $^{12}C^{16}O_2$ laser line. Calculated SMMW emission wavelengths are given schematically and observed emission wavelengths are tabulated. Calculated wavelengths of observed transitions are enclosed in boxes.

In Table III, the observations are compared with the theoretical predictions derived from the spectroscopic constants for $^{32}S^{16}O_2$. The table lists: the frequencies of the $^{12}C^{16}O_2$ pump lines (Petersen et al.); the calculated frequencies and assignments of the nearly coincident IR absorptions; the calculated assignments, transition strengths (S), and frequencies of the SMMW emissions; and the observed SMMW emission frequencies corresponding to the interferometrically measured wavelengths. To guide further investigations, we have listed for each IR absorption the expected SMMW emissions from the primary R- and Q-branch transitions, four cascade transitions, and two ground state refilling transitions. It is possible that additional cascade and refilling transitions as well as P-branch transitions, (as shown by Bugaev and Shliteris) could be induced by higher power laser pumping. The errors in the constants have been propagated through to the predicted SMMW frequencies associated with the ν_1 band, this has not been done for the ground state transitions; the errors in the frequencies are one standard deviation. The calculated frequencies are over three orders of magnitude more accurate than those determined by interferometric measurement, because the molecular constants reflect the accurate results of microwave and infrared heterodyne measurements. The agreement between theory and experiment in regard to frequency is excellent. Only the observed frequency for the $36_{9,27}-35_{8,28}$ transition, 51.79(6) cm^{-1}, does not agree within the stated errors with the predicted frequency, 51.644075(21) cm^{-1}. However, since this cascade line was weak, and since it lies within three standard deviations from the measured value, there is not much reason to doubt the assignment.

In contrast to the work in which only the primary R-branch transitions were observed (Calloway and Danielewicz, 1981A), several cw Q-branch transitions were readily observed and measured by Tobin where a metal waveguide resonator was used; Bugaev and Shliteris observed transitions in the R, P, and Q branch. These observations, and calculations from the spectroscopic constants demonstrate that ground state SMMW absorptions play no significant role in supressing the expected Q-branch transitions as was suggested by Calloway and Danielewicz, 1981A. In retrospect, absence of these lower frequency lines in Calloway and Danielewicz, 1981A has been ascribed to the increased attenuation of the pyrex waveguide at those frequencies (E. J. Danielewicz, private communication).

Cascade and ground state refilling transitions were observed only under TEA-laser pumping by Simonis et al. It is likely that even more transitions would have been seen, had not the extremely strong absorptions of atmospheric water vapor at some frequencies precluded measurement. With the TEA laser on the 9R(28) CO_2 lines, the cascade transition at 58.58(24) cm^{-1} was readily observable, but the parent transition at 62.38 cm^{-1} was not detected. A similar situation occurred with the 9R(40) pump line. The purged optical path used by Calloway and Danielewicz, 1981A permitted detection and measurement of the 62.38 cm^{-1} line and a measurement of the water vapor absorption coefficient at this frequency. Scanning of the resonator cavity was used to determine the wavelengths of the strongly absorbed lines by Tobin (private communication).

The temporal behavior of the SMMW pulse was examined by using the 9R(26) emission for the TEA laser (Simonis et al.). The SMMW emission started 140 ns after the leading edge of the CO_2 laser pulse and continued for about 3 μs along with the trailing tail of the pump pulse; the intensity of the SMMW radiation fluctuated considerably during this time.

All the operating parameters of the optically pumped SO_2 laser have not been measured for each emission line and pumping mode, and so its characterization is still incomplete. However, the situation, as evidenced from Tables II and III, is such that one can reasonably estimate and then check by measurement the values of any unreported characteristics.

Every observed SMMW line has been assigned. Not every line predicted from the molecular constants, however, has been observed. To explain the different emission line patterns corresponding to each pump line, especially for the TEA laser, would require modelling of the type referred to by Temkin and Chang, but this has not been carried out for SO_2.

Theoretical molecular spectroscopy has been able to explain each observed $^{32}S^{16}O_2$ SMMW emission line pumped by the $^{12}C^{16}O_2$ laser. It would be useful to make accurate heterodyne measurements of the SMMW emission frequencies, because such data would improve the knowledge of the ν_1 band constants. The calculated frequencies of Table III should facilitate the search for the heterodyne beats. To conclude this section, the predicted emissions for pumping with the $^{12}C^{18}O_2$ laser (Freed et al.) are given in Table IV, and similar calculations for the $^{12}C^{16}O^{18}$ laser (Freed et al.) are given in Table V.

D. The optically pumped $S^{18}O_2$ SMMW laser

The behavior of $S^{18}O_2$ under the cw pumping of a $^{12}C^{16}O_2$ laser has also been investigated by Calloway and Danielewicz, 1981B. Because the ν_1 band center of $^{32}S^{18}O_2$ is near 1100.65 cm^{-1} (Barbe and Jouve) the overlap of this species with the CO_2 laser should be better than that of the abundant species. Two near-coincidences have been observed and the measured lines are listed in Table VI. At present, assignments for these lines have not been proposed.

A practical problem with this species is its tendency to exchange ^{18}O for ^{16}O from water in the pyrex wall; but from the temporal behavior of the emission, it is not considered probable that the observed SMMW lines originated from $^{32}S^{16}O^{18}O$ which has a band center near 1116(1) cm^{-1} (Polo and Wilson). Calloway and Danielewicz, 1981B, suggest a reexamination of $S^{18}O_2$ in a bakeable high vacuum waveguide system. It would also be of interest to examine other enriched isotopic species, especially $^{34}S^{16}O_2$.

E. The electric discharge laser

Far-infrared or SMMW laser action from sulfur dioxide was first obtained in a cw electric discharge by Dyubko et al. Subsequent work (Hard; Hassler and Coleman; Hubner et al.; Hassler et al.) on pulsed and cw discharges has resulted in a total of 10 reported emission lines. Assignments for these lines have been proposed by Hubner et al.; Hassler et al.; Steenbeckeliers and Bellet. Models similar to those used to explain the HCN laser (Lide and Maki) and the H_2O laser (Hartman and Kleman; Benedict et al.) were invoked to explain the SO_2 discharge laser.

Hubner et al. and Hassler et al. proposed that the laser action could be ascribed to a resonance between the ν_1, $K_p = 14$ levels and $2\nu_2$, $K_p = 16$ levels near $J = 26$. However, because of improved microwave and infrared data, this specific resonance was ruled out by Steenbeckeliers and Bellet; in addition, the results of the recent diode laser heterodyne study of the ν_1 band (Sattler et al., 1981A) also cast doubt on the earlier assignments. Those assignments involved the ν_1, $27_{14,14}$ energy level. This level had been postulated to be perturbed by 0.22 cm^{-1}, but since the IR transition $27_{14,14}-28_{15,13}$, as measured by diode laser heterodyne techniques, is only 1×10^{-4} cm^{-1} from the value calculated from the spectroscopic constants, this level is not involved in a local perturbation. Moreover, no other measured ν_1, $K_p = 14$ levels are perturbed, as would be expected. However, the general mechanism proposed by Hubner et al., and Hassler et al. to explain the discharge laser, namely that of vibrational resonances, is correct, but with a different set of assignments.

Steenbeckeliers and Bellet have explained the observed characteristics of the SO_2 discharge laser on the basis of a local Fermi resonance that mixes the $3\nu_2$, $K_p = 15$ levels and the $\nu_1 + \nu_2$, $K_p = 13$ levels near $J = 30$. (A similar local Fermi resonance has been observed by Pine et al. in the $\nu_1 + \nu_3$, $K_p = 13$ subband near $J = 35$, and this was attributed to interaction with the $2\nu_2 + \nu_3$, $K_p = 15$ levels.) The assignments of Steenbeckeliers and Bellet comprehensively account for the frequencies, intensities, polarizations, temporal behavior, interactions, and cw lasing of the SMMW emissions, and are spectroscopically definitive. The experimental data from Hassler et al. and the assignments from Steenbeckeliers and Bellet are listed for the SO_2 electric discharge laser in Table VII.

Acknowledgements

The authors wish to thank George J. Simonis, Gary L. Wood, Mary S. Tobin, A. R. Calloway, and Edward J. Danielewicz for the use of their data prior to publication. Thanks are also given to Jessica Putnam for her cheerful assistance in preparing the manuscript.

TABLE I.[†] SPECTROSCOPIC CONSTANTS OF THE ν_1 BAND OF $^{32}S^{16}O_2$ IN CM^{-1}

	ν_1 State	Ground State
$\mathscr{A}$	2.0284336(18)[a]	2.02735367(20)[a]
$\mathscr{B}$	0.34251193(31)	0.344173732(37)
$\mathscr{C}$	0.29211359(29)	0.293526384(32)
$\Delta_K \times 10^4$	0.876882(69)	0.863783(20)
$\Delta_{JK} \times 10^5$	-0.37715(17)	-0.390109(61)
$\Delta_J \times 10^6$	0.21960(19)	0.220343(41)
$\delta_K \times 10^6$	0.9301(50)	0.84571(50)
$\delta_J \times 10^7$	0.56736(76)	0.567273(45)
$H_K \times 10^7$	0.12618(12)	0.11786(13)
$H_{KJ} \times 10^9$	-0.7049(44)	-0.6108(14)
H_{JK}	~0[b]	~0[b]
$H_J \times 10^{12}$	0.242(44)	0.3354(62)
$h_J \times 10^{12}$	0.372(58)	0.1754(12)
h_{JK}	~0[b]	~0[b]
$h_K \times 10^9$	0.507(32)	0.5446(88)
ν_o	1151.71352(31)	

[†]TABLE I is taken from Sattler et al., 1981A, Copyright © 1981 by Academic Press, with permission.

[a]Uncertainties cited correspond to one standard deviation.

[b]Not significantly determined.

Table II. Observed Optically Pumped

SMMW Emissions from SO_2

λ(μm) Measured	Polarization	Rel. or Abs. Power (mW) or Energy (μJ)	Opt. Pressure (m Torr)	$^{12}C^{16}O_2$ Line	Pump Power (W) or Energy(J)	Ref.
684.8(21)	∥			9R(28)	7.0W	d
349.1(3)		m		9R(40)		b
311.8(6)				9R(40)		c
311.7(3)		s (61 μJ)	2500	9R(40)	0.5J	b
312.1(9)	⊥	0.02 mW		9R(40)	4.0W	d
258.6(10)				9R(28)		c
258.0(8)	⊥	0.33 mW	~70	9R(28)	7.0W	d
205.3(4)				9R(14)		c
205.3(3)		s (70 μJ)	3000	9R(14)	1.0J	b
208.0(6)	⊥	0.43 mW	~90	9R(14)	9.0W	d
193.1(2)		w		9R(40)		b
180.5(26)		w		9R(18)		b
182.0(5)	⊥			9R(18)	10.0W	d
180.0(10)				9R(40)		c
170.7(7)		s (220 μJ)	4800	9R(28)	1.0J	b
165.2(2)		m		9R(26)		b
160.5(10)				9R(28)		c
160.0(5)	∥	0.9 mW	100	9R(28)	11.3W	a
159.5(5)	∥	0.26 mW		9R(28)	7.0W	d
155.2(2)		s (230 μJ)	3500	9R(26)	1.0J	b
149.7(9)		m		9R(18)		b
146.2(7)		vw		9R(14)		b
142.2(5)	∥	2.9 mW	160	9R(18)	16.0W	a
142.2(3)				9R(18)		c
141.9(6)		s (87 μJ)	2500	9R(18)	1.0J	b
142.0(4)	∥	0.55W		9R(18)	10.0W	d
142.1(12)		w		9R(28)		b
139.9(3)				9R(14)		c
139.7(5)	∥	3.5 mW	250	9R(14)	16.0W	a
139.5(4)	∥	0.37 mW	~70	9R(14)	9.0W	d
128.1(6)		w		9R(18)		b
126.4(1)		51 μJ	4500	9R(34)	0.7J	b

[a]Calloway and Danielewicz, 1981A; [b]Simonis et al.; [c]Tobin; [d]Bugaev and Shliteris. The symbols m(medium) and w(weak) indicate relative energy referred to the SMMW line with the s(strongest) energy pulse for a given pump line. The output pulse energy listed in parentheses is the total energy listed for <u>all</u> SMMW lines (s + m + w) corresponding to the listed pump line.

Table III. Assignments for $^{32}S^{16}O_2$ SMMW Emissions

$^{12}C^{16}O_2$ Laser		Calculated IR absorption		Calculated SMMW Emission			Observed SMMW Emission
Line	Frequency[e] (cm^{-1})	Frequency (cm^{-1})	Assignment	Assignment	S	Frequency (cm^{-1})	Frequency (cm^{-1})
9R(14)	1074.64649	1074.64571(4)	$36_{15,21}-37_{16,22}$	$36_{15,21}-35_{14,22}$	17.17	71.583447(23)	71.58(26)[a], 71.48(15)[c], 71.69(21)[d]
			cascade	$35_{14,22}-34_{13,21}$	16.24	67.696033(25)	
			cascade	$35_{14,22}-35_{13,23}$	15.74	45.425696(20)	
9R(14)	1074.64649	1074.64571(4)	$36_{15,21}-37_{16,22}$	$36_{15,21}-36_{14,22}$	15.90	48.676559(19)	48.72(7)[b], 48.71(10)[c], 48.08(14)[d]
			cascade	$36_{14,22}-36_{13,23}$	16.35	45.425470(19)	
			cascade	$36_{14,22}-35_{13,23}$	16.42	68.332583(24)	68.40(33)[b]
			g.s. refill	$38_{17,21}-37_{16,22}$	19.04	79.3413	
			g.s. refill	$37_{17,21}-37_{16,22}$	15.58	55.0335	
9R(18)	1077.30252	1077.30429(4)	$24_{17,7}-25_{18,8}$	$24_{17,7}-23_{16,8}$	16.95	70.315161(23)	70.32(31)[a], 70.47(29)[b], 70.32(14)[c], 70.42(21)[d]
			cascade	$23_{16,8}-22_{15,7}$	15.98	66.489282(28)	66.80(40)[b]?
			cascade	$23_{16,8}-23_{15,9}$	6.77	51.852009(24)	
9R(18)	1077.30252	1077.30429(4)	$24_{17,7}-25_{18,8}$	$24_{17,7}-24_{16,8}$	6.83	55.037842(19)	55.4(8)[b], 54.95(16)[d]
			cascade	$24_{16,8}-24_{15,9}$	7.50	51.855727(23)	
			cascade	$24_{16,8}-23_{15,9}$	16.10	67.129327(28)	66.80(40)[b]?
			g.s. refill	$26_{19,7}-25_{18,8}$	18.91	77.8990	78.06(36)[b]
			g.s. refill	$25_{19,7}-25_{18,8}$	6.15	61.2514	

(continued)

Table III. (continued)

$^{12}C^{16}O_2$ Laser		Calculated IR Absorption		Calculated SMMW Emission			Observed SMMW Emission
Line	Frequency[e] (cm^{-1})	Frequency (cm^{-1})	Assignment	Assignment	S	Frequency (cm^{-1})	Frequency (cm^{-1})
9R(26)	1082.29624	1082.29325(4)	$30_{14,16}-31_{15,17}$	$30_{14,16}-29_{13,17}$	15.39	64.508745(29)	64.43(8)[b]
			cascade	$29_{13,17}-28_{12,16}$	14.46	60.602394(29)	60.53(7)[b]
			cascade	$29_{13,17}-29_{12,18}$	12.49	42.152455(23)	
9R(26)	1082.29624	1082.29325(4)	$30_{14,16}-31_{15,17}$	$30_{14,16}-30_{13,17}$	12.63	45.421765(23)	
			cascade	$30_{13,17}-30_{12,18}$	13.11	42.152605(23)	
			cascade	$30_{13,17}-29_{12,18}$	14.63	61.239435(29)	
			g.s. refill	$32_{16,16}-31_{15,17}$	17.27	72.2892	
			g.s. refill	$31_{16,16}-31_{15,17}$	12.26	51.8208	
9R(28)	1083.47878	1083.47927(5)	$37_{12,26}-38_{13,25}$	$37_{12,26}-36_{11,25}$	15.08	62.380686(23)	62.50(31)[a], 62.3(6)[c], 62.70(19)[d]
			cascade	$36_{11,25}-35_{10,26}$	14.14	58.414712(23)	58.58(24)[b]
			cascade	$36_{11,25}-36_{10,26}$	17.67	35.489747(16)	
9R(28)	1083.47878	1083.47927(5)	$37_{12,26}-38_{13,25}$	$37_{12,26}-37_{11,17}$	17.84	38.827268(17)	38.67(15)[c], 38.76(12)[d]
			cascade	$37_{11,27}-37_{10,28}$	18.28	35.478826(15)	
			cascade	$37_{11,27}-36_{10,26}$	14.30	59.0432	
			g.s. refill	$39_{14,26}-38_{13,25}$	16.93	70.3081	70.37(59)[b]
			g.s. refill	$38_{14,24}-38_{13,25}$	17.58	45.3694	
9R(28)	1083.47878	1083.47927(5)	$37_{12,26}-38_{13,25}$	$37_{12,26}-38_{11,27}$	4.60	14.6355	14.60(4)[d]

$^{12}C^{16}O_2$ Laser		Calculated IR Absorption		Calculated SMMW Emission			Observed SMMW Emission
Line	Frequency[e] (cm^{-1})	Frequency (cm^{-1})	Assignment	Assignment	S	Frequency (cm^{-1})	Frequency (cm^{-1})
9R(34)	1086.86979	1086.86647(4)	$28_{19,9}-28_{20,8}$	$28_{19,9}-27_{18,10}$	19.14	79.169544(23)	79.11(6)[b]
			cascade	$27_{18,10}-26_{17,9}$	18.17	75.396792(16)	
			cascade	$27_{18,10}-27_{17,11}$	8.35	58.206104(14)	
9R(34)	1086.86979	1086.86647(4)	$28_{19,9}-28_{20,8}$	$28_{19,9}-28_{18,10}$	8.41	61.337536(23)	
			cascade	$28_{18,10}-28_{17,11}$	9.07	58.211123(13)	
			cascade	$28_{18,10}-27_{17,11}$	18.29	76.038112(15)	
			g.s. refill	$29_{21,9}-28_{20,8}$	20.98	86.0070	
			g.s. refill	$28_{21,7}-28_{20,8}$	7.01	67.4268	
9R(40)	1090.02837	1090.02971(7)	$37_{10,28}-38_{11,27}$	$37_{10,28}-36_{9,27}$	13.51	55.672402(22)	55.56(31)[c]
			cascade	$36_{9,27}-35_{8,28}$	12.53	51.644075(21)	51.79(6)[b]
			cascade	$36_{9,27}-36_{8,28}$	18.62	28.677637(12)	28.65(3)[b]?
9R(40)	1090.02837	1090.02971(7)	$37_{10,28}-38_{11,27}$	$37_{10,28}-37_{9,29}$	18.74	32.090581(13)	32.08(3)[b], 32.07(7)[c], 32.04(10)[d]
			cascade	$37_{9,29}-37_{8,30}$	19.25	28.650598(12)	28.65(3)[b]?
			cascade	$37_{9,29}-36_{8,28}$	12.68	52.259458(21)	
			g.s. refill	$39_{12,28}-38_{11,27}$	15.41	63.7148	
			g.s. refill	$38_{12,26}-38_{11,27}$	18.45	38.7705	

[a]Calloway and Danielewicz, 1981A; [b]Simonis et al.; [c]Tobin; [d]Bugaev and Shliteris; [e]Petersen et al.; ? Indicates ambiguity in theoretical assignment.

Table IV. Predicted SMMW Emissions from $^{32}S^{16}O_2$ Pumped by the $^{12}C^{18}O_2$ Laser

$^{12}C^{18}O_2$ LASER		CALCULATED IR ABSORPTION		CALCULATED SMMW EMISSION		
Line	Frequency[a]	Frequency (cm^{-1})	Assignment	Assignment	S	Frequency (cm^{-1})
IIP(20)	1068.94248	1068.9440	$23_{20,4}-24_{21,3}$	$23_{20,4}-22_{19,3}$	19.58	79.0601
				$23_{20,4}-23_{19,5}$	3.72	64.4056
IIP(6)	1079.49163	1079.4936	$38_{13,25}-39_{14,26}$	$38_{13,25}-37_{12,26}$	16.01	66.3225
				$38_{13,25}-38_{12,26}$	18.01	42.1380
IIR(28)	1101.70143	1101.6996	$34_{14,20}-34_{15,19}$	$34_{14,20}-33_{13,21}$	16.07	67.0591
				$34_{14,20}-34_{13,21}$	15.12	45.4255
IIR(46)	1110.81729	1110.8158	$39_{11,29}-39_{12,28}$	$39_{11,29}-38_{10,28}$	14.63	60.2969
				$39_{11,29}-39_{10,30}$	19.51	35.4532

[a]Freed et al.

Table V. Predicted SMMW Emissions from $^{32}S^{16}O_2$ Pumped by the $^{12}C^{16}O^{18}O$ Laser

$^{12}C^{16}O^{18}O$ Line	LASER Frequency[a] (cm^{-1})	CALCULATED Frequency (cm^{-1})	IR ABSORPTION Assignment	CALCULATED Assignment	SMMW S	EMISSION Frequency (cm^{-1})
IIR(19)	1086.22023	1086.2210	$34_{19,15}-34_{20,14}$	$34_{19,15}-33_{18,16}$	19.96	83.0214
				$34_{19,15}-34_{18,16}$	12.59	61.3724
IIR(24)	1089.24975	1089.2527	$34_{18,16}-34_{19,15}$	$34_{18,16}-33_{17,17}$	19.16	79.8860
				$34_{18,16}-34_{17,17}$	13.14	58.2422
IIR(28)	1091.57233	1091.5699	$40_{17,23}-40_{18,22}$	$40_{17,23}-39_{16,24}$	19.38	80.5609
				$40_{17,23}-40_{16,24}$	17.45	55.1059
IIR(30)	1092.70012	1092.6970	$20_{13,7}-21_{14,8}$	$20_{13,7}-19_{12,8}$	13.06	54.8610
				$20_{13,7}-20_{12,8}$	6.58	42.1404
IIR(37)	1096.47281	1096.4727	$23_{16,8}-23_{17,7}$	$23_{16,8}-22_{15,7}$	15.98	66.4893
				$23_{16,8}-23_{15,9}$	6.77	51.8520
IIR(41)	1098.50779	1098.5066	$30_{9,21}-31_{10,22}$	$30_{9,21}-29_{8,22}$	11.58	47.9105
				$30_{9,21}-30_{8,22}$	14.98	28.7966

[a] Freed et al.

Table VI. Observed[a] Optically Pumped SMMW Emission from $S^{18}O_2$

λ(μm) Measured	Polarization	Relative Power	Pressure (m Torr)	$^{12}C^{16}O_2$ Line	Pump Power (W)
471.8	‖	0.1	102	9R(18)	12.0
148.2	⊥	1.0	110	9P(16)	12.6

[a]Calloway and Danielewicz, 1981B.

Table VII. SMMW Emissions from the SO_2 Electric Discharge Laser

Frequency (cm^{-1}) Measured[a]	Calculated[b]	Relative Intensity[c] Pulsed	cw	Assignment[d] $v',J'_{K_p',K_o'}$		$v'',J''_{K_p'',K_o''}$
46.44	46.426	3	1	$s,30_{14,16}$	–	$s^*,30_{13,17}$
48.44	48.435	1	...	$b,29_{14,16}$	–	$b,29_{13,17}$
51.89	51.879	10	5	$b^*,30_{15,15}$	–	$b,30_{14,16}$
66.09	66.093			$s,31_{14,18}$	–	$b^*,30_{15,15}$
66.14	66.139	40	2	$s,31_{14,18}$	–	$s^*,30_{13,17}$
66.67	66.666	10	...	$s,32_{14,18}$	–	$b,31_{15,17}$
70.43	70.436	30	...	$s,29_{13,17}$	–	$b,28_{14,14}$
70.98	90.980			$s^*,30_{13,17}$	–	$b,29_{14,16}$
71.03	71.026	100	15	$b^*,30_{15,15}$	–	$b,29_{14,16}$
71.53	71.556	1	...	$s,31_{13,19}$	–	$b,30_{14,16}$

[a] Hassler et al. Stated uncertainty is 0.02 cm^{-1}.
[b] Steenbeckeliers and Bellet.
[c] Hassler et al.
[d] Steenbeckeliers and Bellet. The letter s indicates the $\nu_1 + \nu_2$ state having a stretching component and the letter b indicates the $3\nu_2$ bending overtone. The asterisk indicates the principal component of the mixed states with components $s,30_{13,17}$ and $b,30_{15,15}$.

References

F. Allario, C. H. Blair, and J. F. Butler, "High Resolution Spectral Measurements of Sulfur Dioxide from 1176.0 to 1265.8 cm^{-1} Using a Single Lead Selenide Laser with Magnetic and Current Tuning," IEEE J. Quantum Electron. QE-11, 205-209 (1975).

A. Barbe and P. Jouve "Force Constants and Form of Vibration of Sulfur Dioxide from Infrared Spectrum of $S^{18}O_2$," J. Mol. Spectrosc. 38, 273-280 (1971).

William S. Benedict, Martin A. Pollack, and W. John Tomlinson, III, "The Water Vapor Laser," IEEE J. Quantum Electron. QE-5, 108-124 (1969).

J. C. D. Brand, D. R. Humphrey, A. E. Douglas, and I. Zanon, "The Resonance Fluorescence Spectrum of Sulfur Dioxide," Can. J. Phys. 51, 530-536 (1973).

R. D. Brown, F. R. Burden, and G. M. Mohay, "Dipole Moment of Sulfur Dioxide," Aust. J. Chem. 22, 251-253 (1969).

V. A. Bugaev and E. P. Shliteris "Sulfur Dioxide Laser Pumped by CO_2 Laser Radiation and Identification of the Transitions," Sov. J. Quantum Electron. 11, 742-744 (1981); Kvant. Elektron. (Moscow) 8, 1241-1248 (June 1981).

A. R. Calloway and E. J. Danielewicz, "Far Infrared Optically Pumped SO_2 Laser," IEEE J. Quantum Electron. QE-17, 579-581 (1981A).

A. R. Calloway and E. J. Danielewicz, "Predicted New Optically Pumped FIR Molecules," Int. J. Infrared and Millimeter Waves 2, 933-942 (1981B).

T. Y. Chang, "Dynamic Stark Effect in CH_3F and Other Optically Pumped Lasers", J. Quantum Electron. QE-13, 937-942 (1977).

R. J. Corice, K. Fox, and G. D. T. Tejwani, "Experimental and Theoretical Studies of the Fundamental Bands of Sulfur Dioxide," J. Chem. Phys. 58, 265-270 (1973).

George F. Crable and William V. Smith, "The Structure and Dipole Moment of SO_2 from Microwave Spectra," J. Chem. Phys. 19, 502 (1951).

B. P. Dailey, S. Golden and E. Bright Wilson, Jr., "Preliminary Analysis of the Microwave Spectrum of SO_2," Phys. Rev. 72, 871 (1947).

V. Dana and J. C. Fontenella, "Etude de la Structure Rotationelle de la Band ν_3 de $^{32}S^{16}O_2$ de 1334 a 1382 cm^{-1}," Mol. Phys. 30, 1473-1479 (1975).

E. J. Danielewicz, Laser Power Optics, San Diego CA 92121 (private communication).

E. J. Danielewicz, "Molecular Parameters Determining the Performance of cw Optically Pumped FIR Lasers," in Digest of the Fourth International Conference on Millimeter Waves and Their Applications, Miami Beach, 10-15 Dec. 1979, IEEE Cat. No. 79 CH1384-7 MTT, pp. 203-204, (1979).

S. F. Dyubko, V. A. Svich, and R. A. Valitov, "SO_2 Submillimeter Laser Generating at Wavelengths 0.141 and 0.193 mm," J.E.T.P. Lett. 7, 320 (1968).

C. Freed, L. C. Bradley and R. G. O'Donnell, "Absolute Frequencies of Lasing Transitions in Seven CO_2 Isotopic Species," J. Quantum Electron. QE-16, 1195-1206 (1980).

Thomas M. Hard, "Sulfur Dioxide Submillimeter Laser," Appl. Phys. Lett. 14, 130 (1969).

B. Hartman and B. Kleman, "On the Origin of the Water-Vapor Laser Lines," Appl. Phys. Lett. 12, 168-170 (1968).

J. C. Hassler and P. D. Coleman, "Far-Infrared Lasing in H_2S, OCS, and SO_2," Appl. Phys. Lett. 14, 135-136 (1969).

J. C. Hassler, G. Hubner, and P. D. Coleman, "Excitation Mechanism of the Far-Infrared Sulfur Dioxide Molecular Laser," J. Appl. Phys. 44, 795-801 (1973).

F. Herlemont, M. Lyszyk, and J. Lemaire, "Infrared Spectroscopy of OCS, SO_2, O_3 with a CO_2 Waveguide Laser," J. Mol. Spectrosc. 77, 69-75 (1979).

Gerhard Herzberg, Molecular Spectra and Molecular Structure II. Infrared and Raman Spectra of Polyatomic Molecules (D. Van Nostrand Company, Inc., New York, 1951).

E. D. Hinkley, A. R. Calawa, P. L. Kelley, and S. A. Clough, "Tunable Laser Spectroscopy of the ν_1 Band of SO_2," J. Appl. Phys. 43, 3222-3225 (1972).

Gunter Hubner, J. C. Hassler, P. D. Coleman, and Guy Steenbeckeliers, "Assignments of the Far-Infrared SO_2 Laser Lines," Appl. Phys. Lett. 18, 511-513 (1971).

David R. Lide, Jr., and Arthur G. Maki, "On the Explanation of the So-Called CN Laser," Appl. Phys. Lett. 11, 62-64 (1967).

Frank J. Lovas, "Microwave Spectral Tables, II. Triatomic Molecules," J. Phys. Chem. Ref. Data 7, 1639-1712 (1978).

D. Patel, D. Margolese, and T. R. Dyke, "Electric Dipole Moment of SO_2 in Ground and Excited Vibrational States," J. Chem. Phys. 70, 2740-2748 (1979).

F. R. Petersen, D. G. McDonald, J. D. Cupp and B. L. Danielson "Accurate Rotational Constants, Frequencies, and Wavelengths from $^{12}C^{16}O_2$ Lasers Stabilized by Saturated Absorption," in Laser Spectroscopy (R. G. Brewer and A. Mooradian, Eds.) pp. 555-569, Plenum, New York, 1979.

A. S. Pine, G. Dresselhaus, B. Palm, R. W. Davies, and S. A. Clough, "Analysis of the 4-μm $\nu_1+\nu_3$ Combination Band of SO_2," J. Mol. Spectrosc. 67, 386-415 (1977).

Shuji Saito, "Microwave Spectrum of Sulfur Dioxide in Doubly Excited Vibration States and Determination of the γ Constants," J. Mol. Spectrosc. 30, 1-16 (1969).

Santiago R. Polo and M. Kent Wilson, "Infrared Spectrum of $S^{16}O^{18}O$ and the Potential Constants of SO_2," J. Chem. Phys. 22, 900-903 (1954).

J. P. Sattler, T. L. Worchesky, and Walter J. Lafferty, "Diode Laser Heterodyne Spectroscopy on the ν_1 Band of Sulfur Dioxide," J. Mol. Spectrosc. 88, 364-371 (1981A).

Joseph P. Sattler, George J. Simonis, Gary L. Wood, Mary S. Tobin, and Walter J. Lafferty, "The Sulfur Dioxide SMMW Lasers," in Digest of the Sixth International Conference on Infrared and Millimeter Waves, Miami Beach, 7-12 Dec. 1981, IEEE Cat. No. 81 CH1645-1 MTT, (1981B).

R. D. Shelton, A. H. Nielsen, and W. H. Fletcher, "The Infrared Spectrum and Molecular Constants of Sulfur Dioxide," J. Chem. Phys. 21, 2178-2183 (1953).

Donald Siebert and George Flynn, "Vibration-vibration energy transfer in laser excited SO_2: Further evidence for a slow V-V step," J. Chem. Phys. 62, 1212-1220 (1975).

G. J. Simonis, G. L. Wood, and J. P. Sattler, "Sulfur Dioxide Pulsed Laser Operation," Int. Conf. on Lasers '80, New Orleans, 15-19 Dec. 1980 (unpublished).

M. H. Sirvetz, "The Microwave Spectrum of Sulfur Dioxide," J. Chem. Phys. 19, 938-941 (1951).

Guy Steenbeckeliers and Jean Bellet, "New Interpretation of the Far-Infrared SO_2 Laser Spectrum," J. Appl. Phys. 46, 2620-2626 (1975).

R. J. Temkin "Theory of Optically Pumped Submillimeter Lasers," IEEE J. Quantum Electron. QE-13, 450-454 (1977).

Mary S. Tobin, Harry Diamond Laboratories, Adelphi MD 20783 (private communication).

C. H. Townes and A. Schawlow, Microwave Spectroscopy (McGraw Hill, New York, 1955).

OPTICALLY PUMPED PH_3 LASER OPERATING IN THE 83-223 μm REGION

Paul D. Coleman

Electro-Physics Laboratory
Department of Electrical Engineering
University of Illinois
Urbana, Illinois 61801

Introduction

Since the first reported generation of far IR laser radiation by Chang and Bridges in 1970, every spectroscopic type of molecule has been made to lase but the two most important types have been symmetric tops having inversion like NH_3 and asymmetric tops with internal rotation like CH_3OH. These symmetric and asymmetric tops have few atoms in the molecule but the inversion and internal rotation modes yield a rich molecular spectrum with many laser transitions possible.

Phosphine (PH_3), a pyramidal molecule with point group symmetry C_{3v} is similar to NH_3. It has a rather dense absorption spectra in the 8-12 μm region which yields many coincident frequency matches for a CO_2 laser pump. Like NH_3 these absorption matches are associated with parallel band transitions $G \rightarrow \nu_2$ but unlike NH_3 the perpendicular band transitions $G - \nu_4$ are also accessible.

In order to make definite assignments of pump laser absorption transitions and laser line transitions, it is necessary to have good spectroscopic data on a molecule. Fortunately, PH_3 has been studied in recent years by Yin and Rao, (1974), Maki et al. (1973),

Tarrago et al. (1981), and by Shimizu (1975). Thus one has available or can calculate energy levels for G, ν_2 and $\nu_4^{\pm\ell}$ to the order of ±0.010 cm^{-1} or less, along with measured G → ν_2 and G - $\nu_4^{\pm\ell}$ transition frequencies.

Tables of data on PH_3 obtained and calculated from the references (Lin and Rao, 1974; Maki et al., 1973; Tarrago et al., 1981; Shimizu, 1975) are included in this paper to aid in illustrating how rather straightforward it becomes to make absorption and laser transition assignments with complete spectroscopic data. Also, these tables will be of help in any future work on optically pumping this PH_3 molecule.

Experimentally the PH_3 molecule was optically pumped with a grating-tuned CO_2 TEA laser which was scanned through its 41 highest power laser lines. Absorption of the CO_2 lines could be anticipated from the spectroscopic data on PH_3. Sixteen different CO_2 laser lines were absorbed by PH_3 resulting in 44 distinct rotational laser transitions. Three of the 44 laser lines could not be assigned to PH_3 and are believed to be due to an impurity gas or a gas formed by the dissociation of PH_3.

Three different laser configurations were used in the study: 1) an open Fabry-Perot resonator, 2) a waveguide cell, and 3) a single pass gas cell. Four lines were found to be generated single pass: the 111.409 cm^{-1} transition from $\nu_2(13,2) \rightarrow \nu_2(12,2)$, the 53.691 cm^{-1} transition from $\nu_4^0(6,0) \rightarrow \nu_4^0(5,0)$, the 59.954 cm^{-1} transition from $\nu_4^{+\ell}(7,1) \rightarrow \nu_4^{+\ell}(6,1)$ and the 82.339 cm^{-1} transition from $\nu_4^{+\ell}(9,3) \rightarrow \nu_4^{+\ell}(8,3)$. The 82.339 cm^{-1} (121.45 μm) line yielded the highest output energy (∿10μJ) for any of the 44 laser transitions obtained.

The laser data for this review paper was taken from the MS thesis work of Niesen (1978) and the paper by Malk et al. (1978).

Experimental Arrangement

The experimental apparatus, shown schematically in Fig. 1, is typical for far IR optical pumped molecular laser work. A CW CO_2

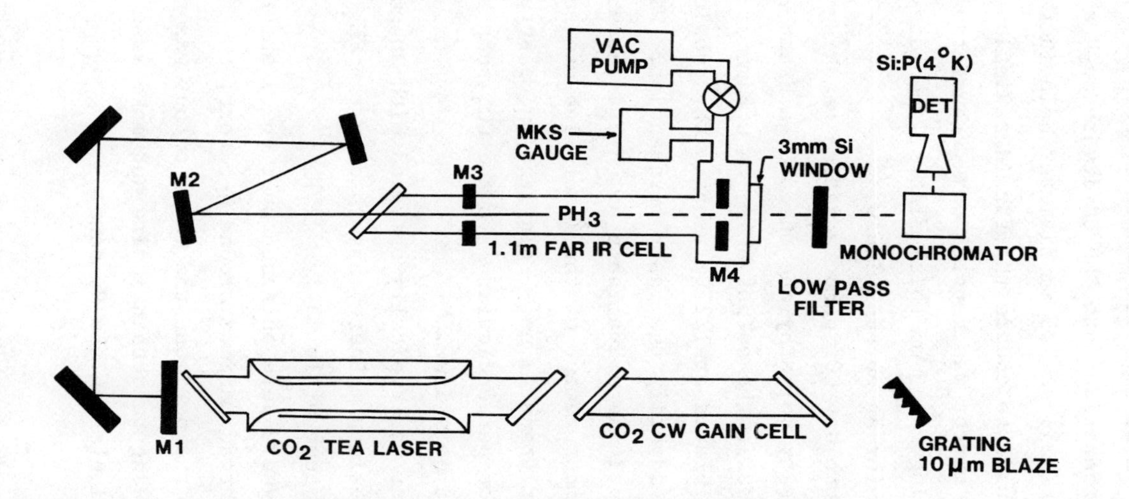

Fig. 1. Schematic of experimental optical pumping configuration. M1: Ge, 10m R, 80% reflectivity; M2: Gold-coated glass, 0.53m R; M3: Gold-coated glass, 4m R, 2 mm diameter hole; M4: Gold-coated glass, flat, 0.5 mm diameter hole.

gain cell is used in the optical cavity of the CO_2 TEA laser to prevent mode locking. The far IR absorption cell containing the PH_3 is 110 cm long. It is a semi-confocal cavity with internal gold-coated mirrors. A 53 cm focal length mirror (M2) was used to focus the CO_2 laser radiation into the cavity through a hole (2 mm) coupled input mirror (M3). A 0.5 mm diameter hole in the flat output mirror (M4) was used to couple the far IR radiation out of the system. This output mirror was mounted on a translatable stage for cavity frequency tuning. A 3 mm thick flat silicon disk (40% reflectivity) was used for an output window.

In order to absorb any CO_2 pump radiation reaching the far IR detector, a high density polyethylene sheet or a Grubb-Parsons low pass filter was placed in the output beam. A 0.5 meter monochromator having a grating blazed at 122 μm along with a cooled Si:P detector was used to identify the laser frequencies. With the excellent spectroscoptic data available on PH_3, an accuracy of ±0.5 percent on the measurement of the laser line was sufficient to identify the laser transitions. Polarization measurements (either parallel $||$ or perpendicular $\perp$ to the CO_2 pump radiation) were made with a grid polarizer.

The CO_2 TEA laser was typically operated with an output of 50 mJ and a 1 μsec pulsewidth. Energy absorbed by the PH_3 gas was estimated to be the order of 10 mJ.

Far IR laser signals were obtained over the pressure range 1-16 torr and had a broad maximum around 4-6 torr.

In this PH_3 laser experiment the laser output was not optimized, the objective being to obtain lasing and make laser line assignments, in particular in the ν_4 band.

PH_3 Spectroscopy

The PH_3 molecule is an oblate symmetric top with four normal modes of vibration, two of interest are shown in Fig. 2. These two modes possess many absorptions in the 8-12 μm range which can be accessed by a CO_2 laser.

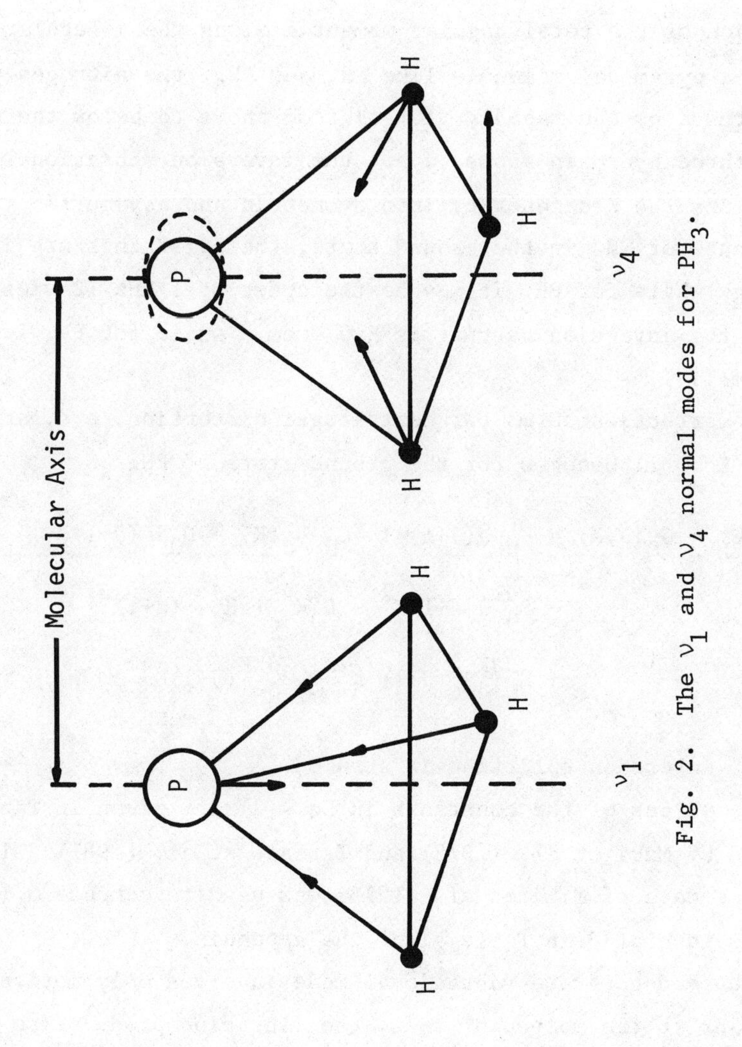

Fig. 2. The ν_1 and ν_4 normal modes for PH_3.

Neglecting distortion terms, the rotational energies W_R of a symmetric top molecule can be expressed as

$$W_R = B_J(J+1) + (C-B)K^2 \quad (1)$$

where J is the total angular momentum quantum number and $K \leq J$ is the projection of the total angular momentum along the molecular axis.

In a pyramidal molecule like NH_3 and PH_3, the nitrogen or phosphorus atom can rapidly vibrate from above to below the plane of the three hydrogen atoms, i.e., the inversion vibrational motion. This splits the K degeneracy into symmetric and asymmetric states. In the case of NH_3 in the ground state, the splittings are in the GHz range while for PH_3 it may be the order of 1 KHz (Davies, 1971). For NH_3 the inversion barrier is 2,072 cm^{-1} while for PH_3 it is 6,085 cm^{-1}.

If correction terms for centrifugal distortion, etc. are included, Eq. (1) becomes for the ground state of PH_3

$$\begin{aligned} W_R = G_0(J,K) &= B_0 J(J+1) + (C_0-B_0)K^2 - D_0^J J^2(J+1)^2 \\ &- D_0^{JK} J(J+1)K^2 - D_0^K K^4 + H_0^J J^3(J+1)^3 \\ &+ H_0^{JJK} J^2(J+1)^2 K^2 + H_0^{JKK}(J+1)K^4 \end{aligned} \quad (2)$$

where no inversion splitting is assumed.

The values of the constants in Eq. (2) are given in Table I as reported by Maki et al. (1973) and Tarrago et al. (1981). In this paper the data of Maki et al. (1973) was used to compute $G_0(J,K)$ which is tabulated in Table VI of the appendix.

When a degenerate vibrational mode involves only motion perpendicular to the molecular axis, the vibration produces an angular momentum ℓh about the molecular axis where ℓ is an integer.

Rotational energy levels for which $|K-\ell|=3n$ where n=0,1,2,3,... belong to molecular symmetry species A, while energy levels for which $|K-\ell| \neq 3n$ belong to specie E. The ground state has $\ell=0$, hence for example

$$G_0(J,K) \quad \begin{array}{l} \text{Specie A if } K=3n \\ \text{Specie E if } K\neq 3n \end{array}$$

The selection rules are in all cases A ↔ A and E ↔ E. One can also state the selection rules for the G → ν_2 parallel band as

$$K = 0 \qquad \Delta J = \pm 1 \qquad \Delta K = 0$$

$$K \neq 0 \qquad \Delta J = 0,\pm 1 \qquad \Delta K = 0$$

or as

$$QP(J,K) \qquad QQ(J,K) \qquad QR(J,K)$$

transitions, the first Q meaning ΔK = 0, while the second letters P,Q,R mean ΔJ = -1,0 and +1.

The perpendicular band G → ν_4 transition selection rules are

$$K = 0 \qquad \Delta J = \pm 1 \qquad \Delta K = +1 \qquad G \rightarrow \nu_4^{+\ell}$$

and

$$K \neq 0 \qquad \Delta J = 0,\pm 1 \qquad \Delta K = \pm 1 \qquad G \rightarrow \nu_4^{\pm\ell}$$

or as

$$PP(J,K) \qquad PQ(J,K) \qquad PR(J,K)$$

$$RP(J,K) \qquad RQ(J,K) \qquad RR(J,K)$$

transitions. Here the first letters P and R indicate ΔK = -1 or +1 while the second letters P,Q and R indicate ΔJ = -1,0 and +1.

The pure rotational selection rules are for the parallel bands

$$K = 0 \qquad \Delta J = \pm 1 \qquad \Delta K = 0$$

and

$$K \neq 0 \qquad \Delta J = \pm 1 \qquad \Delta K = 0$$

$$G(J,K) \rightarrow G(J',K)$$

$$\nu_2(J,K) \rightarrow \nu_2(J',K)$$

while for the perpendicular band

$$K = 0 \qquad \Delta J = \pm 1 \qquad \Delta K = 0 \qquad \nu_4^0 \rightarrow \nu_4^0$$

$$\Delta J = 0,\pm 1 \qquad \Delta K = +1 \qquad \nu_4^0 \rightarrow \nu_4^{-\ell}$$

$$K = 0 \qquad \Delta J = \pm 1 \qquad \Delta K = 0 \qquad \nu_4^{\pm\ell} \rightarrow \nu_4^{\pm\ell}$$

$$\Delta J = 0,\pm 1 \qquad \Delta K = \pm 1 \qquad \nu_4^{\pm\ell} \rightarrow \nu_4^{\mp\ell}$$

Graphic examples of vibration-rotation and pure rotational transitions for G and ν_2 states are shown in Fig. 3 while transitions for G and ν_4 are shown in Figs. 4 and 5.

These diagrams are not to scale in order to illustrate features of the energy level diagrams. For the ν_4 mode, ℓ-doubling is to be observed for K > 0 and K doubling for the A species for K = 2 for $\nu_4^{-\ell}$ and K = 4 for $\nu_4^{+\ell}$.

Absorption and Laser Line Data and Assignments

Using Tables IV and V for absorption transitions for G → ν_2 and G → ν_4 from the appendix or the tables of Tarrago et al. (1981), it is an easy task to locate possible frequency matches between the pump CO_2 laser lines and the PH_2 absorption lines. This data is shown in Table II where the 16 CO_2 pump lines that yielded lasing are marked with an asterisk.

Many of the CO_2 pump lines had a frequency match as close or closer to PH_3 absorption lines as the 16 that produced lasing, assuming the spectroscopic data is accurate. There would appear to be good prospects of achieving further lasing lines using other CO_2 pump lines with additional research work.

Once an absorption match is located, the next step is to use the energy level Table VI of the appendix and seek to find the lasing transitions.

For example, take the QQ(8,4) absorption match which will lead to the excitation of the $\nu_2(8,4)$ level. The allowed pure rotational transitions are $\nu_2(8,4) \rightarrow \nu_2(7,4)$ followed by $\nu_4(7,4) \rightarrow \nu_4(6,4)$ and then $\nu_4(6,4) \rightarrow \nu_4(5,4)$, i.e., a cascade sequence. Energy data is taken from Table VI using TDG.

The numerical energies are seen to be

$$\nu_2(8,4) \rightarrow \nu_2(7,4),\ 1294.769 - 1226.269 = 68.500\ \text{cm}^{-1}$$

$$\nu_2(7,4) \rightarrow \nu_2(6,4),\ 1226.269 - 1166.363 = 59.906\ \text{cm}^{-1}$$

$$\nu_2(6,4) \rightarrow \nu_2(5,4),\ 1166.363 - 1115.043 = 51.320\ \text{cm}^{-1}$$

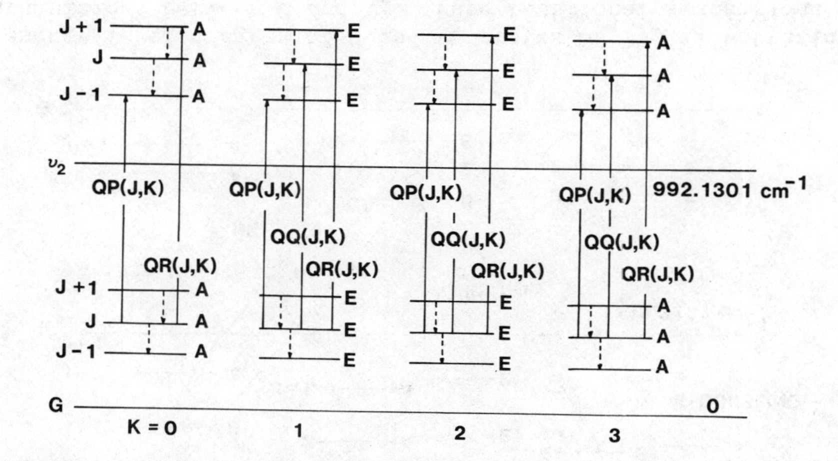

Fig. 3. Partial energy level diagram of G and ν_2 states of PH_3 illustrating vibration-rotation transitions (——) and pure rotation transitions (----). Inversion splitting too small to be seen.

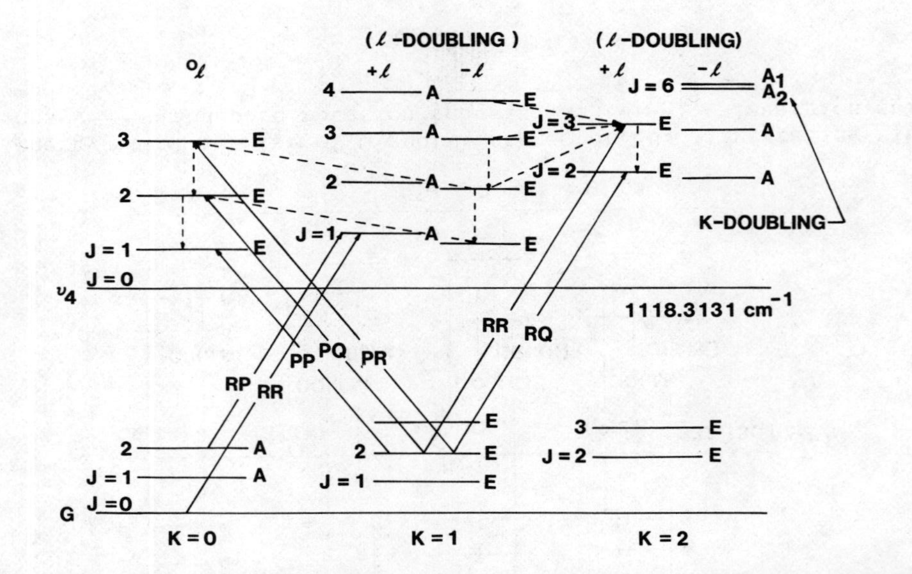

Fig. 4. Partial energy level diagram of G and ν_4 states of PH_3 illustrating PP, PQ, PR, RQ, and RR transitions between G and ν_4. Pure rotational transitions in ν_4 are shown as (----) lines.

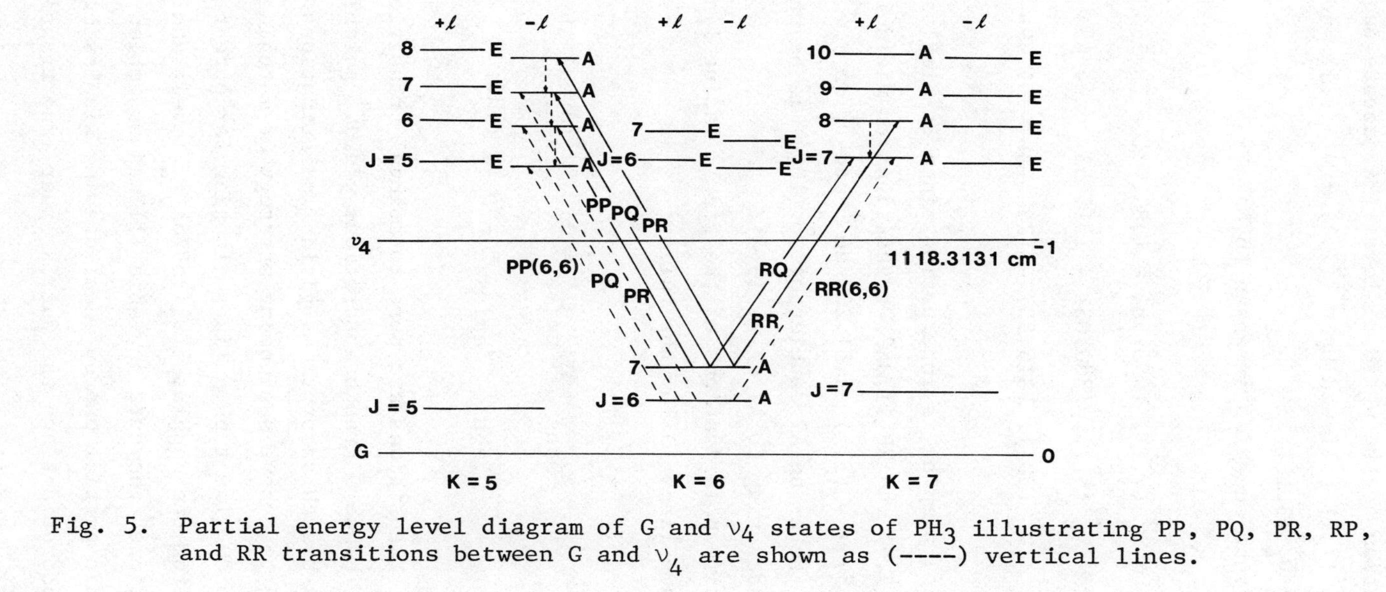

Fig. 5. Partial energy level diagram of G and ν_4 states of PH_3 illustrating PP, PQ, PR, RP, and RR transitions between G and ν_4 are shown as (----) vertical lines.

A second example is the RP(9,0) absorption match which leads to the excitation of the $\nu_4^{+\ell}(8,1)$ level. A cascade sequence is $\nu_4^{+\ell}(8,1) \rightarrow \nu_4^{+\ell}(7,1)$ followed by $\nu_4^{+\ell}(7,1) \rightarrow \nu_4^{+\ell}(6,1)$ and finally by $\nu_4^{+\ell}(6,1) \rightarrow \nu_4^{+\ell}(5,1)$.

The numerical energies are seen to be

$$\nu_4^{+\ell}(8,1) \rightarrow \nu_4^{+\ell}(7,1),\ 1453.584 - 1380.031 = 73.553\ \text{cm}^{-1}$$
$$\nu_4^{+\ell}(7,1) \rightarrow \nu_4^{+\ell}(6,1),\ 1380.031 - 1315.537 = 64.494\ \text{cm}^{-1}$$
$$\nu_4^{+\ell}(6,1) \rightarrow \nu_4^{+\ell}(5,1),\ 1315.537 - 1260.138 = 55.399\ \text{cm}^{-1}$$

The P_9 (24) pumping is interesting since it simultaneously pumps the QR(5,5) and RP(11,3A2) transitions to yield lasing transitions $\nu_2(6,5) \rightarrow \nu_2(5,5)$ at 51.282 cm^{-1} and $\nu_4^{+\ell}(A_1) \rightarrow \nu_4^{+\ell}(A_2)$ at 91.200 cm^{-1}. Here the A1 doublet of $\nu_4^{+\ell}(10,4)$ is being pumped with lasing occurring to A2 of $\nu_4^{+\ell}(9,4)$.

The relative polarizations of the far IR lasing lines relative to the CO_2 pump lines follow the rules

$$\parallel \rightarrow \Delta J_{abs} + \Delta J_{emiss} = 2n \qquad \text{(even)}$$

$$\perp \rightarrow \Delta J_{abs} + \Delta J_{emiss} = 2n+1 \qquad \text{(odd)}$$

All laser transitions are pure rotational transitions where $\Delta J_{emiss} = -1$, hence Q branch absorptions yield $\perp$ polarizations while P and R branch absorptions yield $\parallel$ polarizations.

Following the above assignment strategy of matching PH_2 absorption lines to CO_2 laser pump lines, looking for pure rotational transitions from the populated ν_2 and/or ν_4 level, and correlating the polarizations, the data in Table III was assembled.

A schematic partial energy level drawing displaying the CO_2 laser line absorptions and the far IR laser line transitions is shown in Figs. 6 and 7.

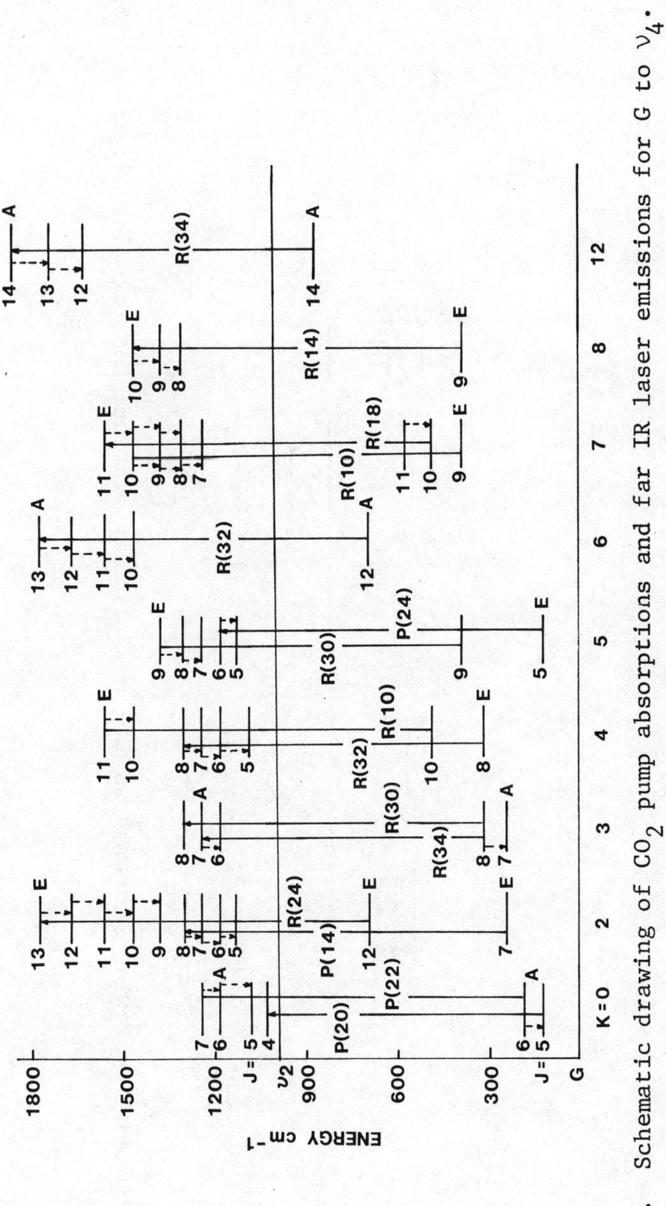

Fig. 6. Schematic drawing of CO_2 pump absorptions and far IR laser emissions for G to ν_4.

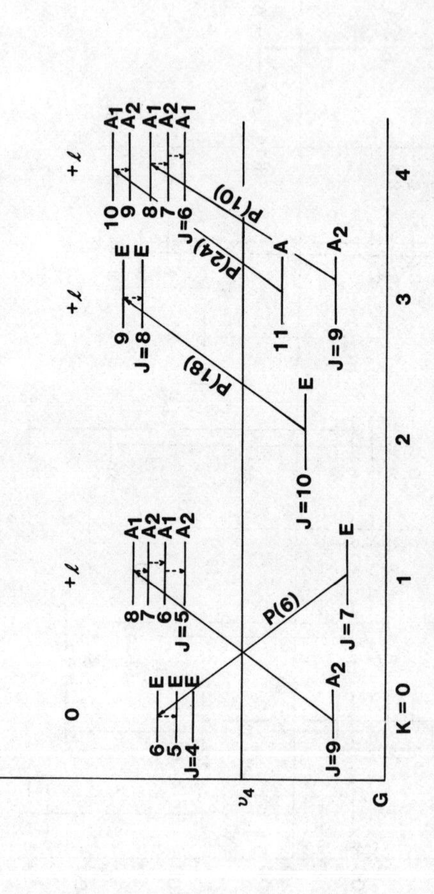

Fig. 7. Schematic drawing illustrating CO_2 pump line absorptions and far IR laser transitions in ν_4 band.

Discussion

The experimental measurement of the frequency of the laser lines was for identification and not for accuracy. Comparing the spectroscopic data of Yin and Tarrago, it would appear that the calculated laser frequencies are accurate to ±0.01 cm^{-1} although on many transitions they agree to ±0.001 cm^{-1}.

Lasing resulting from pumping with CO_2 lines $R(10)_9$ and $R(18)_9$ is of interest since R(10) pumps $\nu_2(10,7)$ and R(18) pumps $\nu_2(11,7)$. This coincidence could be used to study rotational relaxation in PH_3 by using R(18) as a pump and R(10) as a probe. Also the four levels G(10,7), G(9,7), $\nu_2(11,7)$, and $\nu_2(10,7)$ might be used to study four-wave interactions.

The laser transitions

$$\nu_2(13,2) \rightarrow \nu_2(12,2) \text{ at } 111.409 \text{ cm}^{-1}$$
$$\nu_4^0(6,0) \rightarrow \nu_4^0(5,0) \text{ at } 53.691$$
$$\nu_4^{+\ell}(7,1) \rightarrow \nu_4^{+\ell}(6,1) \text{ at } 59.954$$
$$\nu_4^{+\ell}(9,3) \rightarrow \nu_4^{+\ell}(8,3) \text{ at } 82.339$$

were found to have sufficient gain to lase single pass using a far IR cell with no feedback.

The strongest laser signal obtained was the 82.339 cm^{-1} line of the $\nu_4^{+\ell}(9,3) \rightarrow \nu_4^{+\ell}(8,3)$ transition. However, there is no cascading from $\nu_4^{+\ell}(8,3) \rightarrow \nu_4^{+\ell}(7,3)$ at 74.153 cm^{-1} indicating a possible absorption in PH_3 around 74 cm^{-1}.

All laser transitions in the ν_4 band were $\Delta K = 0$ transitions even though $\Delta K = \pm 1$ transitions are allowed.

Acknowledgement

The material in this paper arose through the joint efforts of Edward Malk, Joseph Niesen, Donald Parsons and the author. Our interest in PH_3 as a far IR lasing molecule was whetted through a spirited discussion of PH_3 with Professors K. N. Rao, A. Baldacci and Y. S. Hoh. The author has also had many discussions with Professor J. Henningsen on the general problem of laser assignments in far IR lasers. In this assignment problem one always must rely on the availability of good spectroscopic data, which for PH_3 were the papers of Maki, Yin, and Tarrago. The author gratefully acknowledges the patience of Mrs. J. Smith for typing the extensive tables in this paper.

Table I. Ground State Constants for PH_3.

	MAKI et al.	TARRAGO et al.
B_0	4.4524183 cm^{-1}	4.45241753 cm^{-1}
C_0	3.91894	3.9190139
D_0^J	1.319×10^{-4}	1.3137×10^{-4}
D_0^{JK}	-1.737×10^{-4}	-1.725565×10^{-4}
D_0^K	1.366×10^{-4}	1.3998×10^{-4}
H_0^J	1.83×10^{-8}	1.205×10^{-8}
H_0^{JJK}	-6.2×10^{-8}	-2.352×10^{-8}
H_0^{JKK}	8.0×10^{-8}	-

Table II. Possible Frequency Matches for CO_2 Laser Lines and PH_3 Absorption Lines. (Δ is the frequency offset).

$CO_2(cm^{-1})$	$PH_3(cm^{-1})$	Δ	$CO_2(cm^{-1})$	$PH_3(cm^{-1})$	Δ
R(52) 1095.6663	RP(3,0) 1095.669	-0.0027	*P(22) 1045.0217	QR(6,0) RP(11,7) 1045.037	-0.0153
R(46) 1092.9593	QR(12,9) 1092.922	0.0373	*P(24) 1043.1633	RP(11,3) QR(5,5) 1043.142	0.0213
R(42) 1091.0307	QR(14,0) 1091.045	-0.0143	P(36) 1031.4775	QR(4,0) 1031.479	-0.0015
*R(32) 1085.7655	QR(13,0) 1085.810	-0.0445	P(46) 1021.0578	RP(15,3) 1021.070	-0.0122
*R(24) 1081.0874	QR(12,2) 1081.075	0.0124	P(50) 1016.7227	QR(2,0) RP(16,3) 1016.728	0.0053
*R(18) 1077.3025	QR(10,7) 1077.345	-0.0425	R(52) 993.3792	QQ(12,12) 993.406	-0.0268
*R(14) 1074.6465	QR(9,8) 1074.626	0.0205	R(50) 992.4870	QQ(9,9) 992.507	-0.0200
*R(10) 1071.8837	QR(10,4) 1071.903	-0.0193	*R(34) 984.3840	QQ(14,2) 984.382	0.0020
*P(6) 1058.9487	PP(7,1) 1058.955	-0.0063	*R(32) 983.2530	QQ(8,4) 983.227	0.0300
*P(10) 1055.6251	RP(9,3) 1055.651	-0.0259	*R(30) 982.0962	QQ(9,5) PP(13,11) 982.089	0.0072
*P(12) 1053.9235	RP(9,0) 1053.920	0.0035	R(8) 967.7079	QQ(15,8) 967.726	-0.0181
*P(14) 1052.1956	QR(7,2) 1052.144	0.0516	P(18) 945.9810	QP(5,3) 945.950	0.031
P(16) 1050.4413	PP(8,1) 1050.475	-0.0337	*P(20) 944.1948	QP(5,0) 944.210	-0.0152
*P(18) 1048.6608	PP(7,5) RP(10,2) 1048.684	-0.0232	P(42) 922.9153	QP(7,0) 922.916	-0.0007

Table III. Laser Data and Line Assignments.

CO_2 LINE	ν_{CO_2} cm^{-1}	ABS. TRANS.	ν_{ABS} cm^{-1}	LASER TRANSITION	$(\nu_L)_{CAL}$ (cm^{-1})	$(\nu)_{L_{EXP}}$ (cm^{-1})	POL	OPT PRESS TORR
P(20)	944.194	QP(5,0)	944.212	$G(6,0)\rightarrow G(5,0)$	53.316	53.39	∥	1.6
R(30)	982.095	QQ(9,5)	982.055	$\nu_2(9,5)\rightarrow\nu_2(8,5)$	77.052	77.16	⊥	5.3
		QQ(8,3)	982.055	$\nu_2(8,5)\rightarrow\nu_2(7,5)$	68.461	68.45	⊥	
R(32)	983.252	QQ(8,4)	983.223	$\nu_2(8,4)\rightarrow\nu_2(7,4)$	68.458	68.54	⊥	
				$\nu_2(7,4)\rightarrow\nu_2(6,4)$	59.927	59.99	⊥	2.0
				$\nu_2(6,4)\rightarrow\nu_2(5,4)$	51.312	51.41	⊥	
R(34)	984.383	QQ(7,3)	984.479	$G(8,3)\rightarrow G(7,3)$	70.996	71.69	⊥	3.5
				$\nu_2(7,3)\rightarrow\nu_2(6,3)$	59.937	59.77	⊥	
		QQ(14,12)	984.386	$\nu_2(14,12)\rightarrow\nu_2(13,12)$	119.374	119.20	⊥	
				$\nu_2(13,12)\rightarrow\nu_2(12,12)$	110.795	110.90	⊥	
P(24)	1043.163	QR(5,5)	1043.143	$\nu_2(6,5)\rightarrow\nu_2(5,5)$	51.235	51.36	∥	
		RP(11,3,A2)	1043.142	$\nu_4^{+\ell}(10,4)\rightarrow\nu_4^{+\ell}(9,4)$	91.200	91.16		
P(22)	1045.021	QR(6,0)	1045.044	$\nu_2(7,0)\rightarrow\nu_2(6,0)$	59.977	59.77	∥	2.8
				$\nu_2(6,0)\rightarrow\nu_2(5,0)$	51.422	51.41	∥	
P(18)	1048.661	RP(10,2,E)	1048.648	$\nu_4^{+\ell}(9,3)\rightarrow\nu_4^{+\ell}(8,3)$	82.339	82.44	∥	7.0

(continued)

Table III. (Continued)

CO_2 LINE	ν_{CO_2} cm^{-1}	ABS. TRANS.	ν_{ABS} cm^{-1}	LASER TRANSITION	(ν_L) CAL (cm^{-1})	(ν_L) EXP (cm^{-1})	POL	OPT PRESSS TORR
P(14)	1052.195	QR(7,2)	1052.137	$\nu_2(8,2)\rightarrow\nu_2(7,2)$	68.551	68.68	‖	
				$\nu_2(7,2)\rightarrow\nu_2(6,2)$	59.954	59.70	‖	2.2
				$\nu_2(6,2)\rightarrow\nu_2(5,2)$	51.360	51.44	‖	
P(12)	1053.923	RP(9,0)	1093.925	$\nu_4^{+\ell}(8,1)\rightarrow\nu_4^{+\ell}(7,1)$	73.559	73.64	‖	
				$\nu_4^{+\ell}(7,1)\rightarrow\nu_4^{+\ell}(6,1)$	64.486	64.47	‖	1.7
				$\nu_4^{+\ell}(6,1)\rightarrow\nu_4^{+\ell}(5,1)$	55.389	55.28	‖	
P(10)	1055.625	RP(9,3,A2)	1055.608	$\nu_4^{+\ell}(8,4,A1)\rightarrow\nu_4^{+\ell}(7,4,A2)$	73.148	73.21	‖	4.5
				$\nu_4^{+\ell}(7,4,A2)\rightarrow\nu_4^{+\ell}(6,4,A1)$	63.964	64.23	‖	
P(6)	1058.948	PP(7,1,E)	1058.949	$\nu_4^{0}(6,0)\rightarrow\nu_4^{0}(5,0)$	53.691	53.59	‖	4.1
				$\nu_4^{0}(5,0)\rightarrow\nu_4^{0}(4,0)$	44.828	44.90	‖	
R(10)	1071.884	QR(9,7,E)	1071.894	$\nu_2(10,7)\rightarrow\nu_2(9,7)$	85.556	85.62	‖	
				$\nu_2(9,7)\rightarrow\nu_2(8,7)$	76.933	76.98	‖	2.5
				$\nu_2(8,7)\rightarrow\nu_2(7,7)$	68.332	68.73	‖	
		QR(10,4,E)	1071.904	$\nu_2(11,4)\rightarrow\nu_2(10,4)$	94.259	93.90	‖	2
R(14)	1074.646	QR(9,8)	1074.626	$\nu_2(10,8)\rightarrow\nu_2(9,8)$	85.464	85.54	‖	6.2
				$\nu_2(9,8)\rightarrow\nu_2(8,8)$	76.839	76.75		

CO_2 LINE	ν_{CO_2} cm^{-1}	ABS. TRANS.	ν_{ABS} cm^{-1}	LASER TRANSITION	$(\nu_L)_{CAL}$ (cm^{-1})	$(\nu_L)_{EXP}$ (cm^{-1})	POL	OPT PRESS TORR
R(18)	1077.302	QR(10,7)	1077.358	$G(11,7) \rightarrow G(10,7)$	97.444	97.56	‖	
				$\nu_2(11,7) \rightarrow \nu_2(10,7)$	94.134	94.25	‖	3.0
				$\nu_2(10,7) \rightarrow \nu_2(9,7)$	85.556	85.62	‖	
				$\nu_2(9,7) \rightarrow \nu_2(8,7)$	76.933	76.86	‖	
R(24)	1081.087	QR(12,2)	1081.067	$\nu_2(13,2) \rightarrow \nu_2(12,2)$	111.409	111.60	‖	
				$\nu_2(12,2) \rightarrow \nu_2(11,2)$	102.883	103.30	‖	2.5
				$\nu_2(11,2) \rightarrow \nu_2(10,2)$	94.304	94.10	‖	
R(32)	1085.799	QR(12,6)	1085.80	$\nu_2(13,6) \rightarrow \nu_2(12,6)$	111.359	110.90	‖	
				$\nu_2(12,6) \rightarrow \nu_2(11,6)$	102.773	102.40	‖	3.5
				$\nu_2(11,6) \rightarrow \nu_2(10,6)$	94.293	94.30	‖	

Table IV. Absorption Transitions from G → ν_2 for PH_3. Data without ( ) from TDG. Data with ( ) from YR.

J	K	QP(J,K;A/E)	QQ(J,K;A/E)	QR(J,K;A/E)
0	0	-	992.131A	1000.683A (1000.681)
1	0	983.760A	991.071A	1008.877A2 (1008.873)
1	1	-	991.980E (991.848)	1009.080E (1009.076)
2	0	973.972A (973.976)	-	1016.728A (1016.728)
2	1	974.173E (974.174)	991.274E (991.274)	1016.924E (1016.924)
2	2	-	991.848E (991.848)	1017.527E (1017.528)
3	0	964.372A2 (964.372)	990.072A (990.072)	1024.250A2 (1024.246)
3	1	964.575E (964.707)	990.230E (990.224)	1024.448E (1024.439)
3	2	965.174E (965.173)	990.824E (990.823)	1025.032E (1025.035)
3	3	-	991.798E (991.848)	1026.024A (1026.018)
4	0	954.445A1 (954.446)	988.701A (988.701)	1031.479A1 (1031.458)
4	1	954.639E (954.642)	988.956E (988.940)	1031.659E (1031.654)
4	2	955.230E (955.230)	989.454E (989.468)	1032.224E (1032.224)
4	3	956.228A (956.229)	990.428E (990.420)	1033.192E (1033.190)
4	4	-	991.798E (991.848)	1034.564 (1034.562)
5	0	944.210A2 (944.212)	986.867A (986.867)	1038.399A2 (1038.378)
5	1	944.398E (944.403)	987.195E (987.195)	1038.565E (1038.572)

J	K	QP(J,K;A/E)	QQ(J,K;A/E)	QR(J,K;A/E)
5	2	944.976E (944.978)	987.761E (987.759)	1039.127E (1039.126)
5	3	945.950A (945.952)	988.702A (988.701)	1040.062A (1040.063)
5	4	947.336E (947.339)	990.079E (990.072)	1041.399E (1041.387)
5	5	-	991.798E (991.951)	1043.142E (1043.143)
6	0	933.692A1 (933.694)	985.066A (985.066)	1045.037A (1045.044)
6	1	933.875E (933.882)	985.252E (985.261)	1045.188E (1045.203)
6	2	934.437E (934.435)	985.776E (985.780)	1045.759E (1045.761)
6	3	935.385A (935.385)	986.729A (986.728)	1046.639A (1046.677)
6	4	936.727E (936.731)	988.048E (988.051)	1047.954E (1047.957)
6	5	938.499E (938.508)	989.780E (989.782)	1049.648E (1049.646)
6	6	-	991.980A (991.951)	1051.765A (1051.765)
7	0	922.916A2 (922.912)	982.887A	1051.452A2 (1051.446)
7	1	923.095E (923.096)	983.081E (983.076)	1051.628E (1051.602)
7	2	923.638E (923.641)	983.558E (983.563)	1052.144E (1052.137)
7	3	924.553A (924.557)	984.484A (984.479)	1053.020A (1053.042)
7	4	925.854E (925.856)	985.776E (985.780)	1054.267E (1054.255)
7	5	927.566E (927.570)	987.434E (987.433)	1055.888E (1055.888)
7	6	929.719A (929.720)	989.532A (989.536)	1057.937A (1057.938)
7	7	-	992.094E (992.103)	1060.435E

Table IV. (Continued)

J	K	QP(J,K;A/E)	QQ(J,K;A/E)	QR(J,K;A/E)
8	0	911.913A (911.913)	980.483A (980.492)	1057.633A (1057.633)
8	1	912.087E (912.088)	980.642E (980.643)	1057.801E (1057.803)
8	2	912.610 (912.613)	981.160E (981.142)	1058.301E (1058.302)
8	3	913.491A (913.495)	982.035A (982.055)	1059.152A (1059.140)
8	4	914.745E (914.749)	983.227E (983.223)	1060.341E (1060.344)
8	5	916.392E (916.388)	984.854E (984.855)	1061.911E (1061.909)
8	6	918.461A (918.460)	986.865A (986.867)	1063.876A (1063.860)
8	7	920.994E (920.998)	989.328E (989.327)	1066.244E (1066.241)
8	8	-	992.278E (992.281)	1069.129E (1069.122)
9	0	900.709A2 (900.718)	977.867A	1063.611A2 (1063.607)
9	1	900.877E (900.885)	978.033E (978.149)	1063.768E (1063.778)
9	2	901.381E (901.385)	978.511E (978.510)	1064.260E (1064.281)
9	4	902.228A (902.232)	979.353A (979.354)	1065.070E (1065.062)
9	4	903.432E (903.442)	980.526E (980.492)	1066.219E (1066.241)
9	5	905.010E (905.016)	982.089E (982.055)	1067.718E (1067.711)
9	6	906.994A (907.001)	983.998A (984.011)	1069.592A (1069.379)
9	7	909.413E (909.419)	986.357E (986.360)	1071.903E (1071.904)
9	8	912.317E (912.322)	989.163E (989.152)	1074.626E (1074.629)
9	9	-	992.507A (992.509)	1077.851A (1077.851)

J	K	QP(J,K;A/E)	QQ(J,K;A/E)	QR(J,K;A/E)
10	0	889.335A1 (889.335)	975.079A	1069.412A (1069.422)
10	1	889.492E (889.497)	975.246E (975.269)	1069.592E (1069.579)
10	2	889.978E (889.981)	975.708E (975.720)	1070.023E (1070.022)
10	3	890.791A (890.793)	976.493A (976.497)	1070.776A (1070.788)
10	4	891.945E (891.948)	977.633E (977.636)	1071.903E (1071.904)
10	5	893.455E (893.458)	979.113E (979.118)	1073.367E (1073.368)
10	6	895.347A (895.350)	980.957A (980.958)	1075.165A (1075.167)
10	7	897.657E (897.664)	983.204E (983.223)	1077.345E (1077.358E)
10	8	900.421E (900.425)	985.878E (985.876)	1079.949E (1079.948)
10	9	903.698A (903.703)	989.051A (989.021)	1083.023A (1083.020)
10	10	-	992.767E (992.762)	1086.604E (1086.599)
11	0	877.811A2 (877.812)	974.145A	1075.029A2 (1075.030)
11	1	877.966E (977.964)	972.299E (971.979)	1075.165E (1075.167)
11	2	878.426E (878.433)	972.751E (972.752)	1075.627E (1075.629)
11	3	879.193A (879.200)	973.503E (973.504)	1076.375A (1076.376)
11	4	880.311E (880.315)	974.586E (974.584)	1077.437E (1077.432)
11	5	881.753E (881.758)	975.996E (976.002)	1078.827E (1078.839)
11	6	883.558A (993.562)	977.760A (977.765)	1080.541A (1080.506)
11	7	885.756E (885.759)	979.891E (979.880)	1082.638E (1082.640)

Table IV. (Continued)

J	K	QP(J,K;A/E)	QQ(J,K;A/E)	QR(J,K;A/E)
11	8	888.380E (888.383)	982.448E (982.451)	1085.116E (1085.115)
11	9	891.485A (891.490)	985.450A (985.454)	1088.038A (1088.038)
11	10	895.131E	988.956E (988.940)	1091.426E (1091.422)
11	11	-	993.071E (993.064)	1095.378E
12	0	866.165A (866.163)	969.031A (969.031)	1080.478A (1080.506)
12	1	866.313E (866.315)	969.137E (969.137)	1080.641E (1080.628)
12	2	866.758E (866.756)	969.637E (969.637)	1081.075E (1081.067)
12	3	867.499A (867.494)	970.364A (970.372)	1081.792A (1081.779)
12	4	868.553E (868.555)	971.406E (971.222)	1082.814E (1082.814)
12	5	869.929E (869.944)	972.751E (972.752)	1084.143E (1084.267)
12	6	871.650A (871.690)	974.443A (974.450)	1085.810A
12	7	873.746E (873.750)	976.493E (976.497)	1087.792E (1087.803)
12	8	876.229E (876.228)	978.905E (978.910)	1090.144E (1090.116)
12	9	879.193A (879.166)	981.739A (981.738)	1092.922A
12	10	882.599E (882.599)	985.062E (985.066)	1096.138E
12	11	886.614E (886.617)	988.956E (988.940)	1099.870E
12	12	-	993.406A (993.404)	1104.178A
13	0	854.418A2 (854.309)	965.843A2	1085.810A2
13	1	854.557E (854.425)	966.387E	1085.972E

J	K	QP(J,K;A/E)	QQ(J,K;A/E)	QR(J,K;A/E)
13	2	854.983E (854.981)	966.407E (966.387)	1086.387E
13	3	855.693A (855.708)	967.070A1 (967.074) 967.162A2	1087.042A2 1087.136A1
13	4	856.702E (856.707)	968.107E (968.106)	1089.038E
13	5	858.014E (858.019)	969.403E (969.389)	1089.325E
13	6	859.649A (859.689)	971.011A (971.009)	1090.927A
13	7	861.635E (861.642)	972.941E (972.944)	1092.817E
13	8	863.990E (863.989)	975.246E (975.269)	1095.071E

Table V. Absorption Transitions from G → ν_4 for PH_3. Data without ( ) from TDG. Data with ( ) from YR.

J	K	PP(J,K)	PQ(J,K)	PR(J,K)	RP(J,K)	RQ(J,K)	RR(J,K)
0	0	-	-	-	-	-	(1130.522A1)
1	0	-	-	-	-	(1121.274A2)	1140.198 (1140.174A2)
1	1	-	(1119.045E)	(1137.195E)	-	-	(1141.843E)
2	0	-	(1121.338A1)	-	(1103.814A1)	-	1150.169 (1150.176A)
2	1	1101.239 (1101.236E)	1119.404 (1119.408E)	(1146.559E)	- -	1124.014 (1124.034E)	1151.495
2	2	1098.582 (1098.585E)	1116.852 (1116.852E)	- -	- -	- -	1153.257 (1153.248E)
3	0	-	-	-	(1095.669A2)	(1121.424A2)	1160.521 (1160.512A)
3	1	1092.704 (1092.814E)	1119.878 (1119.876E)	(1155.903E)	(1097.341E)	1124.791 (1124.790E)	1161.501 (1161.493E)
3	2	1090.116 (1090.144E)	1117.344 (1117.349E)	(1153.574E)	-	1126.550 (1126.550E)	1163.090
3	3	1087.040 (1087.205A)	1114.533 (1114.521A)	-	-	-	1164.603 (1164.600A)
4	0	-	-	-	(1087.894A1)	(1121.527A1)	1171.158 (1171.162A1)
4	1	1084.262 (1084.270E)	1120.433 (1120.321E)	(1165.174E)	(1089.202E)	1125.900 (1125.903E)	1171.873 (1171.865E)

J	K	PP(J,K)	PQ(J,K)	PR(J,K)	RP(J,K)	RQ(J,K)	RR(J,K)
4	2	1081.770 (1081.768E)	1117.974 (1117.993E)		(1090.927E)	1127.483 (1127.497E)	1173.236 (1173.231E)
4	3	1078.915 (1078.927A)	1115.246 (1115.249A)	(1160.512A)	-	1129.002 (1129.007A)	1174.632 (1174.640A)
4	4	1075.776 (1075.768E)	1112.182	-	-	-	1175.902 (1175.900E)
5	0	-	-	-	(1080.476A2)	(1121.685A2)	1182.089 (1182.085A2)
5	1	1075.847 (1075.847E)	1120.743 (1120.745E)	(1174.440E)	(1081.441E)	1127.403 (1127.402E)	1182.594 (1182.591E)
5	2	1073.522 (1073.524E)	1118.643 (1118.653E)	-	(1083.023E)	1128.762 (1128.765E)	1183.718 (1183.711E)
5	3	1070.788 (1070.776A)	1116.026 (1116.065A)	-	(1084.533A)	1130.161 (1130.162A)	1184.956 (1184.947A2) (1184.970A1)
5	4	1067.711 (1067.718E)	1113.129 (1113.132E)	-	-	1131.426 (1131.416E)	1186.125 (1186.123E)
5	5	1064.281 (1064.305E)	1109.830 (1109.832E)	-	-	-	1187.134 (1187.116E)
6	0	-	-	-	(1073.367A1)	(1121.843A2)	1193.259 (1193.260A1)
6	1	1067.410 (1067.414E)	1121.118 (1121.121E)	-	(1074.086E)	1129.268 (1129.272E)	1193.625 (1193.627E)
6	2	1065.325 (1065.327E)	1119.404 (1119.283E)	-	(1075.441E)	1130.391 (1130.385E)	1194.512 (1194.513E)

J	K	PP(J,K)	PQ(J,K)	PR(J,K)	RP(J,K)	RQ(J,K)	RR(J,K)
6	3	1062.721 (1062.723A)	1116.852 (1116.852A1) (1116.899A2)	-	(1076.827A)	1131.617 (1131.628A)	1195.578 (1195.543A1) (1195.618A2)
6	4	1059.761 (1059.761E)	1114.073 (1114.081E)	-	(1078.052E)	1132.780 (1132.775E)	1196.628 (1196.626E)
6	5	1056.469 (1056.465E)	1110.921 (1110.924E)	-	-	1133.745 (1133.752E)	1197.544 (1197.541E)
6	6	1052.825 (1052.829A)	1107.445 (1107.439A)	-	-	-	1198.274 (1198.287A)
7	0	-	-	-	(1066.611A2)	(1122.091A2)	1204.663 (1204.657A2)
7	1	1058.947 (1058.955E)	1121.503 (1121.424E)	-	(1067.116E)	1131.426 (1131.416E)	1204.924 (1204.927E)
7	2	1057.123 (1057.124E)	1119.878 (1119.876E)	-	(1068.221E)	1132.343 (1132.344E)	1205.613 (1205.615E)
7	3	1054.706 (1054.685A1) (1054.738A2)	1117.700 (1117.835A2)	-	(1069.437A1) (1069.412A2)	1133.355 (1133.364A1) (1133.440A2)	1206.431 (1206.434A2) (1206.560A1)
7	4	1051.864 (1051.877E)	1115.029 (1115.043E)	-	(1070.582E)	1133.407 (1134.433E)	1206.550 (1207.404E)
7	5	1048.687 (1048.684E)	1112.037 (1112.045E)	(1184.179E)	(1071.540E)	1135.325 (1135.331E)	1208.222 (1208.226E)
7	6	1045.203 (1045.188A)	1108.696 (1108.704A)	(1181.049A)	-	1136.036 (1136.035A)	1208.884 (1208.895A)
7	7	1041.340 (1041.320E)	1105.029 (1105.020E)	(1177.568E)	-	-	1209.354 (1209.355E)

J	K	PP(J,K)	PQ(J,K)	PR(J,K)	RP(J,K)	RQ(J,K)	RR(J,K)
8	0	-	-	-	(1060.133A1)	(1122.272A1)	1216.250 (1216.250A1)
8	1	1050.478 (1050.475E)	1121.801 (1121.817E)	-	(1060.486E)	1133.951 (1133.951E)	1216.450 (1216.447E)
8	2	1048.902 (1048.902E)	1120.524 (1120.434E)	(1144.053E)	(1061.364E)	1134.637	1216.974 (1216.976E)
8	3	1046.667 (1046.723A1) (1046.639A2)	1119.260 (1118.874A1) (1118.511A2)	-	(1062.369A1) (1062.444A2)	1135.434 (1135.437A2) 1135.559 (1135.563A1)	1217.597 (1217.595A1) 1217.792 (1217.784A2)
8	4	1044.028 (1044.030E)	1116.026 (1116.026E)	-	(1063.418E)	1136.393 (1136.389E)	1218.451 (1218.450E)
8	5	1040.996 (1040.999E)	1113.129 (1113.132E)	-	(1064.305E)	1137.195 (1137.195E)	1219.172 (1219.166E)
8	6	1037.637 (1037.636A)	(1109.953A)	(1191.095A)	(1064.986A)	1137.820 (1137.815A)	1219.750 (1219.750A)
8	7	1033.912 (1033.914E)	1106.713 (1106.466E)	(1187.785E)	-	1138.251 (1138.251E)	1220.164 (1220.165E)
8	8	1029.781 (1029.785E)	1102.553 (1102.551E)	-	-	-	1220.357 (1220.363E)
9	0	-	-	-	(1053.920A2)	(1122.523A2)	1228.013 (1228.012A2)
9	1	1041.989 (1042.000E)	1122.070 (1122.091E)	-	(1054.186E)	1136.670 (1136.680E)	1228.162 (1228.159E)
9	2	1040.654 (1040.652E)	1120.931 (1120.931E)	-	(1054.852E)	1137.195 (1137.195E)	1228.564 (1228.564E)

J	K	PP(J,K)	PQ(J,K)	PR(J,K)	RP(J,K)	RQ(J,K)	RR(J,K)
9	3	1038.630 (1038.565A1) 1038.833 (1038.712A2)	1119.260 (1119.283A1)	–	(1055.561A2) (1055.768A1)	1137.820 (1137.815A1) 1137.991 (1137.992A2)	1228.991 (1228.990A2) 1229.257 (1229.255A1)
9	4	1036.178 (1036.175E)	1116.852 (1116.899E)	–	(1056.575E)	1138.643 (1138.637E)	1229.754 (1229.743E)
9	5	1033.303 (1033.312E)	1114.226 (1114.227E)	–	(1057.348E)	1139.330 (1139.324E)	1230.343 (1230.347E)
9	6	1030.101 (1030.107A)	1111.221 (1111.220A)	(1201.062A)	(1057.937A)	1139.873 (1139.877A)	1230.864 (1230.867A)
9	7	1026.549 (1026.547E)	1107.881 (1107.880E)	(1197.997E)	(1058.301)	1140.198 (1140.248E)	1231.211 (1231.206E)
9	8	1022.594 (1022.588E)	1104.198 (1104.178E)	–	–	1140.417 (1140.413E)	1231.308 (1231.361E)
9	9	1018.239 (1018.240A)	1099.866 (1100.106A)	–	–	–	1231.308 (1231.291A)
10	0	–	–	–	(1047.954A1)	(1122.728A1)	(1239.921A1)
10	1	1033.549 (1033.563E)	(1122.416E)	–	(1048.144E)	(1139.624E)	(1240.028E)
10	2	1032.376 (1032.386E)	(1121.424E)	–	(1048.684E)	(1140.008E)	(1240.333E)
10	3	1030.479 (1030.482A2) 1030.689 (1030.690A1)	(1119.664A1) (1119.914A2)	–	(1049.241A1) (1049.431A2)	(1140.413A2) (1140.695A1)	(1240.917A2) (1240.590A1)

J	K	PP(J,K)	PQ(J,K)	PR(J,K)	RP(J,K)	RQ(J,K)	RR(J,K)
10	4	1028.298 (1028.297E)	(1117.730E)	-	(1050.052E)	(1141.160E)	(1241.257E)
10	5	1025.614 (1025.612E)	(1115.249E)	(1213.549E)	(1050.711E)	(1141.733E)	(1241.750E)
10	6	1022.594 (1022.588A)	(1112.407A)	(1210.932A)	(1051.227)	(1142.216A2)	(1242.158A)
10	7	1019.193 (1019.185E)	(1109.238E)	(1207.994E)	(1051.628E)	(1142.495E)	(1242.461E)
10	8	1015.444 (1015.446E)	(1105.754E)	(1204.747E)	(1051.628E)	(1142.616E)	(1242.567E)
10	9	1011.301 (1101.304A)	(1101.897A)	(1201.165A)	-	(1142.495A)	(1242.661A)
10	10	1006.678 (1006.681E)	-	(1197.223E)	-	-	(1242.158E)
11	0	-	-	-	(1042.211A2)	(1123.033A2)	(1251.919A2)
11	1	1025.035 (1025.148E)	(1122.805E)	-	(1042.352E)	(1142.752E)	(1252.027E)
11	2	1024.116 (1024.118E)	(1121.843E)	-	(1042.733E)	(1143.050E)	(1252.255E)
11	3	1022.355 (1022.358A1) 1022.594 (1022.588A2)	(1120.245A2) (1120.544A1)	(1227.169A2)	(1043.142A2) (1043.391A1)	(1143.284A1) (1143.933A2)	(1252.358A2) (1252.802A1)
11	4	1020.390 (1020.392E)	(1118.511E)	-	(1043.836E)	(1143.933E)	(1252.956E)

J	K	PP(J,K)	PQ(J,K)	PR(J,K)	RP(J,K)	RQ(J,K)	RR(J,K)
11	5	1017.885 (1017.884E)	(1116.189E)	(1223.137E)	(1044.375E)	(1144.392E)	(1253.352E)
11	6	1015.011 (1015.014A)	(1113.535A)	(1220.694A)	(1044.820A)	(1144.777A)	(1253.679A)
11	7	1011.804 (1011.804)	(1110.551E)	(1217.932E)	(1045.037E)	(1144.995E)	(1253.679E)
11	8	1008.253 (1008.254E)	(1107.248E)	(1214.871E)	(1045.188E)	(1145.070E)	(1253.965E)
11	9	1004.345 (1004.349A)	(1103.601A)	(1211.498A)	(1044.902A)	(1144.907A)	(1253.855A)
11	10	999.989 (999.990)	(1099.579E)	(1207.797E)	–	(1144.493E)	–
11	11	995.108 (995.114)	(1095.155E)	–	–	–	(1252.884E)
12	0	–	–	–	(1036.674A1)	(1123.339A1)	(1264.072A1)
12	1	–	–	–	(1036.776E)	(1146.047E)	(1264.127E)
12	2	(1015.863E)	(1122.272E)	–	(1037.063E)	(1146.267E)	–
12	3	(1014.562A1) (1014.221A2)	(1120.745A1) (1121.121A2)	–	(1037.288E)	(1146.806A1) (1146.344A2)	–
12	4	(1012.458E)	(1119.283E)	–	(1037.900E)	(1146.925E)	–
12	5	(1010.123E)	(1117.072E)	–	(1038.399E)	(1147.282E)	(1265.118E)
12	6	(1007.425A)	(1114.589A)	–	(1038.712A)	(1147.575A)	–
12	7	(1004.349E)	(1111.777E)	(1227.761E)	(1038.838E)	(1147.575E)	–
12	8	(1001.022E)	–	(1224.883E)	(1038.838E)	(1147.745E)	–

J	K	PP(J,K)	PQ(J,K)	PR(J,K)	RP(J,K)	RQ(J,K)	RR(J,K)
12	9	(997.313A)	(1105.213A)	(1221.695A)	(1038.565A)	(1147.575A)	(1265.390A)
12	10	(993.222E)	(1101.432E)	(1218.214E)	–	(1147.139E)	(1265.049E)
12	11	(988.702E)	–	(1214.402E)	–	(1146.432E)	(1264.445E)
12	12	(983.558A)	–	–	–	–	(1263.548A)
13	0	–	–	–	(1031.336A2)	(1123.698A2)	(1276.279A2)
13	1	–	(1123.339E)	–	(1031.438E)	(1149.484E)	–
13	2	(1007.615E)	(1122.805E)	–	(1031.659E)	(1149.632E)	–
13	3	(1006.489A2)	(1121.817A1) (1121.338A2)	–	(1031.659A1) (1032.114A2)	–	–
13	4	(1004.494E)	(1119.914E)	–	–	(1150.176E)	–
13	5	(1002.332E)	(1117.993E)	–	(1032.556E)	(1150.412E)	–
13	6	(999.979A)	(1115.565A)	–	(1032.798A)	(1150.597A)	–
13	7	(996.925E)	(1112.969E)	–	(1032.666E)	(1150.725E)	–
13	8	(993.765E)	(1109.958E)	–	(1032.798E)	–	(1277.630E)
13	9	(990.230A)	(1106.714A)	–	(1032.556A)	–	–
13	10	(986.357E)	(1103.158E)	–	(1032.114E)	(1149.988E)	(1276.751E)
13	11	(982.089E)	(1099.247E)	(1224.883E)	(1031.336E)	(1149.267E)	–
13	12	(977.365A)	(1094.940A)	(1220.969A)	–	(1148.287A)	–
13	13	(971.974E)	–	–	–	–	–
14	0	–	–	–	(1206.177A1)	(1124.034A1)	–
14	1	–	(1123.770E)	–	(1026.177E)	–	(1288.670E)

J	K	PP(J,K)	PQ(J,K)	PR(J,K)	RP(J,K)	RQ(J,K)	RR(J,K)
14	2	-	(1122.805E)	-	(1026.334E)	-	-
14	3	(997.944A2)	(1121.843A1) (1122.46 A2	-	(1026.279A1)	(1153.661A1)	-
14	4	(996.615E)	(1120.544E)	-	(1026.795E)	(1153.459E)	-
14	5	(994.606E)	(1118.653E)	-	(1027.005E)	(1153.661E)	-
14	6	(992.278A)	(1116.461A)	-	(1027.157A)	(1153.791A)	(1289.174A)
14	7	(989.454E)	(1114.081E)	-	(1027.241E)	(1153.791E)	-
14	8	(986.357E)	(1111.220E)	-	(1027.005E)	(1154.077E)	-
14	9	(983.081A)	(1108.123A)	-	(1026.795A)	-	-
14	10	(979.440E)	(1104.756E)	-	(1026.279E)	-	-
14	11	(975.432E)	(1101.116E)	-	(1025.475E)	-	(1287.998E)
14	12	(971.011A)	(1097.057A)	-	-	-	-
14	13	(966.071E)	-	-	-	-	-
14	14	-	-	-	-	-	(1284.599E)
15	0	-	-	-	(1021.170A2)	(1124.444A2)	-
15	1	-	(1124.145E)	-	(1021.170E)	-	(1300.961E)
15	2	-	(1123.528E)	-	(1021.281E)	-	-
15	3	-	(1123.033A1) (1122.416A2)	-	(1021.070A2)	(1158.023A)	-
15	4	-	(1121.274E)	-	(1021.530E)	(1156.960E)	-
15	5	(986.729E)	(1119.530E)	-	(1021.753E)	(1157.099E)	-
15	6	(984.484A)	-	-	(1021.753A)	(1157.099A)	-

J	K	PP(J,K)	PQ(J,K)	PR(J,K)	RP(J,K)	RQ(J,K)	RR(J,K)
15	7	(982.035E)	(1115.150E)	-	(1021.753E)	(1157.150E)	(1301.274E)
15	8	(979.113E)	(1112.407E)	-	(1021.910E)	-	(1301.206E)
15	9	(975.881A)	(1109.431A)	-	(1021.170A)	-	(1301.274A)
15	10	(972.417E)	(1106.339E)	-	(1020.704E)	-	(1300.445E)

Table VI. Energy Levels for G_0, ν_2, $\nu_4^{+\ell}$, and $\nu_4^{-\ell}$ for PH_3. Data without ( ) from TDG. Data with ( ) from YR.

J	K	$G_0(J,K)$	$\nu_2(J,K)$	$\nu_4^{+\ell}(J,K)$	$\nu_4^{-\ell}(J,K)$
0	0		992.131A	1118.313	1118.313
1	0	8.90A	1000.682A (1000.683)	-	1127.413E (1127.413)
1	1	8.371E	1000.350E (1000.285)	1130.524A2	1123.163E (1123.282)
2	0	26.710A	1017.781A2 (1017.779)	-	1145.586E (1145.582)
2	1	26.177E	1017.451E (1017.494)	1149.079A1 (1149.102)	1141.430E (1141.414)
2	2	24.578E	1016.376E (1016.440)	1150.220E (1150.201)	1135.822A
3	0	53.410A	1043.441A (1043.454)	-	1172.736E (1172.742)
3	1	52.878E	1043.109E (1043.104)	1176.890A1 (1176.879)	1168.642E (1168.635)
3	2	51.282E	1042.107A (1042.105)	1177.668E (1177.671)	1163.138A (1163.140)
3	3	48.616A	1040.415A (1040.454)	1177.801E (1177.834)	1156.249E (1156.257)
4	0	88.996A	1077.665A2 (1077.673)	-	1208.787E (1208.836)
4	1	88.466E	1077.324E (1077.350)	1213.933A1 (1213.931)	1204.867E (1204.855)
4	2	86.874E	1076.315E (1076.325)	1214.367E (1214.373)	1199.464A (1199.469)
4	3	84.215A	1074.643A (1074.636)	1214.362A (1214.365)	1192.685E (1192.670)
4	4	80.481E	1072.279E (1072.317)	1213.222A (1213.218)	1184.466E (1184.443)
5	0	133.454A	1120.463A2 (1120.413)	-	1253.671E (1253.664)
5	1	132.926E	1120.120E (1120.122)	1260.138A2 (1260.154)	1255.140A1 (1120.122)
5	2	131.339E	1119.101E (1119.098)	1260.331E (1260.334)	1244.754A (1244.729)

J	K	$G_0(J,K)$	$\nu_2(J,K)$	$\nu_4^{+\ell}(J,K)$	$\nu_4^{-\ell}(J,K)$
5	3	128.688A	117.408A (117.400)	1260.105E (1260.106)	1238.077E (1238.087)
5	4	124.967E	1115.043E (1115.043)	1258.850A (1258.848)	1229.993E (1229.996)
5	5	120.162E	1112.027E (1112.075)	1256.368E (1256.388)	1220.475 (1220.476)
6	0	186.770A	1171.842A1 (1171.835)	-	1307.365E (1307.355)
6	1	186.244A	1171.497E (1171.500)	1315.537A1 (1315.543)	1308.514A2 (1304.008)
6	2	184.663E	1170.466E (1170.458)	1315.518E (1315.516)	1298.885A2 (1298.890) 1298.938A1
6	3	182.023A	1168.753A (1168.752)	1315.049E (1315.055)	1292.385E (1292.380)
6	4	178.316E	1166.363E (1166.361)	1313.612A1 1313.637A2 (1313.642)	1284.426E (1284.441)
6	5	173.530E	1163.309E (1163.310)	1311.090E (1311.094)	1275.075A (1275.096)
6	6	167.651A	1159.606A (1159.607)	1307.283E (1307.286)	1264.243E (1264.274)
7	0	248.925A	1231.811A2 (1231.812)	-	1369.826E (1369.879)
7	1	248.401E	1231.464E (1231.462)	1380.031A2 (1380.029)	1371.017A1 (1366.709)
7	2	246.827E	1230.421E (1230.412)	1379.863E (1379.848)	1361.836A1) (1361.881) 1361.919A2
7	3	244.199A	1228.687A (1228.689)	1379.175E (1379.173)	1355.554E (1355.544)
7	4	240.508E	1226.269E (1226.278)	1377.641A1 (1377.606) 1377.565A2 (1377.554)	1347.781E (1347.780)
7	5	235.744E	1223.174E (1223.175)	1374.942E (1374.943)	1338.592A (1338.593)
7	6	229.892A	1219.417A (1219.421)	1371.087E (1371.072)	1327.941E (1327.957)

J	K	$G_0(J,K)$	$\nu_2(J,K)$	$\nu_4^{+\ell}(J,K)$	$\nu_4^{-\ell}(J,K)$
7	7	222.934E	1215.021E (1215.037)	1365.944A (1365.927)	1315.759E (1315.779)
8	0	319.897A	1300.373A1 (1300.380)	-	1441.146E (1441.155)
8	1	319.376E	1300.023E (1300.017)	1453.584A1 (1453.588)	1438.241E (1438.288)
8	2	317.810E	1298.970E (1298.963)	1453.332E (1453.326)	1433.555A2 (1433.617) 1433.700A1 (1433.820)
8	3	315.195A	1297.216A (1297.237)	1452.441E (1452.444)	1427.512E (1427.532)
8	4	311.523E	1294.769E (1294.762)	1450.758A2 (1450.754) 1450.639A1 (1450.629)	1419.934E (1419.919)
8	5	306.783E	1291.633E (1291.636)	1447.912E (1447.915)	1410.937A (1410.925)
8	6	300.961A	1287.824A (1287.831)	1443.971E (1443.972)	1400.491E (1400.628)
8	7	294.040E	1283.354E (1283.369)	1438.767A (1438.779)	1388.521E (1388.552)
8	8	285.998E	1278.250E (1278.280)	1432.242E (1432.290)	1375.021A (1375.067)
9	0	399.663A	1377.531A2 (1377.530)	-	1521.237E (1521.221)
9	1	399.144E	1377.173E (1377.215)	1536.150A2 (1536.347)	1518.516E (1518.513)
9	2	397.587E	1376.112E (1376.100)	1535.825E (1535.814)	1514.034A1 (1514.027) 1514.238A2 (1514.237)
9	3	394.987A	1374.339A (1374.338)	1534.818E (1534.783)	1508.218E (1508.203)
9	4	391.336E	1371.866E (1371.854)	1532.983A1 (1532.978) 1532.789A2 (1532.807)	1500.848E (1500.850)
9	5	386.623E	1368.691E (1368.688)	1529.973E (1529.977)	1492.051A (1492.067)

J	K	$G_0(J,K)$	$\nu_2(J,K)$	$\nu_4^{+\ell}(J,K)$	$\nu_4^{-\ell}(J,K)$
9	6	380.834A	1364.827A (1364.833)	1525.947E (1525.954)	1481.812E (1481.837)
9	7	373.953E	1360.292E (1360.302)	1520.707A (1520.709)	1470.124E (1470.151)
9	8	365.958E	1355.099E (1355.119)	1514.263E (1514.178)	1456.888A (1456.813)
9	9	356.828A	1349.282A (1349.335)	1506.306E (1506.365)	1442.007E (1442.082)
10	0	488.194A	1463.275A1 (1463.273)	-	1610.100E (1609.967)
10	1	487.679E	1462.918E (1462.928)	1627.675A1 (1627.676)	1607.534E (1607.488)
10	2	486.132A	1461.842E (1461.855)	1627.304E (1627.317)	1603.211A2 (1603.203) 1603.435A1 (1603.442)
10	3	483.548A	1460.040A (1460.047)	1626.149E (1626.145)	1597.639E (1597.627)
10	4	479.920E	1457.557E (1457.564)	1624.244A2 (1624.253) 1623.989A1 (1623.965)	1590.478E (1590.481)
10	5	475.236E	1454.347E (1454.346)	1621.083E (1621.087)	1581.892A (1581.889)
10	6	469.484A	1450.436A (1450.366)	1616.969E (1616.973)	1571.683E (1571.883)
10	7	462.647E	1445.835E (1445.858)	1611.698A (1611.700)	1560.431E (1560.464)
10	8	454.703E	1440.557E (1440.583)	1605.116E (1605.157)	1547.495A (1547.534)
10	9	445.631A	1434.631A (1434.670)	1597.365E (1597.286)	1532.947E (1532.888)
10	10	435.404E	1428.088E (1428.166)	1588.048A (1588.185)	1516.691E (1616.818)
11	0	585.463A	1557.601A2 (1559.608)	-	(1707.290)
11	1	584.951E	1557.241E (1557.143)	1728.110A2 (1728.116)	1705.266E (1705.345)

J	K	$G_0(J,K)$	$\nu_2(J,K)$	$\nu_4^{+\ell}(J,K)$	$\nu_4^{-\ell}(J,K)$
11	2	583.414E	1556.161E (1556.159)	1727.704E (1727.703)	1701.083A1 (1701.250) 1701.418A2
11	3	580.848	1554.350A (1554.345)	1726.466E (1726.471)	1695.738E (1695.514)
11	4	577.245E	1551.833E (1551.828)	1724.498A1 (1724.501) 1724.139A2 (1724.137)	1688.785E (1688.775)
11	5	572.594E	1548.591E (1548.618)	1721.180E (1721.179)	1680.414A (1680.410)
11	6	566.881A	1544.639A (1544.659)	1717.061E (1716.986)	1670.590E (1670.626)
11	7	560.091E	1539.988E (1539.992)	1711.701A (1711.653)	1659.421E (1659.444)
11	8	552.202E	1534.628E (1534.651)	1705.079E (1705.062)	1646.749A (1646.796)
11	9	543.194A	1528.629A (1528.648)	1697.237E (1697.177)	1632.545 (1632.430)
11	10	533.040	1521.922E (1521.983)	1688.001A (1688.201)	1616.731E (1616.670)
11	11	521.710E	1514.643E (1514.778)	1677.482E (1677.692)	1599.074A
12	0	691.437A	1660.495A1 (1660.486)	-	1814.278E (1813.433)
12	1	690.929E	1660.131E (1660.060)	1837.413A1 (1837.405)	1811.676E (1810.906)
12	2	689.403E	1659.040E (1659.042)	1837.012E (1837.117)	1807.627A2 1808.016A1
12	3	686.856A	1657.226A (1657.232)	1835.720E (1835.700)	1802.481E (1802.143)
12	4	683.279E	1654.689E (1654.624)	1833.657A2 (1834.002) 1833.186A1 (1833.419)	1759.738E (1795.446)
12	5	678.662E	1651.413E (1651.423)	1830.211E (1830.171)	1787.576A (1787.218)
12	6	672.992A	1647.432A (1647.432)	1825.962E (1825.949)	1778.011E (1777.887)

J	K	$G_0(J,K)$	$\nu_2(J,K)$	$\nu_4^{+\ell}(J,K)$	$\nu_4^{-\ell}(J,K)$
12	7	666.252E	1642.734E (1642.738)	1820.577A (1820.556)	1767.072E (1767.043)
12	8	658.423E	1637.304E (1637.323)	1813.752E (1814.023)	1754.647A (1754.732)
12	9	649.483A	1631.175A (1631.226)	1806.105E (1806.261)	1740.743E (1740.646)
12	10	639.407E	1624.394E (1624.467)	1796.973A (1797.044)	1725.272E
12	11	628.165E	1616.929E (1617.105)	1786.500E (1786.793)	1708.897A (1707.897)
12	12	615.727A	1608.926A (1609.131)	1774.519E (1774.663)	1689.085E
13	0	806.081A	1771.944A2	-	1928.913E
13	1	805.577E	1771.577E	1955.528A2 (1955.714)	1926.866E
13	2	804.064E	1770.468E	1955.029E (1955.183)	1922.804A1 1923.255A2
13	3	801.536A	1768.613A2 (1768.610) 1768.689A1	1953,686E (1953.745)	1917.965E
13	4	797.987E	1766.094E (1766.093)	1951.677A1 (1951.918) 1951.130A2	1911.395E
13	5	793.407E	1762.809E (1762.796)	1948.125E (1948.123)	1903.488A
13	6	787.782A	1758.790A (1758.791)	1943.794E (1943.773)	1894.029E
13	7	781.096E	1754.027E (1754.040)	1938.367A (1938.380)	1883.221E
13	8	773.331E	1748.553E (1748.600)	1931.816E (1931.745)	1871.132A
13	9	764.463A	1742.362A (1742.408)	1923.869E (1923.947)	1857.548E
13	10	754.469	1735.465E (1735.563)	1914.846A (1914.842)	1842.434E
13	11	743.320E	1727.890E (1728.030)	1904.387E (1904.447)	1825.709A

J	K	$G_0(J,K)$	$\nu_2(J,K)$	$\nu_4^{+\ell}(J,K)$	$\nu_4^{-\ell}(J,K)$
13	12	730.986A	1719.727A (1719.926)	1892.477E (1892.609)	1807.226E
13	13	717.434E	1710.876E (1711.196)	1879.064E (1879.275)	1786.737E
14	0	-	1891.924A1	-	2052.621E
14	1	-	1891.559E	2082.387A1	2050.157E
14	2	-	1890.423E	2081.892E	2046.678A2 2047.276A1
14	3	-	1888.710A2	2080.517E	2041.874E
14	4	-	1886.025E	2077.806A1 2078.485A2	2035.499E
14	5	-	1882.740E	2074.799E	2027.727A
14	6	-	1876.695A	2070.523E	2018.704E
14	7	-	1873.902E	2064.996A	2008.143E
14	8	-	1868.382E	2058.422E	1996.179A
14	9	-	1862.125A	2050.940E	1982.865E
14	10	-	1885.145E	2041.468A	1968.080E
14	11	867.138E	1847.461E (1847.402)	2031.152E	1951.830A
14	12	854.914A	1839.080A (1839.300)	2019.304E	1933.765E
14	13	841.484E	1830.111E (1830.424)	2006.072A	1913.939A
14	14	827.343E	1820.509E (1821.516)	1991.211E	1891.998A

References

Chang, T.Y. and Bridges, T.J., 1970: "Laser Action at 452, 496 and 541 μm in Optically Pumped CH_3F", Opt. Comm. 1, 423-425.

Davies, P.B., Neumann, R.M., Wofsy, S.C., and Klemperer, W., 1981: "Radio Frequency of Phosphine", J. Chem. Phys. 55, 3564-3568.

Malk, E.G., Niesen, J.W., Parsons, D.F., and Coleman, P.D., 1978: "Laser Emission in the 83-223 μm Region from PH_3 with Laser Line Assignments", IEEE J. Quan. Elect. QE-14, 544-550.

Maki, A.G., Sáms, R.L., and Olson, W.B., 1973: "Infrared Determination of C_0 for Phosphine via Perturbation-Allowed $\Delta|K-\ell| = \pm 3$ Transitions in the $3\nu_2$ Band", J. Chem. Phys. 58, 4502-4512.

Niesen, J.W., 1978: MS Thesis, Department of Electrical Engineering, University of Illinois, "A Phosphine Far IR Laser".

Shimizu, F., 1975: "Stark Spectroscopy of PH_3 by 10 μm CO_2 and N_2O Lasers", J. Phys. Soc. Japan, 38, 293.

Tarrago, G., Dang-Nhu, M., and Goldman, A., 1981: "Analysis of Phosphine Absorption in the Region 9-10 μm and High-Resolution Line-by-Line Simulation of the ν_2 and ν_4 Bands", J. Mol. Spectrosc. 88, 311-322.

See for example, Townes, C.H. and Schawlow, A.L., 1955: Microwave Spectroscopy, McGraw Hill, New York, Chapter 12.

Yin, K.L. and Rao, K.N., 1974: "ν_2 and ν_4 Fundamentals of Phosphine Occurring at 8-12 μm", J. Mol. Spectrosc. 51, 199-207.

THE OPTICALLY PUMPED FORMIC ACID LASER

D. Dangoisse and P. Glorieux

Laboratoire de Spectroscopie Hertzienne,
Associé au CNRS - Université de Lille I
59655 - Villeneuve d'Ascq Cedex - France

I - INTRODUCTION

Since the advent of optically pumped FIR lasers, the formic acid (HCOOH) has appeared as one of the most promising candidates for high efficiency FIR laser action. In fact some of its emission lines are among the most intense laser lines available in the far infrared region. Up to now more than 225 lines have been observed from HCOOH and its isotopic species that cover the region from 133.9 μm to 1730.8 μm [1A-13A]. A remarkable point about the study of this laser is that it illustrates the original contribution of the spectroscopy of FIR lasers to the obtention of accurate molecular parameters. When the first observation of FIR laser action in formic acid was made, the only high resolution data available were provided by microwave spectroscopy of the ground and some excited states [1C-3C]. From these data and some low resolution infrared spectra, it has been possible to assign some HCOOH emission lines and thus to get additional spectroscopic parameters [4C]. Thereafter a high resolution infrared spectrum of the infrared bands pumped by the CO_2 laser has been recorded and assigned [1B]. The high accuracy of the information concerning the main isotope

$H^{12}COOH$ made it possible to extend the assignment to $H^{13}COOH$ laser lines [12A]. It may be emphasized that all the results have been obtained on an asymmetric top molecule whose spectrum is by far more complicated than that of most commonly assigned symmetric tops as for instance the CH_3X series (X = F, Cl, Br, I).

II - EXPERIMENTAL RESULTS AND MEASUREMENTS

The wavelengths of the FIR lines are measured either directly by scanning one of the laser cavity mirrors through a number of wavelengths or by directing the FIR beam into an external Fabry-Perot resonator. The accuracy of the measurements is a part of 10^{-3}.

Most of the lines are now measured with an accuracy better than 10^{-6} by mixing FIR output with harmonics of various phase locked millimeter sources. Silicon tungsten, In Sb or antimony-bronze point contact diodes, or Schottky diodes are generally used as mixers. For the other lines which are often the weakest, only a wavelength measurement is available.

Except for a few cases, the frequency measurements are generally in good agreement with the given error range. The different cases of discrepancy will be mentioned below. The scatter in these measurements mostly originates from the inaccuracy in tuning the resonator exactly to the center of the emission line.

1. Submillimeter laser lines from $H^{12}COOH$

A great number of FIR laser lines has been observed by several authors and it is now absolutely confirmed that they originate from the $H^{12}COOH$ species. However for a few of them, as discussed below, their belonging to formic acid is not clearly stated.

The first submillimeter laser action in formic acid pumped by a pulsed CO_2 laser which provided peak power of 100-150 W in 80 µsec pulses, was obtained in 1973 by Wagner et al. [2A] with an open resonator. The FIR wavelengths were measured by means of an external Ni wire mesh Fabry-Perot interferometer, but the accuracy of the measurements was not clearly mentioned. Some of these oscillations have not been obtained until now with a C.W. pumping (see first column of Table I). Pulsed operation in formic acid has been reinvestigated by Plant et al. [3A] and their measurements are reported in the second column of Table I.

Then, new experiments using a C.W. CO_2 laser delivering a power of 10 to 30 watts and either a Fabry-Perot resonator or a waveguide as FIR cavity have been performed [4A-10A]. A total of fifty nine distinct lines have been reported in the literature (see table I). The frequency of twenty eight of these lines is now accurately measured. The average difference between the measurements of various authors is 0.8 MHz. We will now discuss for several pump lines the various difficulties encountered :

<u>9R26</u> : The existence of the two lines observed by Plant et al. should be verified since their wavelengths are exactly the same as for the 9R28.

<u>9R22</u> : It is not sure that the 414 μm line mentioned by Plant is exactly the same as the 419.5 μm line measured by the other authors. The line located at 693 788.5 MHz is observed by Dyubko et al. [6A] in $H^{12}COOD$ instead of $H^{12}COOH$ in [8A]. This disagreement should be checked. It can be noticed that this line is assigned in $H^{12}COOH$.

<u>9R20</u> : Is the 414 μm line reported by Plant the same as the one pumped by the 9R22 ?

The FIR frequency is measured in [8A] at 672 335.5 MHz instead of 672 331.8 MHz in [6A]. This discrepancy greater than the experimental uncertainty is not clearly explained and this line should be remeasured.

<u>9R18</u> : Among the six lines reported by Plant only three seem to correspond to observations made by other authors.

The emission line of 761 607.7 MHz frequency in [8A] is observed in [6A] at 761 610.2 MHz. The agreement is however good between the three other measurements.

Table I Submillimeter emissions from the $H^{12}COOH$ molecule.

$H^{12}COOH$	Wavelength (μm)					Frequency (MHz)				Polarization	offset
	Wagner et al [2A] pulsed	PLant et al [3A] pulsed	Dyubko et al [4A] C.W.	Radford [5A] C.W.	Landsberg [11A] C.W.	Dyubko et al [6A] C.W.	Dangoisse et al [7A - 8A] C.W.	Weiss et al [10A] C.W.	Radford et al [9A] C.W.		$\nu_{HCOOH} - \nu_{CO_2}$ (±0.1 MHz)
9R40			743.0 (*)	785		403 721.3	403 721.6 381 336.9			II II C	- 21.00
9R38			458.6 (*)			653 821.5	653 822.2		653821.4 ± 0.5	⊥	19.58
9R36	458.43					379 561.2				? II	
9R34	359.81					833 000 (a)				II	
9R32	229.39					380 958.8				? II	
9R30			670.0 (*)			447 764.9	447 765.0		447 766.0 ± 0.8	II	- 30 (± 20)
9R28	512.88	512 530	513.2 (*) 534.8 (*)			584 388 584 372.9 561 724	584 388.2		584 386.9 ± 0.7	II II II C	+ 41.45
9R26		512 530								II II	
9R24	418.51		745.0 (*)	761		717 000 (a) 402 920.5	402 919.6			II II ?	6.70
9R22	419.55 433.10 447.58	414	420.0 (*) 433.0 (*)	433	133.9 196.5 418.6 432.1	716 157.4	716 155.8 693 788.5 516 538.7		716 156.4 ± 0.5 671 419.5 (b)	II ⊥ II II C ? ?	+ 19.67
9R20	432.50 445.81	414 428				692 951 692 895 672 331.8	692 949.5 672 335.5	692 951.4 ± 0.2	692 950.5 ± 0.5	II II II II C	27.84

$H^{12}COOH$ (2)	Wavelength (μm) Wagner et al [2A] pulsed	Plant et al [3A] pulsed	Dyubko et al [4A] C.W.	Radford [5A] C.W.	Landsberg [11A] C.W.	Frequency (MHz) Dyubko et al [6A] C.W.	Dangoisse et al [7A] - [8A] C.W.	Weiss et al [10A] C.W.	Radford et al [9A] C.W.	Polarization	Offset $\nu_{HCOOH}-\nu_{CO_2}$ (± 0/1 MHz)
9R18	393.62 405.55 421	368 392 403 428 441 496	394.2 (*) 406.0 (*)			761 610 739 160.3	761 607.6 739 161.0	761 607.7 ± 0.3	761 606.5 ± 0.5 739 161.0 ± 0.7	II II C II ? ?	+ 32.42
9R16	446.75	388, 401, 413	447.0 (*)			670 867.2	670 867.2			II,II,II II	+ 10(±20)
9R14	334.91 342.74		336.3 (*)			892 000 (a)				II ?	
9R10	460.51									?	
9R8	420.26					713 127.6				II	
9R6			421.0			425 000				II II	
9R4	309.23		303.0 (*)		302.2 309.5	991 775.8	991 776.9		991 777.8 ± 1.0	II II C	- 1.25
9P8	302.08									?	
9P14	493.28									?	
9P16	437.70 518.83	435	437.6 534.5 (*)			685 316.6 581 929.7 561 747.5	581 930.3 561 748.6			⊥ ⊥ II	64.46
9P18	334.82									?	
9P20	254.80		254.5			1 178 000 (a)				II	
9P26	405.75		405.0			740 000 (a)				II	

(Continued)

Table I Submillimeter emissions from the $H^{12}COOH$ molecule (Continued).

$H^{12}COOH$ (3)	Wavelength (μm)					Frequency (MHz)				Polarization	Offset
	Wagner et al [2A] pulsed	Plant et al [3A] pulsed	Dyubko et al [4A] C.W.	Radford [5A] C..W.	Landsberg [11A] C.W.	Dyubko et al [6A] C.W.	Dangoisse et al [7A] - [8A] C.W.	Epton et al [13A] C.W.	Radford et al [9A] C.W.		$\nu_{HCOOH}-\nu CO_2$ (± 0.1 MHz)
9P28						247 075.8				II	
9P30	278.61		278.5			1076 000 (a)				II	
9P38	580.52	577	582.0			516 170.7				⊥	
9P42	492									?	
10R42	404.1					742 000 (a)				⊥	
10R24						937 000 (a)				II	
10R22	311.45 319.48	309	311.0			964 000 (a)		962 250 ± 5		II ?	
10P14	445.21					674 000 (a)				⊥	
10R14				930 (c)							
10R32				930 (c)							

(a) Calculated from a wavelength measurement, (b) see Ref. [16A], (c) Taken from Ref [1A]

c : cascade

(*) In this publication pump lines were in doubt. All the lines followed by (*) have been replaced with the supposed right pump line

9R16 : Three lines observed by Plant have not been observed in C.W. operation.

10R22 : The frequency of the first FIR line has been measured with an accuracy of 5 MHz by mixing the laser output on a point - contact diode with the 311 μm line from the HCN discharge laser [13A] .

2. Submillimeter laser lines from $H^{13}COOH$

Formic acid has been extensively studied during the past few years by many authors, but however its ^{13}C isotopic species has not often been used [8A] .

In table II are reported new lines which have been recently measured with $H^{13}COOH$ [12A] . As in the case of $H^{12}COOH$, when the main emission is a strong one, cascade emissions have been observed.

Some lines require a special attention as explained hereafter : The emission generated by the 9P6 seems to have two components separated by only 3 MHz. These components have been observed simultaneously. This could originate from the fact that :

i) there is only one rotational transition and the emission is not single mode.

ii) the emission corresponds to two near degenerate transitions.

In order to avoid all ambiguity, pump lines as 9R18, 9R20 and 9R28 have not been used with the $H^{13}COOH$ (90 % purity in ^{13}C sample obtained from the CEA*) because the most intense FIR lines of $H^{12}COOH$ are still observable in the spectrum.

3. Submillimeter laser lines from the deuterated analogs of formic acid

The deuterated species of formic acid have been mainly studied in [6A] by Dyubko et al. and give rise to a great number of oscillations which are reported in table III, IV and V.

---

* CEA : Commissariat à l'énergie atomique, France.

Table II Submillimeter emissions from the $H^{13}COOH$ molecule (12A).

CO_2 LASER	Wavelength (μm)	Frequency (MHz)	Polarization	Pressure (mTorr)	Intensity (Arb. Unit)
9 R 32	572.330	523810.4	⊥	40	5
9 R 30	1030.378	290953.9	//	50	50
	1116.483	268514.8	cascade		
9 R 26	448.533	668383.0	⊥	50	25
	464.627	645231.8	cascade		2
9 P 6	313.797	955370.3	//	50	17
		955368.1			
9 P 12	788.919	380004.0	//	50	67
	838.369	357589.7	cascade		7
9 P 14	255	-	-	30	1
	1491.846	200953.9	//	50	33
9 P 16	258.425	1160071.8	//	50	100
9 P 20	548.843	546225.3	//	25	3
9 P 24	536.096	559214.1	//	50	23
	381.615	785587.2	//	30	7
9 P 26	382.357	784063.1	//	50	7
9 P 30	477.963	627229.1	//	50	7
9 P 32	393.485	761888.8	//	50	23
10 R 46	480	-	-	40	-
10 R 32	891.087	336434.4	//	40	5
10 R 28	310	-	-	40	1

Table III Submillimeter emissions from the H COOD molecule (6A).

CO_2 LASER	Wavelength (μm)	Frequency (MHz)	Polarization	Intensity (Arb. Unit)
9 R 40	668		//	5
9 R 38	472.9		⊥	2
9 R 30	594		⊥	3
9 R 22	417.0		//	30
9 R 16	392.9		//	5
9 R 6	304.1		//	1.5
9 P 10	395.0		//	5
9 P 12	813.7572	368405.3	//	3
	826		//	5
9 P 14	477.4		//	1
9 P 16	395.0		//	3
9 P 18	582.5536	514617.8	⊥	10
9 P 22	657		//	2.5
9 P 28	339.9		//	6
	446.8		//	0.1
9 P 30	1541.750	194449.9	//	30
9 P 34	373.8		//	0.4
9 P 36	351.0		//	3
9 P 38	355.2		//	10
	361.2		//	12
9 P 40	411.2		//	3
	531		//	1
9 P 44	498.0		//	2
10 R 42	727.9491	411831.6	//	13
10 R 40	493.1562	607905.7	//	6
10 R 38	391.6886	765384.6	⊥	7
	1157.318	259040.7	⊥	3
10 R 36	372.0		⊥	2.5
	695.6720	430939.4	//	40
	733		//	5
10 R 32	919.9355	325884.2	//	600
	986.3125	303952.8	//	30
10 R 30	325.9		//	1.5
10 R 28	369.9678	810320.5	//	40
10 R 26	450.1		⊥	3
	689.9981	434483	//	100

(continued)

Table III Submm emissions from the H COOD molecule (6A)(Continued).

CO_2 LASER	Wavelength (μm)	Frequency (MHz)	Polarization	Intensity (Arb. Unit)
10 R 24	1730.833	173207	//	10
10 R 22	351.9		//	10
	398.1		//	2
10 R 20	1161.676	258068.8	//	6
10 R 16	358.2		//	0.2
	433.2		//	0.5
10 R 14	240		⊥	3
	926.2087	323677	⊥	500
10 R 12	395.7124	757601.9	⊥	50
	660		⊥	3
	692		⊥	3
10 R 10	324.1		//	1
	630.1661	475735.6	⊥	30
10 R 8	347.0		⊥	3
10 R 6	353.1		//	50
10 R 4	387.8		⊥	12
10 P 6	430.4380	696482.3	//	30
10 P 12	450.9799	664757.9	⊥	20
10 P 14	567.1065	528635.2	//	50
	590		//	5
10 P 16	461.2610	649941	//	300
	472.1		//	30
10 P 24	429.6898	697695.1	//	30
10 P 26	372.0		//	3
10 P 28	353.1		//	15
10 P 30	356.0		//	8
10 P 32	291.9		//	15
10 P 36	819		//	0.1

Table IV Submm emissions from the DCOOD molecule (6A).

CO_2 LASER	Wavelength (μm)	Frequency (MHz)	Polarization	Intensity (Arb. Unit)
9 R 16	283.1		//	5
9 R 8	335.7087	893013.6	//	20
9 R 4	366.9		//	3
9 P 12	1070.231	280119.5	//	1
	843.2369	355525.8	//	1
	998.5140	300238.6	//	50
9 P 16	276.1		//	10
	935.0095	320630.4	//	100
9 P 38	1281.649	233911.6	//	6
10 R 40	350.2		//	50
10 R 36	241.2		//	1.5
10 R 32	265.1		//	5
10 R 30	323.1		//	20
10 R 28	325.2		//	0.5
	508		//	1
10 R 26	567.8683	527926	//	300
	591.6157	506735.1	//	100
10 R 24	304.0832	985889.7	//	150
	310.0		//	60
	645		//	5
10 R 22	397.1		//	20
10 R 20	218.0		//	100
	396.0		//	50
	789.4203	379762.8	//	150
	835		//	5
10 R 18	1009.409	296997.9	//	100
10 R 14	351.9		//	30
	415.2		//	5
10 R 12	389.9070	768882	//	50
	380.5654	787755.5	//	500
10 R 10	395.1488	758682.5	//	40
	452.2		//	5
10 R 6	298.0		//	5
10 R 4	478.9		//	2
10 P 4	737		//	2
	795		//	3
	812		//	25
10 P 8	491.8906	609469.8	⊥	60
	508.7911	589225	⊥	100
	1158		⊥	3
10 P 10	726.9203	412414.5	⊥	30
	761.7617	393551.5	⊥	10
10 P 12	469.2		⊥	6
10 P 14	442.8		⊥	2
10 P 18	425.2		⊥	15
	593		//	15
10 P 20	561.2939	534109.6	⊥	250
	927.9814	323058.7	⊥	20
10 P 26	877.5481	341625.1	//	300
	936.6023	320085.1	//	10
	779.8744	384411.2	//	10
10 P 30	457.3410	655511.9	//	15
	666		//	10
10 P 34	514.9507	582177	//	200
	526.4856	569421.9	//	150
	527.2146	568634.6	//	150
10 P 44	414.1		//	10

Table V Submm emissions from the DCOOH molecule (6A).

CO_2 LASER	Wavelength (μm)	Frequency (MHz)	Polarization	Intensity (Arb. Unit)
10 R 36	697.4552	429837.6	//	10
10 R 30	647.3485	463108.3	//	1
10 R 34	713.1056	420404	//	80
	752.7485	398263.8	//	50
10 R 28	971.8064	308489.9	//	3
10 R 24	312.0		//	3
	1237.966	242165.4	//	4
10 R 20	265.1		⊥	0.3
10 R 16	341.8		⊥	3
10 R 14	433.2353	691985.3	⊥	15
19 R 12	365.2		//	0.8
	1047.579	286176.6	⊥	20
10 P 6	710		⊥	0.5
10 P 8	639.1282	469064.7	⊥	50
10 P 14	466.5461	642578.4	⊥	15
	479.9040	624692.6	//	30
10 P 16	433.2		//	1
10 P 20	362.1		//	3
10 P 22	328.4570	912729.7	//	30
10 P 30	272.0		//	3

4. Submillimeter gain and output power

Evidence of the gain anisotropy with polarization of the pump radiation has been given by Dyubko et al. [1D]. Gain measurements reported in table VI have been obtained for the parallel and perpendicular polarization of the input signal with the pump radiation.

Table VI Submillimeter gain in the $H^{12}COOH$ medium (1D).

CO_2 LASER	ν(MHz)	Polarization of the laser emission	Gain, dB/m	
			For // polarization	For ⊥ Polarization
9 R 18	761 610	//	3.09	2.97
9 R 20	692 951	//	3.15	3.0
9 R 28	584 388	//	2.63	2.6
9 R 38	653 821.5	⊥	1.46	2.2
9 R 40	403 721.3	//	1.93	1.85

Power measurements have been made by Dyubko et al. in 1975 [6A] and are listed in table VII. New values have been given (see figure 1) for some strong lines in [2D] with an accuracy of 10 %. The FIR laser is a 1.2 m long open resonator with 100 mm internal diameter. The pump power of the CO_2 laser is 30 W for the strong lines. The optimum pressure is generally situated in the 50-200 m Torr range in C.W. operation and depends on the pump lines. The threshold pump power is often less than 5W at optimum pressure [3D] . This threshold could be reduced for example with the 744 μm line of $H^{12}COOH$ (9R40) to less than 50 mW when the pressure is lowered to a few m Torr.

Duxbury [3D] suggests to do temperature studies of this molecule since it is 25 % dimerised at room temperature. Pump absorption properties of some emission lines are given in [4D-5D].

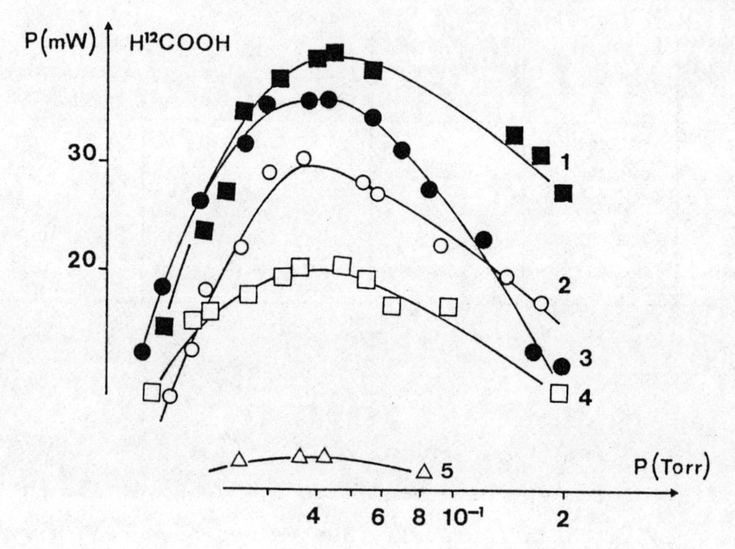

Figure 1 Dependence of the output power on the pressure in the active gas obtained for various wavelengths (2D). (1) 393.6 μm (9R18), (2) 418.6 μm (9R22), (3) 432.6 μm (9R90), (4) 513.0 μm (9R28), (5) 742.6 μm (9R40).

Peak power of 5 Watts have been obtained by Bluyssen et al. [6D]. Pulse shapes are discussed as a function of pressure.

Table VII Output power on the strongest lines of the formic acid laser (6A).

	CO_2 LASER	ν(MHz)	Output power (mW)
HCOOH	9 R 18	761 610	15
	9 R 20	692 951	8
	9 R 22	716 157.4	6
	9 R 28	584 388	10
HCOOD	10 R 14	323 677	8
	10 R 32	325 884.2	10
	10 P 16	649 941	10
DCOOD	10 R 12	787 755.5	12
	10 R 20	1 375 000	6
	10 R 26	527 926	4
	10 P 8	589 225	1
	10 P 20	534 109.6	3
	10 P 26	341 625.1	2
	10 P 34	582 177	2

III - SPECTROSCOPIC MEASUREMENTS AND ASSIGNMENTS IN THE $H^{12}COOH$ AND $H^{13}COOH$ SPECIES

Formic acid is a slightly asymmetric top molecule belonging to the C_s symmetry group ($\kappa = -0.95$). The molecule has nine vibrational states, but efficient pumping concerns mainly the two fundamental bands ν_6 and ν_8 (table VIII). Pumping could also probably occur from hot bands since the lowest excited states ν_7 and ν_9 are

Table VIII Vibrational energies in $H^{12}COOH$ and $H^{13}COOH$ below 1200 cm^{-1}.

Vibration	$H^{12}COOH$ (cm^{-1})	$H^{13}COOH$ [2B] (cm^{-1})	[12A]	Type of IR bands
ν_7 (OCO bend.)	626.158 *	621.4	–	a and b
ν_9 (OH.Tors.)	640.722 *	636.1	–	c
ν_8 (CH Wag.)	1033.467 **	1022.0	–	a and b
ν_6 (CO-(OH def.)	1104.851 **	1093.5	1095.409	c

* from [7B]

** from [14 A]

a little populated at room temperature ($\frac{h\nu_7}{kT}$ # $\frac{h\nu_9}{kT}$ # 3). The principal inertia axis a and b are located in the plane of the molecule. The dipole moment (μ = 1,4 D) is situated in this plane. The ratio of the projection of this dipole moment on the a and b axis of inertia is

$$\left| \frac{\mu_b}{\mu_a} \right|^2 = 0.035$$

Thus, the emission corresponding to the a moment are favored and all emissions but two have been assigned as a-type rotational transitions.

The type of the IR transitions is related the vibration to induced dipole moment. The selection rule of vibrational transitions are governed by $\frac{\partial \mu_a}{\partial Q_6}$ ($\Delta J = 0 \pm 1$, $\Delta Ka = 0$) and $\frac{\partial \mu_b}{\partial Q_6}$ ($\Delta J = 0, \pm 1$, $\Delta K_a = \pm 1$, $\Delta K_c = \pm 1$) for the ν_6 in plane vibration and $\frac{\partial \mu_c}{\partial Q_8}$ ($\Delta J = 0, \pm 1$, $\Delta K_c = 0$) for the ν_8 out of plane vibration.

1. Assignment of FIR laser lines in $H^{12}COOH$

Many infrared and microwave studies are concerned with formic acid spectra.

a) Infrared results. The earliest studies of formic acid are related to the band contour analysis of the IR spectrum recorded by several authors with low resolution techniques [3B-5B]. Accurate values of the ν_6 and ν_8 band centers were then deduced from the assignment of accurately measured FIR laser lines [14A] and from the infrared study of the ν_6 and ν_8 bands [6B]. Deroche et al. [7B] used a detailed analysis of the Coriolis coupled ν_7 and ν_9 bands to obtain molecular parameters for the excited states.
Recently, twenty five frequencies of the infrared spectrum of the ν_6 and ν_8 states have been measured with Sub-Doppler techniques [1B].

b) Microwave results. The microwave spectroscopy of formic acid has been developped for many years, and allows a good knowledge of $H^{12}COOH$ in the ground, ν_6, ν_7, ν_8 and ν_9 states. The most recent parameters of the ground state have been obtained by Willemot et al. [1C] using Watson's theory of centrifugal distortion and a least square method including 284 transitions with very high values of J and Ka, observed between 8 and 300 GHz. The submillimeter rotational spectrum of the ground state has also been studied by Dyubko et al. [15A]. A gas spectrometer fed with a backward wave oscillator operating in the high frequency range has been used to measure the frequencies of 182 Ra-type transitions right up to 560 GHz. To our knowledge, these frequencies have not been published and the accuracy of the measurements is not mentioned. The fifteen parameters of the Watson's theory of centrifugal distortion have been derived from this study.
The ν_6 and ν_8 microwave spectrum has been investigated between 8 and 200 GHz to calculate a set of parameters [2C-4C]. The two states are coupled by a Coriolis resonance but the authors failed

to determine the Coriolis constants from the observed rotational spectra ($J < 20$, $Ka < 7$).

An analysis of the ν_6 and ν_8 excited states has also been carried out by Dyubko et al. [15A] and frequency measurements have been performed in the 300-400 GHz range : 20 and 16 lines in the ν_6 and ν_8 states respectively have been observed. To our knowledge, these frequencies are unpublished. As in the previous study, Coriolis coupling was not considered.

c) FIR assignments. Assignments of the FIR lines have been carried out independently by Dyubko et al. [15A] and by Dangoisse et al. [14A]. Many lines have been assigned in the ν_6 and ν_8 states. Spectroscopic constants have been derived with the aid of simultaneous analysis of the rovibrational and rotational lines in the three states (GS, ν_6, ν_8). The difficulty of these works was mainly due to the Coriolis coupling which was not derivable from the available data and therefore these studies do not describe fully the rotational spectra of the ν_6 and ν_8 states.

Very recently a least square analysis of Landsberg et al. [1B] using Watson's theory of centrifugal distortion with the addition of a Coriolis coupling term confirmed most of the assignments. This analysis including accurately measured IR [1B], microwave [4C] and FIR [8A] frequencies improved the molecular constants.

The molecular constants obtained for the GS, ν_6 and ν_8 states by Dyubko et al. [15A], Willemot et al. [4C] and Landsberg et al. [1B] are given for comparison on table IX.

Finally 19 of the 28 accurately measured frequencies have been assigned (see table X). Three lines (516 538.7 MHz [9R22], 447 767.0 MHz [9R30] and 561 748.6 MHz [9P16] have been misassigned in [15A] and [3C]. The assignment of the 516 170.8 MHz line [9P38] is not fully confirmed and thus not reported here. Neither these emissions nor any of the previous unassigned ones could be assigned as rotational transitions in the ν_6 or ν_8 state. The unassigned transitions could originate from pumping in hot bands of IR absor-

Table IX Comparison of $H^{12}COOH$ molecular parameters.

MHz	Ground state		$v_6 = 1$			$v_8 = 1$		
	[15 A]	[1C] σ	[15 A]	[4 C] σ	[1 B] σ	[15 A]	[4 C] σ	[1 B] σ
			33122309	33122599	33122629.0 (10)	30982778	30982561	30982554 (41)
A	77512.26514	77512.2310 (63)	77863.139	77853.19 (3)	77600.1 (18)	76731.086	76724.20 (6)	76977.5 (18)
B	12055.11105	12055.1045 (8)	12002.97218	12003.043 (8)	12003.1709 (59)	12001.3902	12001.46 (1)	12001.623 (18)
C	10416.12030	10416.1145 (8)	10352.19645	10352.251 (8)	10352.0459 (68)	10419.8642	10419.90 (1)	10419.643 (17)
Δ_J	0.01000035	0.0099894 (15)	0.0100965	0.01026 (9)	0.0102447 (76)	0.0095708	0.0103 (1)	0.009763 (34)
Δ_{JK}	- 0.0861672	- 0.08625 (62)	- 0.0939925	- 0.0922 (2)	- 0.089313 (87)	- 0.063761	- 0.0660 (3)	- 0.0699 (16)
Δ_K	1.702608	1.70229 (19)	2.3022	1.705 (10)	1.7742 (13)	2.446	1.289 (29)	1.260 (31)
δ_J	0.00194850	0.00194920 (24)	0.00199882	0.00207 (1)	0.0020429 (56)	0.0019509	0.001814 (5)	0.001820 (18)
δ_K	0.0428111	0.042600 (29)	0.053542	0.0581 (10)	0.04896 (27)	- 0.00389	0.010 (1)	0.0230 (28)
H_J	0.1323×10^{-7}	0.928×10^{-8} (39)	0	*	0.164×10^{-7} (32)	0	*	0
H_{JK}	0.158×10^{-6}	0	0	0	$- 0.759 \times 10^{-6}$ (57)	0	0	0
H_{KJ}	$- 0.10318 \times 10^{-4}$	$- 0.1026 \times 10^{-4}$ (14)	0	*	$- 0.2567 \times 10^{-4}$ (84)	0	*	0
H_K	0.11812×10^{-3}	0.1195×10^{-3} (18)	0	*	0.302×10^{-4} (38)	0	*	0
h_J	0.5809×10^{-8}	0.6035×10^{-8} (48)	0	*	0.125×10^{-7} (29)	0	*	0
h_{JK}	0.634×10^{-7}	0	0	0	0	0	0	0
h_K	0.1785×10^{-4}	0.1255×10^{-4} (48)	0	*	0	0	*	0

$\Omega_{68} = \frac{1}{2}\left((\nu_6/\nu_8)^{\frac{1}{2}} + (\nu_8/\nu_6)^{\frac{1}{2}}\right)$, $2 A \Omega_{68} \zeta^a_{68} = 23292$ (80) MHz [1 B]

0 : Constrained to 0.0
* : Constrained to G.S value

Table X Assignments of optically pumped transitions in the ν_6 and ν_8 bands of $H^{12}COOH$.

CO_2 LASER Line	CO_2 LASER Freq (MHz) [11 D]	IR ABSORPTION Obs Freq (MHz) [18]	IR ABSORPTION O-C (MHz)	IR ABSORPTION Assignment	SMMW EMISSION Obs Freq (MHz)	SMMW EMISSION O-C (MHz) [18]	SMMW EMISSION Assignment
9 R 40	32678228.36	32678207.36	2.25	(0) $19_{3,17} \rightarrow$ (6) $18_{3,16}$	403721.6	- 0.29	(6) $18_{3,16} \rightarrow$ (6) $17_{3,15}$
				Cascade	381336.9	- 0.08	(6) $17_{3,15} \rightarrow$ (6) $16_{3,14}$
9 R 38	32647432.22	32647451.80	- 0.31	(0) $30_{2,28} \rightarrow$ (6) $30_{1,29}$	653822.2	0.71	(6) $30_{1,29} \rightarrow$ (6) $29_{1,28}$
9 R 28	32481876.59	32481918.04	0.46	(0) $27_{5,23} \rightarrow$ (6) $26_{5,22}$	584388.2	0.71	(6) $26_{5,22} \rightarrow$ (6) $25_{5,21}$
				Cascade	561724	- 1.47	(6) $25_{5,21} \rightarrow$ (6) $24_{5,20}$
9 R 24	32410185.68	32410192.38	0.53	(0) $17_{9,9} \rightarrow$ (6) $18_{8,10}$ (0) $17_{9,8} \rightarrow$ (6) $18_{8,11}$	402919.6	0.87	(6) $18_{8,10} \rightarrow$ (6) $17_{8,9}$ (6) $18_{8,11} \rightarrow$ (6) $17_{8,10}$
9 R 22	32373156.19	32373175.86	- 0.78	(0) $33_{12,21} \rightarrow$ (6) $32_{12,20}$ (0) $33_{12,22} \rightarrow$ (6) $32_{12,21}$	716155.8	- 0.12	{ (6) $32_{12,20} \rightarrow$ (6) $31_{12,19}$ (6) $32_{12,21} \rightarrow$ (6) $31_{12,20}$
				Cascade	693788.5	0.82	{ (6) $31_{12,19} \rightarrow$ (6) $30_{12,18}$ (6) $31_{12,20} \rightarrow$ (6) $30_{12,19}$
				Cascade	671419.5	1.26	{ (6) $30_{12,18} \rightarrow$ (6) $29_{12,17}$ (6) $30_{12,19} \rightarrow$ (6) $29_{12,18}$
					1525600	-	(6) $32_{12} \rightarrow$ (6) 32_{11}
					2238900	-	(6) $32_{12} \rightarrow$ (6) 31_{11}
9 R 20	32335334.03	32335361.87	- 0.02	(0) $34_{0,34} \rightarrow$ (6) $33_{0,33}$	692949.5	- 1.34	(6) $33_{0,33} \rightarrow$ (6) $32_{0,32}$
				Cascade	672335.5	- 0.71	(6) $32_{0,32} \rightarrow$ (6) $31_{0,31}$
		32335393.84	0.06	(0) $34_{1,34} \rightarrow$ (6) $33_{1,33}$	692895.0	-	(6) $33_{1,33} \rightarrow$ (6) $32_{1,32}$
9 R 18	32296717.04	32296749.45	- 0.11	(0) $35_{9,27}$ (6) $34_{9,26}$ (0) $35_{9,26}$ (6) $34_{9,25}$	761607.6	1.64	{ (6) $34_{9,26} \rightarrow$ (6) $33_{9,25}$ (6) $34_{9,25} \rightarrow$ (6) $33_{9,24}$
				Cascade	739161.0	2.65	{ (6) $33_{9,25} \rightarrow$ (6) $32_{9,24}$ (6) $33_{9,24} \rightarrow$ (6) $32_{9,23}$

CO_2 LASER		IR ABSORPTION			SMMW EMISSION		
line	Freq (MHz)	Obs Freq (MHz) [IR]	O–C (MHz)	Assignment	Obs Freq (MHz)	O–C (MHz) [IR]	Assignment
9 R 4	32004017.38	32004016.13	− 0.01	(0) $45_{6,40}$ → (6) $44_{6,39}$	991776.9	0.72	(6) $44_{6,39}$ → (6) $43_{6,38}$
				Cascade	968600.0		(6) $43_{6,38}$ → (6) $42_{6,37}$
9 P 16	31491437.39	31491501.85	− 1.07	(0) $26_{13,13}$ → (6) $26_{12,14}$	581930.3		(6) $26_{12,14}$ → (6) $25_{12,13}$
				(0) $26_{13,14}$ → (6) $26_{12,15}$			(6) $26_{12,15}$ → (6) $25_{12,14}$
		–	11.07	(0) $30_{3,28}$ → (8) $30_{4,26}$	685316.6	− 0.04	(8) $30_{4,26}$ → (8) $29_{4,25}$
9 P 28	31159508.16	–	− 11.07	(0) $12_{3,10}$ → (8) $11_{4,8}$	247075.8	0.06	(8) $11_{4,8}$ → (8) $10_{4,7}$
10 R 22	29296136.37	–	–	(0) $43_{6,37}$ → (8) $42_{5,37}$	962250	–	(8) $42_{5,37}$ → (8) $41_{5,36}$

ption leading to a FIR emission in the excited vibrational states. It has been shown for several molecules that a great number of laser lines could originate from absorption in hot bands (e.g. pumping from the ν_3 state (# 611 cm^{-1}) of CH_3Br [7D]). That is possible in the case of formic acid which has two low vibrational states ν_7 and ν_9.

Let us comment some parts of the assignments for the 9R22 CO_2 line. Three FIR lines are assigned as cascade ones. However the third frequency (671 419.5 MHz) measured by Radford [16A] has not been reported in the final version of his paper [9A]. This transition cannot be in these conditions considered as definitely assigned. In a recent study of formic acid laser, two emissions have been observed by Landsberg [11A] at 133.9 μm and 196,5 μm (0.2 % accuracy). These emissions have the shortest wavelength known with formic acid laser and have been assigned to b-type emissions as reported on figure 2. They are the only case of emissions of this type to be assigned up to now in formic acid.

d) Experimental checking of the assignments. Double resonance is an interesting way to check some of the assignments.

To check the assignment of the 512 μm line [9R28], Dyubko et al. [15A] have performed an IR-FIR double resonance experiment. The gas sample is pumped by the CO_2 laser, the probe radiation delivered by a carcinotron is resonant either with the

(6) $27_{5,23} \rightarrow$ (6) $26_{5,22}$ transition (587 GHz) or the

(6) $28_{5,24} \rightarrow$ (6) $27_{5,23}$ transition (607 GHz) in the ν_6 state.

In each case a change in the probe signal has been observed, indicating that the $27_{5,23}$ level is a common level for the two transitions involved in laser action. This method could be used for the checking of all the assignments, nevertheless a high frequency tunable source is needed. So it is easier to use Q_a-type transitions which can be located in the microwave region.

The introduction of an additional field in an optically pumped cavity where the molecules interact also with two other fields

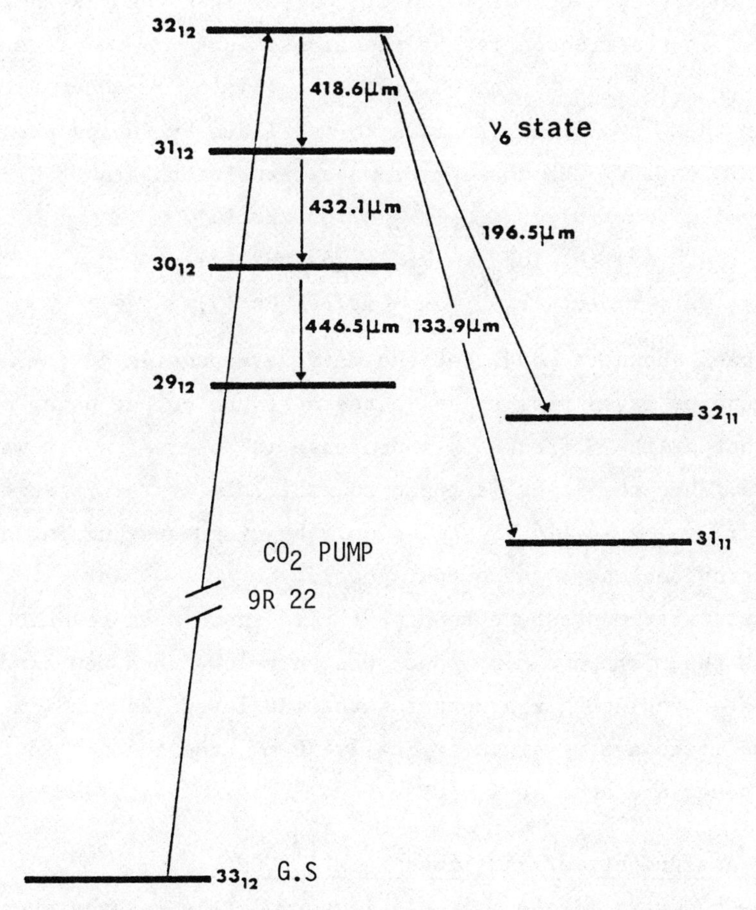

Figure 2 Schematic representation of energy levels in $H^{12}COOH$, showing IR and FIR lines.

gives rise to a great variety of phenomena allowing us to study the laser medium [8D-9D]. Unfortunately this method is limited to a few cases and need a suitable energy level diagram (figure III). The IR pumping [9R40] creates a population inversion between the (6) $18_{3,16}$ and (6) $17_{3,15}$ levels and also between the (6) $17_{3,15}$ and (6) $16_{3,14}$ ones by collision processes. Then the FIR radiation gets rise at the frequencies of the transitions (6) $18_{3,16} \rightarrow$ (6) $17_{3,15}$ and (6) $17_{3,15} \rightarrow$ (6) $16_{3,14}$. One or the other of these two lines are made to oscillate by tuning the length of the FIR cavity. The three microwave transitions are of Q_a-type :

(0) $19_{3,16} \leftrightarrow$ (0) $19_{3,17}$ 22303 MHz (α)
(6) $18_{3,15} \leftrightarrow$ (6) $18_{3,16}$ 17435 MHz (β)
(6) $17_{3,14} \leftrightarrow$ (6) $17_{3,15}$ 12710 MHz (γ)

It has been shown in [8D] that the microwave pumping at the α and γ frequencies gives rise to an increase of the output power of the laser, but at the β frequency a decrease is observed as shown on figure 3. When the laser is tuned to oscillate at the frequency of the (6) $17_{3,15} \rightarrow$ (6) $16_{3,14}$ emission, the microwave pumping at the γ frequency decreases the output power.

The first experiments have been performed in an open resonator in which the microwave pump power was very low. Then new experiments have been done in a copper waveguide laser (25 mm diameter). When the microwave power is typically 50 mW, the changes in laser output power are near 30 % [10D].

2. Assignments of FIR lines in H ^{13}COOH

The geometry of the molecule is very little changed when a ^{13}C atom is substitued for a ^{12}C one. The general shape of the spectrum is thus assumed to be very similar for the two different species. This assumption has been largely taken into account in using results obtained from the spectroscopic studies and the assignments of the FIR emissions of H ^{12}COOH [12A]. Energy level calculations for the H ^{13}COOH species require data coming from previous microwave and infrared spectroscopic studies.

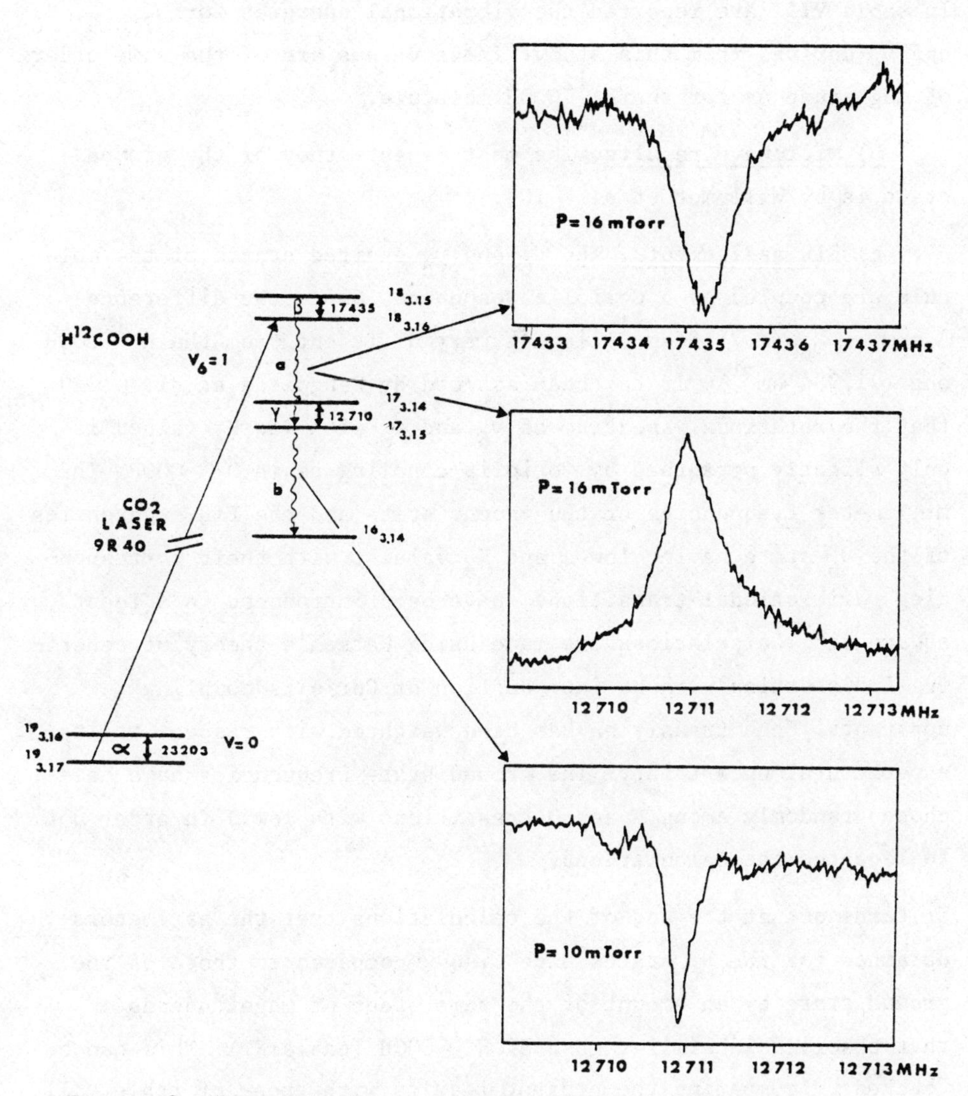

Figure 3 Change in laser output power as a function of the microwave frequency (9D).

a) Infrared results. There are almost no available infrared data for this molecule. R.L. Reddington [2B] has studied the various isotopic species of formic acid isolated in a neon matrix. In table VIII are reported the vibrational energies for ν_7, ν_9, ν_6 and ν_8 derived from this study. These values are of the same order of magnitude as for the $H^{12}COOH$ molecule.

b) Microwave results. The most recent study of the ground state is by Willemot et al. [1C].

c) FIR assignments. The ν_6 and ν_8 excited states of the molecule are coupled by a Coriolis resonance. Since the difference $\Delta = \omega_8 - \omega_6$ (# 71.5 cm^{-1}) is not very different from the $H^{12}COOH$ one (71.384 cm^{-1}), it has been assumed by Dangoisse et al. [12A] that the rotational spectrum of ν_6 and ν_8 with low K_a values is only slightly perturbed by Coriolis coupling as in $H^{12}COOH$. The millimeter frequencies of the ground state and the FIR frequencies of the ν_6 state having low J and K_a values, with their corresponding rovibrational transitions, have been introduced in a least square fit. Calculations are made using Watson's theory of centrifugal distortion without introduction of Coriolis coupling constants. Each transition has been weighted with respect to its experimental uncertainty. The ground state frequencies have been chosen randomly among R and Q transitions with low J in order not to lenghten the calculations.

It turns out at the end of the calculations that the parameters obtained for the ν_6 states have changed compared to those of the ground state by an amount of the same order of magnitude as that observed in similar case of $H^{12}COOH$ (table XI). This can be checked by comparing the present results with those of table VIII. Finally, 12 emissions have been assigned in the ν_6 state (table XII). The emissions which could be assigned in the ν_8 state are generally very weak. The possibility for these emissions to involve another excited state cannot be discarded and the assignments are only given here as tentative ones (table XIII).

Table XI Rotational constants of $H^{13}COOH$ (12A).

Constants (MHz)	$H^{13}COOH$ (a) G.S	$H^{13}COOH$ (b) G.S	$H^{13}COOH$ (b) ν_6
A	75580.800 (69)	75580.744	75907.7
B	12053.5663 (33)	12053.5623	11995.2
C	10378.9988 (33)	10379.0010	10312.3
Δ_J	0.009914 (17)	0.0099	0.0101
Δ_{JK}	- 0.084799 (55)	- 0.0850	- 0.0913
Δ_K	1.655 (13)	1.655 *	1.655 *
δ_J	0.0019784 (13)	0.0020	0.0020
δ_K	0.04229 (26)	0.04229 *	0.04229 *

(a) obtained from [1C]

(b) obtained from IR, FIR and MW transitions [12 A]

The P^6 parameters are fixed at their value in the G.S of $H^{12}COOH$.

* fixed at its value in the G.S

() one standard error. Errors are not indicated in the second and third columns because the small number of available experimental datas make them insignificant.

Table XII Assigned transitions in the ν_6 excited state of $H^{13}COOH$ (12A)

CO_2 laser	cm^{-1}	$H^{13}COOH$ absorption (ν) J K_a K_c	$H^{13}COOH$ emission (ν) J K_a K_c	Obs.Freq. (MHz)	$f_{obs}-f_{calc}$ ** (MHz)
9 R 30	1084.635	(0) 14 3 12 → (6) 13 3 11	(6) 13 3 11 → (6) 12 3 10	290953.8	-0.2
			(6) 12 3 10 → (6) 11 3 9	268514.8	+0.8
9 P 12	1053.923	(0) 18 7 11 → (6) 17 6 12	(6) 17 6 12 → (6) 16 6 11 }	380004.0	-0.4
		(0) 18 7 10 → (6) 17 6 11	(6) 17 6 11 → (6) 16 6 10 }		-1.0
			(6) 16 6 11 → (6) 15 6 10 }	357589.7	+0.5
			(6) 16 6 10 → (6) 15 6 9 }		+0.4
9 R 32	1085.765	(0) 24 2 22 → (6) 24 2 23	(6) 24 2 23 → (6) 23 2 22	523810.4	-0.1
9 R 26	1082.296	(0) 29 4 25 → (6) 29 3 26	(6) 29 3 26 → (6) 28 3 25	668383.0	+0.3
			(6) 28 3 25 → (6) 27 3 24	645231.8	-0.4
9 P 20	1046.854	(0) 27 2 25 → (6) 26 1 26	(6) 26 1 26 → (6) 25 1 25	546285.3	+0.1
9 P 24	1043.183	(0) 26 8 18 → (6) 25 7 19	(6) 25 7 19 → (6) 24 7 18 }	559214.1	+0.4
			(6) 25 7 18 → (6) 24 7 17 }		-0.1
9 P 32	1035.473	(0) 35 8 28 → (6) 34 7 27	(6) 34 7 27 → (6) 33 7 26*	761888.7	+31.7
9 P 14	1052.195	(0) 10 9 2 → (6) 9 8 1	(6) 9 8 1 → (6) 8 8 0 }*	200953.9	+1.4
		(0) 10 9 1 → (6) 9 8 2	(6) 9 8 2 → (6) 8 8 1 }*		
9 P 6	1058.948	(0) 44 4 41 → (6) 43 4 40	(6) 43 4 40 → (6) 42 4 39*	955368.1	-287.6
				955370.3	-285.4

* Tentative assignment

** Calculated with the parameters reported in ref. [12A]

Table XIII Tentative assignments in the ν_8 excited state of $H^{13}COOH$ (12A).

CO_2 laser	cm^{-1}	$H^{13}COOH$ emission $(\nu)J\ K_a$		Obs. Frequency (MHz)
9 P 24	1043.103	$(8)35_8$	$\rightarrow (8)34_8$	785587.2
9 P 26	1041.279	$(8)35_{(K_a \geq 9)}$	$\rightarrow (8)34$	740063.1
9 P 30	1039.369	$(8)28_{(K_a \sim 12)}$	$\rightarrow (8)27$	627229.1
10 R 32	983.252	$(8)15_5$	$\rightarrow (8)14_5$	336434.4

d) Experimental checking of the assignments. Results obtained from assignment of laser emissions have been compared to those given by RF spectroscopy when the sample gas placed in the cavity of a CO_2 laser is irradiated simultaneously by RF and IR fields. The interest of these experiments is to confirm the assignments of the FIR emissions based mainly on energy level calculations. The RF spectrum and the IR transition associated with two pump lines have been assigned and tentatively assigned for a third pump line (table XIV). In this last case the tentative assignment is in good agreement with that obtained from the submillimeter study. The assignment of the 668 383.0 MHz line has been confirmed by a double resonance on the active medium of the laser : the FIR laser output power increases when a microwave field at 42 792.5 MHz is sent into the laser. The frequency is exactly coincident with the predicted frequency of the $29_{4,25} \leftarrow 29_{4,26}$ transition.

Table XIV RF Spectroscopy inside the cavity of a CO_2 laser (12A).

CO_2 laser	Obs.Freq. (MHz)	Assignment (ν) JK_a	Calc.Freq. (MHz) (G.S)	IR transition	Calc.Freq. (cm^{-1})	CO_2 Freq. (cm^{-1})
9 R 4	173.12 *	(0) 14_4	173.25			
	297.81 *	(0) 15_4	297.89			
	493.80 *	(0) 16_4	493.79			
	792.65	(0) 17_4	792.51	$(0)17_{4,14} \rightarrow (6)16_{3,13}$	1067.540	1067.539
9 P 6	188.00 *	(0) 21_5	187.78			
	296.09	(0) 22_5	296.21	$(0)22_{5,17} \rightarrow (6)21_{4,18}$	1058.984	1058.948
	457.13 *	(0) 23_5	457.21			
9 P 32	6.32	(0) 35_8	4.95			
	54.00 *	(6) 33_7	54.64			
(a)	81.76	(6) 34_7	82.20	$(0)34_7 \rightarrow (6)35_8$		
	120.74 *	(6)335_7	122.06			

* Collisional satellites
(a) Tentative assignment

IV - SPECTROSCOPIC MEASUREMENTS IN THE DEUTERATED ANALOGS

A lot of emissions have been observed from the deuterated analogs of formic acid optically pumped by a CO_2 laser and are not assigned until now. The only attempt is from Dyubko et al. for H COOD [17A].

The infrared studies give only approximate values of the vibrational energies of the molecules [2B,8B] and the microwave analysis concern only the ground state and the excited state below 600 cm^{-1}.

The graphical method of assignment used successfully in the two previous cases does not give a reliable solution [9D]. Several hypothesis could explain this failure.

i) the excited states near 1000 cm^{-1} for the various isotopic species do not correspond exactly to the same normal modes as for the main species.

ii) Fermi and Coriolis resonances predicted in previous infrared studies [2B,8B] could strongly perturb the rotational spectrum of the excited states.

iii) Several emissions are very weak and could be pumped via hot band IR transitions.

In order to undertake the assignment of these emissions with any chance of success, a reinvestigation of the FIR spectrum to search for cascade emission should be of great help, and accurate IR data are needed.

V - CONCLUSION

Formic acid is one of the most promising molecules to increase the number of laser emissions since the molecular constants of the ^{12}C and ^{13}C species are now accurate enough to attempt a prediction of coincidences with the various CO_2 isotopes or sequence band lasers. Pumping with more powerful CO_2 lasers could also increase the number of emissions by pumping via hot bands.

Note : New emissions lines from $H^{12}COOH$ and power measurements are also available in a Thesis [18A] sent to us at the end of this work.

References

A - Wavelength and frequency measurements - Assignment of FIR emissions in formic acid

[1A] D.J.E. Knight : "Ordered list of far-infrared lines", NPL Report Qu 45 (1981).

[2A] R.J. Wagner, A.J. Zelano and L.H. Ngai : "New submillimeter laser lines in optically pumped gas molecules", Opt. Commun 8 pp 46-47 (1973).

[3A] T.K. Plant, P.D. Coleman and T.A. Detemple : "New optically pumped far-infrared lasers", IEEE. J. Quant. Electron. QE9, pp 962-963 (1973).

[4A] S.F. Dyubko, V.A. Svich and L.D. Fesenko : "Submillimeter laser using formic acid vapor pumped with carbon dioxide laser radiation", Sov. J. Quant. Electron, 3 p 446 (1974).

[5A] H.E. Radford : "New C.W. lines from a submillimeter waveguide laser", IEEE. J. Quant. Electron. QE 11 pp 213-214 (1975).

[6A] S.F. Dyubko, V.A. Svich and L.D. Fesenko : " Submillimeter HCOOH, DCOOH, HCOOD and DCOOD laser", Sov. Phys. Tech. Phys., 20 pp 1536-1538 (1976).

[7A] D. Dangoisse, A. Deldalle, J.P. Splingard and J. Bellet : "Mesure précise des émissions continues du laser submillimétrique à acide formique", C.R. Acad. Sci. Paris t. 283 série B pp 115-118 (1976).

[8A] A. Deldalle, D. Dangoisse, J.P. Splingard and J. Bellet : "Accurate measurements of C.W. optically pumped FIR laser lines of formic acid molecule and its isotopic species $H^{13}COOH$, HCOOD and DCOOD", Opt. Commun. 22 pp 333-336 (1977).

[9A] H.E. Radford, F.R. Petersen, D.A. Jennings and J.A. Mucha : "Heterodyne measurements of submillimeter laser spectrometer frequencies", IEEE. J. Quant. Electron. 13 pp 92-94 (1977).

[10A] G. Kramer and C.O. Weiss : "Frequencies of some optically pumped submillimeter laser lines", Appl. Phys. 10 187-188 (1976).

[11A] B.M. Landsberg : "New C.W. optically pumped FIR emissions in HCOOH, D_2CO and CD_3Br", Appl. Phys., 23 pp 345-348 (1980).

[12A] D. Dangoisse and P. Glorieux : "Optically pumped C.W. submillimeter emissions from $H^{13}COOH$: measurements and assignments", J. Mol. Spectrosc., 92 pp 283-297 (1982).

[13A] P.J. Epton, W.L. Wilson. Jr, F.K. Tittel and T.A. Rabson : "Frequency measurement of the formic acid laser 311 μm line", Appl. Opt. 18 pp 1704-1705 (1979).

[14A] D. Dangoisse, E. Willemot, A. Deldalle and J. Bellet : "Assignment of the HCOOH C.W. submillimeter laser", Optics Commun. 28 pp 111-116 (1979).

[15A] O.I. Baskakov, S.F. Dyubko, M.V. Moskienko and L.D. Fesenko: "Identification of active transitions in a formic acid vapor laser", Sov. J. Quant. Electron., 7 pp 445-449 (1977).

[16A] H.E. Radford : Private communication (1976).

[17A] S.F. Dyubko, O.I. Baskakov, M.V. Moskienko and L.D. Fesenko: "Assignment of some optically pumped FIR laser lines of $C_2H_2F_2$, HCOOD and CH_3I, Conf. Digest of the third Int. Conf. on submillimeter waves and their applications", Guildford U.K., pp 68-69 (1978).

[18A] H.P. Röser : "Die entwicklung eines optisch gegumpten Submm lasers als Lokaler Oszillator in einem heterodyn - system", Thesis, Bonn (1979).

B - Infrared spectroscopy

[1B] B.M. Landsberg, D. Crocker and R.J. Butcher : "Offset-locked CO_2 waveguide laser study of formic acid. Reassessement of Far-infrared laser assignments", J. Mol. Spectrosc. 92 pp 67-76 (1982).

[2B] R.L. Reddington : "Vibrational spectra and normal coordinate analysis of isotopically labeled Formic acid Monomers", J. Mol. Spectrosc. 65 pp 171-189 (1977).

[3B] V. Williams : "Infrared Spectra of monomeric formic acid and its deuterated forms I high frequency region, II low frequency region (2200-800 cm^{-1})", J. Chem. Phys. 15 pp 232-251 (1947).

[4B] T. Miyazawa and K.S. Pitzer : "Internal rotation and infrared spectra of formic acid monomer and normal coordinate treatment of out of plane vibration of monomer dimer and polymer", J. Chem. Phys. 27 pp 1076-1086 (1959).

[5B] C. Hisatsune and J. Heicklen : "Are there two structural isomers of formic acid", Can. J. Spectrosc. 18 pp 135-142 (1973).

[6B] J.C. Deroche and P. Pinson : "Assignments in infrared spectra of ν_6 and ν_8 of formic acid", Six. Coll. High. res. mol. spectrosc. C-10 (1979).

[7B] J.C. Deroche, J. Kauppinen and E. Kyrö : "ν_7 and ν_9 bands of formic acid near 16 µm", J. Mol. Spectrosc. 78 pp 379-394 (1979).

[8B] R.C. Milliken and K.S. Pitzer : "Infrared spectra and vibrational assignment of monomeric formic acid", J. Chem. Phys. 27 pp 1305-1308 (1957).

C - Microwave spectroscopy

[1C] E. Willemot, D. Dangoisse, N. Monnanteuil and J. Bellet : "Microwave spectra of molecules of astrophysical interest : XVIII : Formic acid", J. Phys. Chem. Ref. data 9 pp 59-160 (1980).

[2C] E. Willemot, D. Dangoisse and J. Bellet : "Etude des spectres de rotation des états vibrationnels excités ν_6 (1105 cm^{-1}) et ν_8 (1033 cm^{-1}) de l'acide formique", C.R. Acad. Sci. Paris t 279 série B pp 247-250 (1974).

[3C] E. Willemot, D. Dangoisse and J. Bellet : "Microwave spectrum of formic acid and its isotopic species in D, ^{13}C and ^{18}O. Study of Coriolis resonances between ν_7 and ν_9 vibrational excited states", J. Mol. Spectrosc. 73 pp 96-119 (1978).

[4C] E. Willemot, D. Dangoisse and J. Bellet : "Microwave spectrum of the vibrational excited states ν_6 and ν_8 of formic acid. Contribution to the assignment of the formic acid submillimeter laser", J. Mol. Spectrosc. 77, 161-168 (1979).

D - Other references

[1D] S.F. Dyubko, L.D. Fesenko and O.I. Baskakov : "Investigation of submillimeter wave amplification in optically pumped molecular gases", Sov. J. Quant. Electron. 7 pp 859-862 (1977).

[2D] L.D. Fesenko and S.F. Dyubko : "Optimization of the parameters of optically pumped submillimeter lasers", Sov. J. Quant. Electron. 6 pp 839-843 (1976).

[3D] G. Duxbury and H. Herman : "Optically pumped millimetre lasers", J. Phys. B : Atom. Molec. Phys. 11 pp 935-949 (1978).

[4D] C.O. Weiss : "Pump saturation in molecular Far-infrared laser", IEEE. J. Quant. Electron. QE12 pp 580-584 (1976).

[5D] C.O. Weiss and G. Kramer : "Vibrational relaxation in the HCOOH Far-infrared laser", Appl. Phys. 9 pp 175-177 (1976).

[6D] H.J.A. Bluyssen, A.F. Van Etteger, T.C. Maan and P. Wyder : "Very short Far-infrared pulses from optically pumped CH_3OH/D, CH_3F and HCOOH lasers using an E.Q.-switched CO_2 laser as a pump source", IEEE. J. Quant. Electron. QE16 pp 1347-1351 (1980).

[7D] J.C. Deroche and C. Betrencourt-Stirnemann : "Rotational analysis of CH_3Br ν_6 perpendicular band through far infrared laser lines", Molecular Physics 32 pp 921-930 (1976).

[8D] D. Dangoisse, A. Deldalle and P. Glorieux : "Double resonance on the active medium of an optically pumped submillimeter laser : HCOOH", J. Chem. Phys. 69 pp 5201-5202 (1978).

[9D] D. Dangoisse : "Lasers moléculaires optiquement pompés - Application à la spectroscopie submillimétrique", Thèse d'Etat, Lille (1980).

[10D] D. Dangoisse, P. Glorieux and M. Lefebvre : "Double resonance on the active medium of a FIR laser : dispersion effects", 7th Coll. on high resol. Mol. Spectrosc. Q9 Reading (1981).

[11D] C. Freed, L.C. Bradley and R.G. O'Donnell : "Absolute frequencies of Lasing transitions in seven CO_2 isotopic species", IEEE. J. Quant. Electron. QE16 pp 1195-1206 (1980).

OPTICALLY PUMPED INFRARED LASER ACTION IN PROPYNE

E. Arimondo and M. Ciocca

Istituto di Fisica Sperimentale and
Unità del GNSM-Università di Napoli, Italy

G. Baldacchini

ENEA, Centro Ricerche Energia
Frascati, Roma
Italy

The propyne, or methylacetilene, molecules (CH_3CCH) have received scarce attention in the development of optically pumped infrared lasers.

The literature reports the investigations by two groups in different conditions, in the far infrared region by Chang and Mc Gee[4] in 1979 and in the near infrared region by Fischer and Wittig [7] in 1981. The first study led to emission in the 400-1000 μm region following pumping by 9 μm CO_2 lasers lines. The second group reported emission at 16 μm through pumping by few 10 μm CO_2 laser lines. As evident in the low resolution spectrum of fig. 1 the propyne vibration produces an absorption spread over all the 9-10 μm regions of the CO_2 laser, so that new emission lines may result in more detailed investigations.

From the spectrum of fig. 1 it results that the 9 μm region is

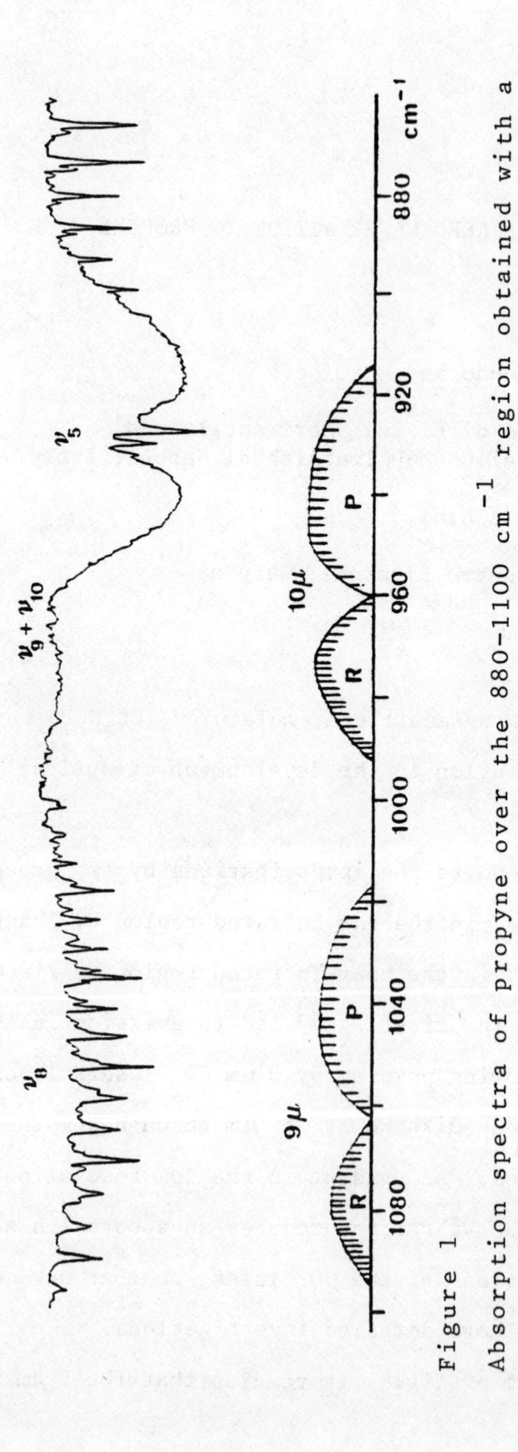

Figure 1

Absorption spectra of propyne over the 880-1100 cm $^{-1}$ region obtained with a Perkin-Elmer 283 B Spectrometer (upper part) and schematic of CO_2 laser lines (lower part). The spectrum was obtained at 150 Torr.

dominated by the ν_8 perpendicular band, with the Q branches and the K = 3n intensity alternances clearly shown. This band has been investigated only in low resolution spectroscopy by Thomas and Thompson[11]. The laser lines lying in this region and having coincidences mainly with the $\nu_8 \leftarrow 0$ transitions have been used by Chang and Mc Gee[4] in their study of optically pumped far infrared lasers (lines reported in Table 1). The 10 μm region presents the absorption of the $\nu_5 \leftarrow 0$ parallel band whose structure is not resolved in the low resolution spectrum of the figure. The laser Stark study by Burrel et al.[3] and the recent combined laser Stark and microwave study of Meyer et al.[10] provided very accurate molecular constants for the ν_5 absorption. The spectrum of fig. 1 presents some features at 970 cm^{-1}, that may be interpreted as components of the $\nu_9+\nu_{10} \leftarrow 0$ combination band.

The coincidences of this band with 10μm CO_2 lines at 945 cm^{-1}, 964 cm^{-1} and 973 cm^{-1} led Fisher and Wittig[7] to obtain laser emission on the $\nu_9+\nu_{10} \rightarrow \nu_{10}$ band. Thus components of the $\nu_9+\nu_{10} \leftarrow 0$ combination band are located also under the high frequency side of the ν_5 absorption.

In table II the vibrational and rotational constants for the states involved in the 9 μm and 10μm absorptions have been reported. For the ground state the quartic and sextic centrifugal constants have been derived in the laser Stark studies (9,10) and microwave investigations (2). It is surprising that quartic

TABLE I - Laser lines in propyne

λ[a] (µm)	$1/\lambda$[a] (cm^{-1})	Ref.	CO_2 pump	Offset (MHz)	Pol.	Out/in	Pressure (Torr)	Thresh	IR Assignment	$\Delta\nu_{IR}$[b]
1097.11(10)	(9.1149)	(4)	$9P_8$	-20.	$\|\|$	0.32/50 mW/W	0.051	9W	${}^pR_1(15)\ \nu_8\leftarrow 0$	0.1
675.29(10)	(14.8085)	(4)	$9P_{40}$	0.	$\|\|$	0.33/55 "	0.039	15W	${}^pP_2(27)\ \nu_8\leftarrow 0$	0.86
583.77(10)	(17.1300)	(4)	$9P_{20}$	-15.	$\|\|$	0.004/50 "	0.047	7W	${}^pR_3(29)\ \nu_8+\nu_{10}\leftarrow\nu_{10}$	0.95
566.41(10)	(17.6951)	(4)	$9P_{18}$	+25.	$\|\|$	0.58/63 "	0.055	34W	${}^pR_4(30)\nu_8\leftarrow 0$	0.79
531.03(10)	(18.8313)	(4)	$9P_6$	0.	$\|\|$	0.004/45 "	0.092	9W	${}^rP_4(34)\nu_8\leftarrow 0$	1.29
516.77(10)	(19.3510)	(4)	$9R_{12}$	+35.	$\|\|$	0.006/47 "	0.039	25W	${}^rR_0(33)\nu_8\leftarrow 0$	0.76
428.87(10)	(23.3171)	(4)	$9R_{38}$	+35.	$\|\|$	0.012/20 "	0.051	9W	${}^rR_2(40)\nu_8\leftarrow 0$	1.11
(16.42)	609. (2)	(7)	$10R_{16}$	-	-	VVW pulsed	0.4	$1.5 J/cm^2$	? $\nu_9+\nu_{10}\leftarrow 0$	-
(16.39)	610. (2)	(7)	$10P_{14}$	-	-	VVW "	0.4	1.5 "	? $\nu_9+\nu_{10}\leftarrow 0$	-
(16.12)	620.3(5)	(7)	$10R_{16}$	-	-	VW "	0.4	1.5 "	? $\nu_9+\nu_{10}\leftarrow 0$	-
(15.72)	636.1(5)	(7)	$10R_{16}$	-	-	M "	0.7	0.5 "	? $\nu_9+\nu_{10}\leftarrow 0$	-
(15.72)	636.3(5)	(7)	$10R_4$	-	-	W "	0.7	0.5 "	? $\nu_9+\nu_{10}\leftarrow 0$	-

λ^a (μm)	$1/\lambda^a$ (cm^{-1})	Ref.	CO_2 pump	Offset (MHz)	Pol.	Out/In	Pressure (Torr)	Thresh	IR	Assignment	$\Delta\nu_{IR}^b$
(15.71)	636.5(5)	(7)	$10P_{14}$	-	-	S pulsed	1.2	0.5	?	$\nu_9+\nu_{10}\leftarrow 0$	-

a) in bracket lines we have reported the wavelengths or wavenumbers derived from the measured ones.

b) $\Delta\nu = \nu_{meas.} - \nu_{calc.}$; evaluated in cm^{-1}

TABLE II Spectroscopic constants of propyne

	Band origin (cm^{-1})	Ref.	α_{ν}^{B} (MHz)	Ref.	$A_{\nu}-A_0-B_{\nu}-B_0$ (MHz)	Ref.	ζ	Ref.
$\nu_5 \leftarrow 0$	930.27540(7)	(2)	+37.757(4)	(3)	-189.3(23)	(3)	-	-
$\nu_8 \leftarrow 0$	1054.16	(11)	+25. (4)	(9)	-689.	(11)	0.397	(11)
$\nu_8+\nu_{10} \leftarrow \nu_{10}$	1047.0	(11)			see note(*)			
$\nu_9 \leftarrow 0$	633.28	(11)	-6.	(11)	+30.	(11)	1.00	(11)
$\nu_{10} \leftarrow 0$	329.2	(1)	-24.110(2)	(2)	+60.	(11)	0.892	(2,6)
B_0	8545.87712(6) MHz	(2)						
A_0	158620. MHz	(6)						

(*) fixed to the values of the $\nu_8 \leftarrow 0$ band

centrifugal constants are not available for the ν_8 band and that the coincidences with the 9 μm CO_2 laser lines have not been applied to make a high resolution laser Stark study of this band.

The apparatus used by Chang and Mc Gee[(4)] to produce optically pumped far infrared laser emission in propyne, described in many papers of this review series, relied on a pulsed CO_2 200W peak power and 150 μs pulses and a mirror cavity in the far infrared laser.

The experiment by Fisher and Wittig [(7)] was based on a single line 7.5 J TEA CO_2 laser, 160 $\pm$ 30 ns full width half maximum pulse. The optically pumped laser is composed by a 6 m cavity containing propyne cooled at the liquid nitrogen temperature. The pump laser was double passed through the propyne cavity. The influence of the propyne pressure and temperature and of the pump fluence on the output power of the 16 μm laser was tested in the experiment. For the strong 636.5 cm^{-1} line (Table I) the optimum operating pressure increased with the pump fluence. The output power of all laser lines decreased dramatically with increasing temperature, as a consequence of the near threshold operation.

On the basis of the vibrational constants of Thomas and Thompson[(11)] and the rotational constants of Meyer et al.[(9)], we have derived tentative assignments for the far infrared laser lines observed by Chang and Mc Gee[(4)] (see Table I). It turned out that nearly all the observed laser lines are produced by the $\nu_8 \leftarrow 0$ pumping. The accuracy of the fit for the infrared pump and for infrared laser wave numbers is very good considering the low accuracy of the available molecular constants. A systematic deviation up to 1 cm^{-1} in the infrared wave numbers of the $\nu_8 \leftarrow 0$ transitions with high J numbers may result by the absence of centrifugal rotational corrections in the calculation. Thus a more accurate analysis of the vibrational and rotational ν_8 constants is required.

For one line, at 583.77 μm pumped by the $9P_{20}$, it turned out that the infrared pumping should occur through the $\nu_8+\nu_{10} \leftarrow \nu_{10}$ hot band. The band origin of this band was derived by Thomas and Thompson[11], while for other vibrational and rotational constants we made use of the constants for the $\nu_8 \leftarrow 0$ band.

We have not reported in Table I the calculated wavelengths of the FIR transitions because of the poor accuracy resulting from the missed quartic centrifugal rotational constants.

The laser lines in the 16 μm region are produced by the CO_2 pumping on the $\nu_9+\nu_{10} \leftarrow 0$ combination band and emission on the $\nu_9+\nu_{10} \rightarrow \nu_{10}$ hot band. By combining two degenerate vibrations (E type) in a C_{3v} molecule, vibrational states with A_1,A_2 and E species are produced[8] (see Fig.2). The combination bands $\nu_9+\nu_{10} \leftarrow 0$ up to the A_1 and E states are dipole allowed, and give rise to absorbtion bands with $\Delta K=0$ and $\Delta K = \pm 1$ respectively. The $A_2 \nu_9+\nu_{10} \leftarrow 0$ band may be dipole allowed through Coriolis coupling with A_1 and E states.

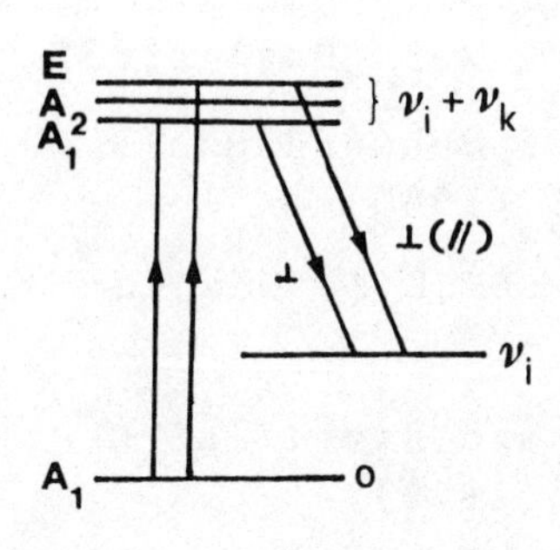

Figure 2

Schematic diagram of the IR pumping on a combination band $\nu_i + \nu_k \leftarrow 0$ and IR emission on a hot band $\nu_i + \nu_k \rightarrow \nu_i$ with ν_i and ν_k degenerate vibrations.

The A_1 parallel band $\Delta K=0$ in no way looks like an ordinary parallel band because the Q branches are equally spaced on either side of the band origin. The hot band $\nu_9 + \nu_{10} \rightarrow \nu_{10}$ from the A_1 species of the upper state $\nu_9 + \nu_{10}$ to the E type ν_{10} states produces a perpendicular band. The $E \rightarrow E$ type $\nu_9 + \nu_{10} \rightarrow \nu_{10}$ subband can occur both as a perpendicular and parallel band, but only the perpendicular component has an intensity comparable with ν_9 since ν_9 is a perpendicular band[8]. As already mentioned, the methylacetilene spectra show up perpendicular bands centered near 952 cm^{-1} and 974 cm^{-1}, that may be attributed to the E and A_1 type $\nu_9 + \nu_{10} \leftarrow 0$ bands, but a more detailed investigation of the spectra is required to make definitive assignments of the IR and FIR lines.

REFERENCES

(1) Anttila R.; private communication reported in ref. (5).

(2) Bauer A., Boucher D., Burie J., Demaison J. and Dubrulle A.; J. Phys. Chem. Ref. Data 8, 537 (1979)

(3) Burrel P.M., Bjarnov E. and Schwendeman R.M.; J. Mol. Spectrosc. 82, 193 (1980)

(4) Chang T.Y. and Mc Gee J.D.;IEEE J. Quantum Electron. QE-9 , 62 (1976)

(5) Duncan J.L., Mc Kean D.C. and Nivellini G.D. J. Mol. Structure 32, 255 (1976).

(6) Duncan J.L., Mc Kean D.C., Mallison P.D. and Mc Culloch R.D.; J. Mol. Spectrosc. 46, 232 (1976)

(7) Fischer T.A. and Wittig C.; Appl. Phys. Lett. 39, 6 (1981).

(8) Herzberg G.; "Molecular Spectra and Molecular Structure" Vol. II, Van Nostrand, 1945, New York, London pag. 266

(9) Meyer C. and Sargent-Rozey M.; J. Mol. Spectrosc. 83, 343 (1980)

(10) Meyer F., Dupré J., Meyer C., Lambeau C., De Vleeschouwer M., Layahe J.G. and Fayt A., Int. J. Inf. Mill. Waves 3, 83 (1982)

(11) Thomas R.K. and Thompson H.V., Spectrochimica Acta, 24 A 1937 (1968).

INDEX